Competency Based Mathematics

for

Secondary Schools

Book 4

(MODULES 15 TO 17)

Nji Emmanuel Ndi
GBHS Mankon - Bamenda
North West Region Cameroon
Tel: (+237) 676 684 050
Email: manuelndike@gmail.com

First Edition

Printed by CreateSpace, an Amazon.com Company

EStore address: www.CreateSpace.com/7056024

Available from Amazon.com, CreateSpace.com, and other retail outlets

Available on Kindle and other retail outlets

Books by Nji Emmanuel Ndi

Complete Ordinary Level Mathematics Passport
Rudiments of Ordinary Level Mathematics (Second Edition)
Advanced Level Mathematics Key Facts
Competency Based Mathematics for Secondary Schools Book 1
Competency Based Mathematics for Secondary Schools Book 2
Competency Based Mathematics for Secondary Schools Book 3
Competency Based Mathematics for Secondary Schools Book 4
Competency Based Mathematics for Secondary Schools Book 5

Copyright © 2017 Nji Emmanuel Ndi
All rights reserved.
ISBN-10: 1545072019
ISBN-13: 978-1545072011

DEDICATION

Dedicated to all emerging and emergent Societies

Competency Based Mathematics for Secondary Schools. Book 4

Table of Contents

DEDICATION I
   Acknowledgement vii
   How to Use this Book viii
   Notations Used in this Book x

MODULE 15: NUMBERS, FUNDAMENTAL OPERATIONS AND RELATIONSHIPS IN THE SETS OF NUMBERS AND BETWEEN ELEMENTS OF SETS 1

TOPIC 1 : ESTIMATIONS, APPROXIMATIONS AND ERRORS 2
   1.1   ROUNDING DOWN AND ROUNDING UP 3
   1.2   THE CONCEPT OF ESTIMATIONS 3
   1.3   ESTIMATING SUMS AND DIFFERENCES 3
   1.4   ESTIMATING PRODUCTS AND QUOTIENTS 4
   1.5   ESTIMATING OF PRODUCTS AND QUOTIENTS 5
   1.6   THE CONCEPT OF APPROXIMATIONS 6
   1.7   DECIMAL PLACES 6
   1.8   SIGNIFICANT FIGURES 6
   1.9   STANDARD FORM 8
   1.10 CALCULATIONS INVOLVING STANDARD FORM 9
   1.11 CALCULATIONS INVOLVING APPROXIMATION 10
   1.12 MAXIMUM AND MINIMUM VALUES FROM A GIVEN APPROXIMATION 11
   1.13 ABSOLUTE ERROR 12

TOPIC 2 : SURDS, INDICES AND LOGARITHMS 16
   1.14 RATIONAL AND IRRATIONAL NUMBERS 17
   SURDS 17
   2.1   CONCEPT OF A SURD 17
   2.2   LAWS OF SURDS 17
   2.3   THE CONJUGATE OF A SURD 19
   2.4   RATIONALIZING THE DENOMINATOR 19
   2.5   SURD EQUATIONS 20
   THEORY OF INDICES 21
   2.6   LAWS OF INDICES 21
   2.7   EXPONENTIAL OR INDEX EQUATIONS 22
   THEORY OF LOGARITHMS 23
   2.8   LAWS OF LOGARITHMS 23
   2.9   CHANGE OF BASE FORMULA 27

TOPIC 3 : MATRICES 33
   3.1   REVIEW AND REVISION 34
   3.2   THE DETERMINANT OF A 2×2 SINGULAR MATRIX 34
   3.3   THE ADJOINT OF A 2×2 MATRIX 35
   3.4   THE INVERSE OF A 2×2 MATRIX 36
   3.5   SIMULTANEOUS EQUATIONS – THE MATRIX METHOD 40
   3.6   INFORMATION MATRICES 42

Table of Contents

MODULE 16: PLANE GEOMETRY 48

TOPIC 4 : VECTORS IN 2-DIMENSIONS 49

- 4.1 Vectors as Coordinates in Two Dimensions 50
- 4.2 Parallel and Perpendicular Vectors 50
- 4.3 Position Vectors 52

Vector Geometry 55

- 4.4 The Position Vector of the Midpoint 55
- 4.5 Proportional Division of a vector 55
- 4.6 The Scalar or Dot Product 60
- 4.7 Angle between two vectors 61

TOPIC 5 : SIMPLE TRANSFORMATIONS 68

TRANSFORMATIONS WITHOUT MATRICES 69

- 5.1 Translation, Reflection, Rotation 69
- 5.2 Enlargements 70
- 5.3 Shear 71
- 5.4 Stretch 72
- 5.5 Invariant Points and Invariant Lines 73

TRANSFORMATIONS USING MATRICES 74

- 5.6 Transformation Matrix (Matrix Operator) 74
- 5.7 Transformations Involving Many Points 77
- 5.8 Describing Transformation 78
- 5.9 Finding Transformation Matrices 83
- 5.10 Inverse Transformations 86
- 5.11 Transformations by Singular Matrices 87
- 5.12 Transformations involving Change of Area 87
- 5.13 Composite or Successive Transformations 89
- 5.14 Translations and Translation Vectors 90

SUMMARY 91

TOPIC 6 : CONSTRUCTIONS AND LOCI 99

CONSTRUCTIONS 100

- 6.1 Constructing a Line of Given Length 100
- 6.2 Constructing a Triangle ABC with Sides of Given Length 100
- 6.3 Height, Median and Perpendicular Bisector 101
- 6.4 Constructing a Perpendicular Bisector (or Mediator) of a Given Line Segment 102
- 6.5 Constructing a Perpendicular to a Given Line Segment, from a Given Point . 102
- 6.6 Bisecting a Given Angle 103
- 6.7 Construction of Some Special Angles 104
- 6.8 Trisecting a Right Angle 105
- 6.9 Constructing an Inscribed Circle of a Given Triangle 106
- 6.10 Constructing a Circumscribed Circle of a Given Triangle 107
- 6.11 Dividing a Given Line Segment into Congruent Segments 108
- 6.12 Construction of Polygons 109

	LOCI	111
6.13	LOCI IN 2-DIMENSIONS	111

TOPIC 7 : TRIGONOMETRY 119

7.1	REVIEW AND REVISION	120
7.2	RECIPROCAL TRIGONOMETRIC RATIOS	121
7.3	RECIPROCAL TRIG RATIOS FROM TABLES	123
7.4	THE GENERAL ANGLE	124
7.5	THE MEANING OF NEGATIVE ANGLES	126
7.6	RADIAN MEASURE	128
7.7	BASIC TRIGONOMETRIC IDENTITIES	129
7.8	GRAPHS OF TRIGONOMETRIC FUNCTIONS	130
7.9	SIMPLE TRIGONOMETRIC EQUATIONS	132

TOPIC 8 136

TOPIC 9 : APPLICATIONS 136

8.1	SOLUTIONS TO TRIANGLES	137
8.2	ANGLES OF ELEVATION AND DEPRESSION	143
8.3	BEARINGS IN TWO DIMENSIONS	146
8.4	SINE AND COSINE FORMULAE	150

TOPIC 10 : CIRCLES 159

9.1	VOCABULARIES ASSOCIATED WITH CIRCLES	160
9.2	MENSURATION OF THE CIRCLE	160
9.3	ARC LENGTH	162
9.4	AREA OF A SECTOR	163
9.5	AREA OF A SEGMENT	165
9.6	CIRCLE THEOREMS	168

TOPIC 11 : THE EARTH AS A SPHERE 191

10.1	SURFACE AREA AND VOLUME OF A SPHERE	192
10.2	VOLUME AND SURFACE AREA OF A HEMISPHERE	193
10.3	POSITION OF A PLACE ON THE SURFACE OF THE EARTH	193
10.4	GREAT CIRCLES	194
10.5	SMALL CIRCLES	195
10.6	TIME ZONES IN THE WORLD	196

TOPIC 12 : NETWORKS 199

11.1	NETWORK TERMINOLOGY	200
11.2	EULER'S FORMULA	200
11.3	ORDERED LISTS AND UNORDERED LISTS	201
11.4	ODD AND EVEN VERTICES	202
11.5	TRAVERSABLE NETWORKS	203
11.6	CONDITIONS FOR A NETWORK TO BE TRAVERSABLE	205
11.7	NETWORKS IN REAL LIFE	207

11.8	NETWORK GRAPHS	210
11.9	TYPES OF NETWORK GRAPHS	213
11.10	NETWORK TREES	217

TOPIC 13 : FLOW DIAGRAMS 224

12.1	THE CONCEPT OF A FLOW DIAGRAM	225
12.2	TYPES OF FLOW DIAGRAMS	225
12.3	BASIC FLOW DIAGRAM SYMBOLS	230
12.4	INSTRUCTION BOXES	231
12.5	DECISION BOXES	233
12.6	DATA AND STORES	236

MODULE 17: 249

ALGEBRA AND LOGIC 249

TOPIC 14 : ALGEBRAIC PROCESSES 250

13.1	REVIEW AND REVISION	251
13.2	QUADRATIC IDENTITIES	252
13.3	THE TRINOMIAL PERFECT SQUARE TEST	252
13.4	QUADRATIC EXPRESSIONS IN TWO VARIABLES	253
13.5	METHOD OF COMPLETING THE SQUARE	254
13.6	THE QUADRATIC FORMULA	256
13.7	DERIVATION OF THE QUADRATIC FORMULA	256
13.8	NON-STANDARD QUADRATIC EQUATIONS	258
13.9	INCOMPLETE QUADRATIC EQUATIONS	259
13.10	WORDED PROBLEMS THAT LEAD TO QUADRATIC EQUATIONS	261
13.11	SIMULTANEOUS EQUATIONS, ONE LINEAR-ONE QUADRATIC	263

TOPIC 15 : POLYNOMIALS 267

14.1	CONCEPT OF REMAINDER AND FACTORS	268
14.2	LONG DIVISION OF POLYNOMIALS	269
14.3	VARIABLE SUBSTITUTION IN POLYNOMIALS	270
14.4	REMAINDER THEOREM	271
14.5	FACTOR THEOREM	271
14.6	TRIAL AND ERROR METHOD FOR FINDING ROOTS OF POLYNOMIALS	272
14.7	SUM AND DIFFERENCE OF TWO CUBES	273

TOPIC 16 : INEQUALITIES AND INEQUATIONS 276

15.1	REPRESENTATION OF INEQUALITIES	277
15.2	CONDITIONAL INEQUALITIES OR INEQUATIONS	278
15.3	BUILDING UP INEQUALITIES	278
15.4	SOLVING INEQUATIONS	279
15.5	LINEAR INEQUALITIES	279
15.6	COMPOUND INEQUALITIES	280
15.7	ABSOLUTE VALUE INEQUALITIES	281
15.8	QUADRATIC INEQUATIONS	282

15.9	Absolute Inequalities	284
15.10	Nature of roots of a Quadratic Equation	284
15.11	Simultaneous Linear Equations (Graphical Method)	285
15.12	Half planes	286
15.13	Vertical and Horizontal Boundary Lines	288
15.14	Simultaneous Linear Inequations	289

TOPIC 17 : SEQUENCES AND SERIES 300

16.1	Number Sequence	301
16.2	The nth Term of a Sequence	301
16.3	Sum of the First N Terms of a Sequence	302
16.4	Arithmetic and Geometric Progressions	303
16.5	Arithmetic Progression (A.P.)	303
16.6	The n^{TH} term of an A.P.	304
16.7	Sum of the first N terms of an A.P.	305
16.8	The Arithmetic Mean	307
16.9	Geometric Progression (G.P.)	308
16.10	The nth term of a G.P.	309
16.11	The Sum of N terms of a G.P.	310
16.12	Sum to infinity, $S\infty$	312
16.13	Geometric Mean	313

Acknowledgement

My deepest gratitude goes to God Almighty for the inspiration and for the strength.
Many thanks go to Mme. Mbuameh Daisy and Mr. Mburubah Walters for their critical proof reading of the typescript and for offering very useful suggestions which went a long way to reshape the work, the North West Regional Pedagogic Inspector for Mathematics Mr. Nfor Samuel Ndi who preview the initial manuscript and gave ample advice, which went a long way to reshape the document. I heartily thank the Former North West Regional Pedagogic Inspector for Mathematics Mr. Nji Samuel Tatah who made a very commendable effort to edit the Mathematics content of the book. I cannot forget the last minute encouragements and advice which the National inspector of Mathematics Mme Babila Emilia inspired me with. I equally pay much tribute to my students on which this material was tested. I cannot end here without thanking my sweet heart Nji Irene Nfih and my Children who encouraged and supported me in one way or the other during the course of the work.
Many thanks go to the WAEC and the CGCE Board for allowing their past questions to be used directly or indirectly.

Nji Emmanuel Ndi
G.B.H.S. Mankon, Bamenda
North West Region
Cameroon
TEL: (+237)76684050
E-mail: manuelndike@gmail.com

Competency Based Mathematics for Secondary Schools. Book 4

How to Use this Book

This book is written in a very special way with different sections boxed and represented by special symbols as follows.

? Brainstorming Exercise

✎ Example

Exercise

💬 Discussion Exercise 💬

👍 Real life Examples

Integration Activity

🔍 Investigative Activity

✍ Multiple Choice Exercise

Review and Revision Exercise

👪 Group Activity

🛠 Skill Building Exercise

How to Use this Book

The various sections represented by different symbols are out to facilitate navigation through the book. By investing enough time and energy in each section both students and teachers will realize that their speed and understanding will be greatly enhanced.

The brain storming exercises are aimed at provoking and invoking the learners' minds to prepare them for the task at hand. The teacher is highly encouraged to orally question the students during lessons using questions under this section.

The investigative exercises are meant to give the learner ample opportunity to experiment and self discover facts and concepts and develop methods and skills without being told.

The group activities and discussion exercises are aimed at developing a team spirit in the learners.

Many well designed examples are vividly used and solved to facilitate the learner's understanding by showing the necessary steps required for a particular solution. There are a good number of real life examples which point out the application of the subject matter in real life situations. The student is advised to study these examples very carefully.

There are many well graded exercises and skill building exercises to test the level of understanding of the learner and to facilitate skill development in the learner. The student is advised to attempt all the questions as each question may have its own technique.

Many integration activities have been designed to unify groups of sub topics, topics or modules in some cases.

Where necessary review exercises have been given to help the learner retain the skills acquired in the earlier sections.

Finally each topic ends with a good number of multiple choice questions. In each question only one of the alternatives is correct. Write down the letter corresponding to the correct answer.

For greatest achievement, the learner is advised to study regularly what he does not know and work without fear of making mistakes whether with the teacher or during group work.

By consistently and systematically going through this course as instructed, the learner will be overwhelmed with the competencies acquired at each level and at the end of the course.

Notations Used in this Book

Notation	Meaning
$\{\ldots\}$	The set of elements…or the unordered list with elements…
$n(A)$	The number of elements in set A
$\{x:\ \ \}$	The set of all x such that
$\in$	Is an element of …
$\notin$	Is not an element of …
$\{\ \}$ or $\emptyset$	The empty set.
$\mathscr{E}$	The universal set.
$\cup$	The union of …
$\cap$	The intersection of…
$\subseteq$	Is a subset of …
$\subset$	Is a proper subset of …
$A \backslash B$	The difference between the sets A and B.
$(a, b, c, \ldots)$	An ordered list of elements a, b, c, …
$\{a, b, c, \ldots\}$	The set or an unordered list of elements a, b, c, …
$\mathbb{Z}$	The set of integers, $\{0, \pm 1, \pm 2, \pm 3, \pm 4, \ldots\}$
$\mathbb{N}$	The set of all positive integers and zero, $\{0, 1, 2, 3, 4, \ldots\}$
$\mathbb{Z}^+$	The set of positive integers $\{+1, +2, +3, +4 \ldots\}$
$\mathbb{Q}$	The set of rational numbers
$\mathbb{Q}^+$	The set of positive rational numbers
$\mathbb{R}$	The set of all real numbers $\{x: x \in \mathbb{R}\}$
$\mathbb{R}^+$	The set of all positive real numbers $\{x \in \mathbb{R}: x > 0\}$
$f(x)$	f of x or the image of x under the function f
f^{-1}	The inverse function of the function f
fg or $f \circ g$	The function f of the function f
$=$	Is equal to
$\neq$	Is not equal to
$\approx$	Is approximately equal to
$<$	Is less than
$>$	Is greater than
$\not<$	Is not less than
$\not>$	Is not greater than

$\leq$	Is less than or equal to
$\geq$	Is greater than or equal to
$a < x < b$ or $]a, b[$ or (a, b)	An open interval on the number line
$a \leq x \leq b$ or $[a, b]$	A closed interval on the number line
$\{x : a < x < b\}$	The set of elements x such that a is less than x and x is less than b
a	The vector **a**
AB	The vector represented in magnitude and direction by AB
$\lvert x \rvert$	The modulus or absolute value of x i.e. $\{x$ for $x > 0, -x$ for $x < 0, x \in \mathbb{R}\}$
$\boldsymbol{a} \cdot \boldsymbol{b}$	The dot or scalar product of the vectors **a** and **b**
A^{-1}	The inverse of the non-singular matrix A
A^T	The transpose of the matrix A
$\lg x$ or $\log x$	The common logarithm of x
x^n	The number x, raised to the power n
$\propto$	Is proportional
∞	Infinity
$\sqrt{\ }$	The positive square root
$-\sqrt{\ }$	The negative square root
$\sqrt[n]{a}$	The nth root of a
p:	The statement or preposition p
T or 1 in truth tables	True
F or 0 in truth tables	False
$\sim p$ or p' or $\neg p$	The negation of a statement p
$p \wedge q$	The conjunction of the statements p and q
$p \wedge q$	The disjunction of the statements p and q
$A \cap B$ or $\{x : x \in A \wedge x \in B\}$	The intersection of sets A and B.
$A \cup B$ or $\{x : x \in A \vee x \in B\}$	The union of sets A and B.
$p \Rightarrow q$ or $p \to q$	p implies q or p is sufficient for q or p only if q or q is necessary for p
$p \Leftrightarrow q$ or $p \leftrightarrow q$ or p iff q	p is a necessary and sufficient condition for q or p implies and is implied by q or p if and only if q

$\forall x$	For all or for every element x
$\exists x$	There exists or for at least one or for some element x
$\exists! x$	There exists one and only one element x
$\equiv$	Is equivalent or is congruent to
///	Is similar to
$\perp$	Is perpendicular to
$\parallel$	Is parallel to
$G = (V, E)$	The graph of the set V of vertices together with the set E of edges
$D = (V, A)$	The directed graph (digraph) of the set V of vertices and the set A of ordered edges
$G = (V, E, A)$	The mixed graph of the set of vertices V, unordered edges E and ordered edges A.
$n := $	The store n takes the value…
°	Degree
°C	Degrees Celsius
°F	Degrees Fahrenheit

Module 15

Numbers, Fundamental Operations and Relationships in the Sets of Numbers and between Elements of Sets

Family of Situations
Module 15 is an extension of modules 1, 5 and 10. At the end of this module; the student is expected to have acquired more competencies within the **families of situations** *'Representation, determination of quantities and identification of objects by numbers'*.

Categories of Action
The categories of action for module 15 include:
1. Determination of number, reading and writing information using numbers,
2. Verbal interaction on information containing numbers,
3. Representation and treatment of information and quantities.

Credit
The module is expected to be covered within 6 weeks teaching 4 periods of 50 minutes per week (or within 24 periods of 50

Topic 1

ESTIMATIONS, APPROXIMATIONS AND ERRORS

Objectives

At the end of this topic, the learner should be able to:

1. Make reasonable approximations and estimates of quantities and measures.
2. Approximate a given number to a given number of decimal places or significant figures.
3. Write a normal form number in standard form and vice versa.
4. Perform calculations in standard form.
5. State the value of a digit in a given number.
6. Find the maximum and minimum values.
7. Find the maximum and minimum errors from calculations and measurements.

Module 15, Topic 1: Estimations, Approximations and Errors

1.1 Rounding Down and Rounding Up

Approximation is the act of considering a number as its closest more convenient number, by means of rounding down or rounding up the number so as to handle it more conveniently or ease calculations with it.

To Round down a number we give a less exact smaller value to the number and to round up a number we give a less exact larger value to the number.

When the digit in the place after that required for the given degree of accuracy is any of 5, 6, 7, 8 or 9, we round up the digit required for the degree of accuracy by increasing it by 1 and ignore the rest in the case of decimals or replace them with zeros in the case of whole numbers.

When the digit in the place after that required for the given degree of accuracy is 0, 1, 2, 3 or 4, we round down the digit required for the degree of accuracy by maintaining this digit and ignoring the rest in the case of decimals or replace them with zeros in the case of whole numbers.

To round to the nearest ten, hundred, thousand…, replace the number by the closest multiple of 10, 100, 1000…, respectively.

To round to the nearest tenth, hundredth, thousandth, etc write the number to 1, 2, 3, etc decimal places.

1.2 The Concept of Estimations

An **estimate** is an intelligent guess on the dimensions of an object or the result of a calculation. Estimates are approximate values. To estimate the result of a calculation, round up or round down each number in the calculation to a reasonable or given degree of accuracy.

1.3 Estimating sums and differences

 Example

(a) *In whole numbers*
1. Estimate the following
(b) (i) 245 + 350 +570   (ii) 431 + 53   (iii) 748–394

Solution
(i) 245 + 350 +570 ≈ 200 + 400 + 600 ≈ 1200
(ii) 431 + 53 ≈ 430 + 50 ≈ 480   (iii) 748–394 ≈ 700 – 400 ≈ 300

(b) *In Decimals*

2. Estimate to the nearest whole number.
   (i) 4.68 + 0.71   (ii) 6.7234 − 3.5138

   Solution
   (i)   4.68 ≈ 5   (ii)   6.7234 ≈ 7
       + 0.71 ≈ +1   − 3.5138 ≈ −4
       ─────────   ─────────
           6   3

3. Estimate to the nearest tenth.
   (a) 3.623 + 0.29 + 5.386   (b) 4.86 − 3.456

   Solution
   (a) 3.623 + 0.29 + 5.386 ≈ 3.6 + 0.3 + 5.4 = 9.3
   (b) 4.86 − 3.456 ≈ 4.9 − 3.5 = 1.4

Exercise 1:1

1. Estimate the following sums and differences
   (a) 844 + 239   (b) 464 + 37   (c) 7982 + 1486
   (d) 5391 − 247   (e) 26212 − 14084   (f) 748 − 394
   (g) 509 − 42   (h) 13486 + 4842 + 29072

2. Estimate to the nearest whole number.
   (a) 0.89 + 2.083   (b) 6.4139 + 2.8238 + 3.2500
   (c) 4.9 − 0.87   (d) 8.7234 − 2.6006

3. Estimate to the nearest tenth.

   (a)   0.835   (b)   7.428   (c)   0.82   (d)   12.34
       + 0.27   6.37   − 0.243   − 8.45
                        + 0.843

1.4 Estimating products and quotients

(a) *In whole numbers*

Example

1. Estimate the following products
   (i)   58 × 24   (ii) 653 × 56
   (iii) 48830 × 750   (iv) 275 × 24

Module 15, Topic 1: Estimations, Approximations and Errors

Solution
(i) 58 ×24 ≈ 60 ×20 ≈ 1200
(ii) 653 ×56 ≈ 650 ×60 ≈ 39000
(iii) 48,830 ×750 ≈ 48800 ×800 ≈ 36600,000
(iv) 275 ×24 ≈ 300 ×20 ≈ 6,000

2. Estimate the following quotients
   (i) $\dfrac{562}{18}$ (ii) $\dfrac{68}{27}$ (iii) $\dfrac{62}{24}$

Solution
(i) $\dfrac{562}{18} \approx \dfrac{600}{20} \approx 30$ (ii) $\dfrac{68}{27} \approx \dfrac{70}{30} \approx 2.3$ (iii) $\dfrac{62}{24} \approx \dfrac{60}{20} \approx 3$

1.5 Estimating of Products and Quotients

Example

(b) *In Decimals*
Give an estimate for the following.
(i) 4.755×0.5 (ii) $\dfrac{89.93}{4.1}$

Solution
(i) $4.755 \times 0.5 \approx 5 \times 0.5 \approx 2.5$ (ii) $\dfrac{89.93}{4.1} = \dfrac{899.3}{41} \approx 22.5$

Exercise 1:2

1. Estimate the following products.
   (a) 21 × 45   (b) 763 × 53   (c) 4.8×6
   (d) 0.21×3.81   (e) 8.6×5.9   (f) 3.1×8.2

2. Estimate the following quotients.
   (a) $32\overline{)85}$ (b) $91\overline{)873}$ (c) $3\overline{)6.31}$
   (d) $4.2\overline{)0.439}$ (e) $0.31\overline{)6.125}$ (f) $5.7\overline{)0.483}$

1.6 The Concept of Approximations

An approximation is a value or a quantity that is closed to, but not the same as a desired value or a quantity for a specific purpose.

1.7 Decimal Places

The terms tenth, hundredth, thousandth etc respectively mean one, two, three etc decimal places. Thus rounding off to the nearest tenth, hundredth, thousandth etc is the same as rounding off to 1, 2, 3 etc decimal places.

1.8 Significant Figures

The significant figures of a given number are the figures from the leftmost non-zero digit to the rightmost non-zero digit (or rightmost zero digit in the case where the digit on the required degree of accuracy is zero) required to express the number to a given degree of accuracy. This definition implies that, 4.071 to four significant figures for instance, is the same as 4.0710 to five significant figures and 4068 to three significant figures for instance is the same as 4070 to four significant figures.

 Example

1. Round 6526 to the nearest:
   (a) ten   (b) hundred   (c) thousand.
   In each case, state the number of significant figures.

   Solution
   (a) 6526 to the nearest ten = 6530   [3 s.f.]
   (b) 6526 to the nearest hundred = 6500   [2 s.f.]
   (c) 6526 to the nearest thousand = 7000   [1 s.f.]

2. Round 1.045 to: (a) 2 decimal places   (b) 1 decimal place
   In each case, state the number of significant figures.

   Solution
   (a) 1.045 to 2 d.p.s. = 1.05   [3 s.f.]   (b) 1.045 to 1 d.p. = 1.0 [1 s.f.]

3. Round 0.01027 to:
   (a) 4 decimal places.   (b) 3 decimal places.
   (c) 2 decimal places.   (d) 1 decimal place.
   In each case, state the number of significant figures.

Module 15, Topic 1: Estimations, Approximations and Errors

Solution
(a) 0.01027 to 4 d.p.s. = 0.0103 [3 s.f.]
(b) 0.01027 to 3 d.p.s. = 0.010 [1 s.f.]
(c) 0.01027 to 2 d.p.s. = 0.01 [1 s.f.]
(d) 0.01027 to 1 d.p.s. = 0.0 [no s.f.]

4. Round 0.2007043 to:
   (a) 6 significant figures.
   (b) 5 significant figures.
   (c) 4 significant figures.
   (d) 3 significant figure.
   (e) 2 significant figures.
   (f) 1 significant figure.

Solution
(a) 0.2007043 to 6 significant figures = 0.200704.
(b) 0.2007043 to 5 significant figures = 0.20070.
(c) 0.2007043 to 4 significant figures = 0.2007.
(d) 0.2007043 to 3 significant figures = 0.201.
(e) 0.2007043 to 2 significant figures = 0.20.
(f) 0.2007043 to 1 significant figure = 0.2.

Notice that (b) and (c) are virtually the same and (e) and (f) are virtually the same.

Exercise 1:3

1. Complete the following table

	Number	Number of significant figures			
		1	2	3	4
a	0.0068398				
b	2.0068398				
c	4.69768				
d	1.006127				

2. Complete the following table

	Number	Number of Decimal places			
		1	2	3	4
a	0.0068398				
b	2.0068398				
c	4.69768				
d	1.006127				

3. Express to 3 decimal places.

| (a) 0.003646 | (b) 0.4567 | (c) 0.5046 |

4. Express to 2 decimal places.
   (a) 14.9028   (b) 23.1058   (c) 6.0381
5. Express to 3 significant figures.
   (a) 0.02485   (b) 4.027956
6. Express to 2 significant figures.
   (a) 547.53   (b) 59.81798   (c) 5382
7. Express to 1 significant figure.
   (a) 0.009238   (b) 5.097   (c) 0.2309

1.9 Standard Form

In science, very large or very small numbers occur. For instance, the mass of an election is 0.000,000,000,000,000,000,000,000,000,000,911 g; the velocity of electromagnetic waves in air is 300,000,000 m/s. We call this ordinary way of writing numbers the **normal or decimal form**. Writing numbers in this form, often lead to errors especially when there are so many zeros.

The **standard form** or **scientific notation** provides an easier way of writing such large or small numbers.

To express a number in standard form:

(a) Move the decimal point to the right of the first non-zero digit, counting the number of steps.

(b) Multiply the resulting number by 10 raised to the power corresponding to the number of steps moved.

(c) If the decimal point was originally to the left of its current position the power is negative and if it was originally to the right, the power is positive.

Summarily, a number in standard form or scientific notation is of the form

$$N = A \times 10^n, \text{where } n \in \mathbb{Z} \text{ and } 1 \leq |A| < 10$$

If $|N| < 1, n < 0$

If $|N| = 1, n = 0$

If $|N| > 1, n > 0$

Module 15, Topic 1: Estimations, Approximations and Errors

Example

Express the following in standard form
(a) 0.000,000,000,000,000,000,000,000,000,000,000,911
(b) 300,000,000   (c) 0, 00048

Solution
(a) $0.000,000,000,000,000,000,000,000,000,000,000,911 = 9.11 \times 10^{-34}$
(b) $300,000,000 = 3 \times 10^8$ (c) $0,00048 = 4.8 \times 10^{-4}$

Exercise 1:4

Express the following numbers in standard form.
(a) 5000   (b) 480   (c) 10200   (d) 700000
(e) 0.0032   (f) 0.000073   (g) 0.925   (h) 0.001
(i) 0.5600   (j) 3000×10^{-8} (k) 19.6×10^{-4} (l) 0.034×10^{-8}

1.10 Calculations Involving Standard Form

In performing calculations in standard form the powers of 10 are manipulated using the multiplication and division laws of indices. Thus,

$$10^n \times 10^m = 10^{n+m}$$
$$10^n \div 10^m = 10^{n-m}$$
$$(10^m)^n = 10^{mn}$$

Example

Evaluate the following leaving your answers in index form.
(a) $10^3 \times 10^5$ (b) $10^{-6} \times 10^4$ (c) $10^{-8} \times 10^{-3}$
(d) $10^3 \div 10^5$ (e) $10^{-6} \div 10^4$ (f) $10^{-8} \div 10^{-3}$

Solution
(a) $10^3 \times 10^5 = 10^{3+5} = 10^8$ (b) $10^{-6} \times 10^4 = 10^{-6+4} = 10^{-2}$
(c) $10^{-8} \times 10^{-3} = 10^{-8+(-3)} = 10^{-11}$ (d) $10^3 \div 10^5 = 10^{3-5} = 10^{-2}$
(e) $10^{-6} \div 10^4 = 10^{-6-4} = 10^{-10}$ (f) $10^{-8} \div 10^{-3} = 10^{-8-(-3)} = 10^{-5}$

Exercise 1:5

1. Evaluate giving your result in standard form.
   (a) $9.5 \times 10^7 - 3.08 \times 10^6$
   (b) $\dfrac{0.45 \times 0.91}{0.0117}$
   (c) 0.06×0.09
   (d) $\dfrac{0.24}{0.012}$
   (e) $\dfrac{8.75}{0.025}$
   (f) $\sqrt{\dfrac{0.81 \times 10^5}{2.25 \times 10^7}}$
   (g) $\dfrac{9.6 \times 10^8}{0.24 \times 10^5}$
   (h) $\dfrac{0.9687}{0.001}$
   (i) $\dfrac{0.203 \times 0.55}{3.05}$
   (j) $\sqrt{\dfrac{1.44 \times 10^5}{8.1 \times 10^4}}$
   (k) $\sqrt{\dfrac{0.0016 \times 0.0081}{0.36}}$
   (l) $\sqrt{\dfrac{76.42 \times 10^{-1}}{0.004 \times 10^2}}$

2. Given that, $a = 6 \times 10^3$, $b = 2 \times 10^{-4}$, $c = 4 \times 10^{-5}$. Find the value of $\dfrac{a \times b}{c}$ expressing your answer in the standard form.

3. The weight of a single atom of hydrogen is about 0.00000000000000000000000017 grams. Express this weight in standard form.

4. Evaluate $\dfrac{1946 \times 10^{-1}}{2 \times 10^2}$, giving the answer
   (a) In standard form.
   (b) Correct to 1 decimal place.
   (c) Correct to 2 significant figures.

5. Evaluate $\dfrac{12.78 \times 10^{-3}}{9 \times 10^{-1}}$, expressing your answer
   (a) In standard form
   (b) Correct to two significant figures
   (c) Correct to three decimal places

1.11 Calculations Involving Approximation

In calculations involving approximations, it is advisable to approximate at the end of the calculation. This reduces the amount of error.

 Example

Find $1.25 + 3.17$ correct to 2 significant figures.

Module 15, Topic 1: Estimations, Approximations and Errors

Solution

Correct Solution	Incorrect Solution
1.25	1.25 to 2 sig. fig. = 1.3
3.17	3.17 to 2 sig. fig. = 3.2
4.42	To 2 sig. fig. sum = 4.5
To 2 sig. fig. sum = 4.4	

1.12 Maximum and Minimum Values from a given Approximation

An approximated value has both a maximum and a minimum value.

 Example

1. The length of a line is 3.0 cm to 2 significant figures. State the maximum and minimum values of the length.

   Solution
   The largest number which can be rounded down to 3.0 cm is 3.049 cm.
   Therefore maximum length = 3.049 cm
   The smallest number which can be rounded up to 3.0 cm is 2.95 cm.
   Therefore minimum length = 2.95 cm

2. Given that all numbers are correct to one decimal place. Find the maximum and minimum values of 16.3 × 2.8.

   Solution
   The largest number which can be rounded down to 16.3 is 16.349.
   The largest number which can be rounded down to 2.8 is 2.849.
   Therefore maximum value = 16.349 × 2.849 ≈ 46.578301 = 46.6

   The smallest number which can be rounded up to 16.3 is 16.245.
   The smallest number which can be rounded up to 2.8 is 2.745.
   Therefore minimum value = 16.245 × 2.745 ≈ 44.592525 = 44.5

1.13 Absolute Error

Absolute error is the magnitude of the difference between the approximate value and Real value of a quantity.

$$\text{Absolute error} = |\text{Real value} - \text{Approximate value}|$$

Relative Error

Relative error is the ratio of the absolute error in a measurement to the true value.

$$\text{Relative error} = \frac{\text{Absolute error}}{\text{Real value}}$$

Percentage Error

Percentage error is the relative error expressed as a percentage.

$$\text{Petrcentage error} = \text{Relative error} \times 100\%$$

$$\text{Petrcentage error} = \frac{\text{Absolute error}}{\text{Real value}} \times 100\%$$

Example

The age of a man who is exactly 43 years 9 months old is written to the nearest year.

Calculate

(a) The absolute error in the approximation.

(b) The relative error in the approximation.

(c) The percentage error in the approximation.

Solution
Age of man = 43 years 9 months = 43.75 years

Age of man to the nearest year is 44 years.

(a) Absolute error = 44 − 43.75 = 0.25 years = 3 months.

(b) Relative error = $\frac{\text{absolute error}}{\text{Real value}} = \frac{0.25}{43.75} = 0.0057$

(c) The percentage error = $\frac{0.25}{43.75} \times 100\% = 0.57\%$

Module 15, Topic 1: Estimations, Approximations and Errors

 Exercise 1:6

1. The number of people attending a football match is 27000, correct to 2 significant figures. The greatest possible attendance shown by this figure is:
2. During a Physics practical, a student measured the length of a wire as 18 mm instead of 20 mm. What is the
   (a) Absolute error (b) relative error (c) percentage error, in the measurement of the wire?
3. A student measured the length of a pole as 34.5 m instead of 34.8 m. Calculate
   (a) the absolute error (b) the relative error (c) the percentage error, in the measurement of the pole.
4. To convert 4 km/h to m/s, a student obtained an answer 1 m/s. Calculate his percentage error.

 Multiple Choice Exercise 1

1. Given that $x = 0.0102$, correct to three significant figures. The value which cannot be the actual value of x, is:
   [A] 0.01021   [B] 0.01014   [C] 0.01015   [D] 0.01016
2. The number of people attending a football match is 27000, correct to 2 significant figures. The greatest possible attendance shown by this figure is:
   [A] 27,000   [B] 27499   [C] 27599   [D] 26999
3. After evaluating 2.35×0.48, the answer to 2 decimal places is:
   [A] 11.28   [B] 1.13   [C] 1.128   [D] 1.10
4. The value of $3.769 \div 0.7$ to the nearest tenth is:
   [A] 5.41   [B] 5.0   [C] 10   [D] 5.4
5. To the nearest whole number, the result of $\dfrac{6.6 \times 1.8}{5.4}$ is:
   [A] 2.2   [B] 3   [C] 2   [D] 22
6. $0.000252 \div 0.007$ to two decimal places is:
   [A] 0.04   [B] 0.03   [C] 0.36   [D] 0.40
7. By evaluating $\dfrac{7 + 3.32}{9.91 - 5.11}$, the answer to one decimal place will be:
   [A] 21.5   [B] 2.1   [C] 22.0   [D] 2.2
8. $0.44734 \div 0.01$, evaluated to the nearest hundredth is:
   [A] 44.70   [B] 45   [C] 44.73   [D] 44.00
9. $\dfrac{6.3 \times 60 \times 0.2}{3.6 \times 1.4}$, when simplified, the answer to the nearest ten is:
   [A] 15   [B] 20   [C] 10   [D] 1.5
10. 46×900 expressed in standard form is:
    [A] 4.14×10^3 [B] 4.14×10^5 [C] 4.14×10^4 [D] 4.14×10^6
11. In standard form converting 4 hours to seconds the result is:

[A] 1.44×10^3 [B] 1.44×10^{-3} [C] 1.44×10^4 [D] 1.44×10^{-4}

12. 258 km when expressed to mm and expressed in standard form becomes:
    [A] 2.58×10^8 [B] 2.58×10^7 [C] 2.58×10^6 [D] 2.58×10^5

13. $\dfrac{8.75}{0.025}$, expressed in standard form is:
    [A] 3.5×10^{-3} [B] 3.5×10^{-2} [C] 3.5×10^1 [D] 3.5×10^2

14. Given that $0.000208 = 2.08\times10^x$, the value of x is:
    [A] 4 [B] -4 [C] 5 [D] -5

15. The product of 0.06 and 0.09 in standard form is:
    [A] 5.4×10^2 [B] 5.4×10^{-3} [C] 5.4×10^1 [D] 5.4×10^{-2}

16. Evaluating $0.009 \div 0.012$, the answer in standard form is:
    [A] 7.5×10^2 [B] 7.5×10^{-3} [C] 7.5×10^{-1} [D] 7.5×10^{-2}

17. The sum of 728.93 and 0.46 expressed in standard form to three significant figures is:
    [A] 72.9×10 [B] 7.29×10^2 [C] 728×10^0 [D] 7.28×10^0

18. The sum of 2.48×10^3 and 5.9×10^4 is:
    [A] 6.148×10^1 [B] 6.148×10^4 [C] 6.148×10^3 [D] 6.148×10^2

19. $4\times10^2 \times 2\times10^{-4}$ is equal to:
    [A] 8×10^6 [B] 8×10^{-2} [C] 8×10^{-8} [D] 6×10^{-2}

20. $7.580\times10^9 + 7.677\times10^9$ is equal to:
    [A] 1.5257×10^{10} [B] 1.5257×10^8 [C] 1.5257×10^9 [D] 1.5257×10^7

21. Given that $x=5.7\times10^6, y=1.8\times10^6$, $x-y=$:
    [A] 3.9×10^{-6} [B] 3900000×10^6 [C] 3.9×10^6 [D] 39×10^5

22. The population of two towns A and B are 5.77×10^6 and 3.66×10^6 respectively. The difference in the population of A and B is:
    [A] 2.11×10^0 [B] 2.11×10^{-2} [C] 2.11×10^4 [D] 2.11×10^6

23. The result of $7.42\times10^{-6} - 4.33\times10^{-7}$ is:
    [A] 6.987×10^{-6} [B] 5.987×10^{-6} [C] 4.987×10^{-6} [D] 3.987×10^{-6}

24. $5.72\times10^3 - 2.37\times10^2$ in the normal form, is equal to:
    [A] 4483 [B] 4473 [C] 5483 [D] 3450

25. The product of 0.012 and 0.0008 in standard form is:
    [A] 9.6×10^6 [B] 9.6×10^{-6} [C] 9.6×10^5 [D] 9.6×10^{-5}

26. When simplified, $(2\times10^{-5})\times(4\times10^{-2})$ becomes:
    [A] 8.0×10^{-7} [B] 8.0×10^7 [C] 8.0×10^3 [D] 8.0×10^{-3}

27. On simplification $8\times10^5 + 2\times10^3$ gives:

[A] 4×10^8 [B] 4×10^2 [C] $4 \times 10^{\frac{5}{3}}$ [D] $4 \times 10^{-\frac{5}{3}}$

28. 450×70 is:

[A] 3.15×10^5 [B] 3.15×10^4 [C] 3.15×10^3 [D] 3.15×10^2

29. The result of $\dfrac{0.126}{36}$ is:

[A] 3.5×10^4 [B] 3.5×10^{-4} [C] 3.5×10^3 [D] 3.5×10^{-3}

30. The simplified value of $\dfrac{(12 \times 10^8)(16 \times 10^{-6})}{1 \times 10^{-2}}$ in standard form is:

[A] 1.92×10^6 [B] 192×10^4 [C] 1.92×10^4 [D] 1.92×10^2

31. On evaluation $\sqrt{\dfrac{0.81 \times 10^{-5}}{2.25 \times 10^7}}$ equals:

[A] 3.6×10^{-13} [B] 6.0×10^{-7} [C] 3.6×10^{13} [D] 6.0×10^7

32. Given that $p = 3.6 \times 10^{-3}$ and $q = 2.25 \times 10^6$, the value of $\sqrt{\dfrac{p}{q}}$ is:

[A] 1.6×10^{-9} [B] 4.0×10^{-5} [C] 1.6×10^9 [D] 4.0×10^5

33. While doing his Physics practical, Che recorded a reading as 1.12 cm, instead of 1.21 cm. The percentage error he made is:
    [A] 1.17% [B] 6.38% [C] 7.44% [D] 8.5%

34. Miss Yaje sold an article for 7500 FRS instead of 12750 FRS. Her percentage error correct to one decimal place is:
    [A] 41.2% [B] 5.3% [C] 1.7% [D] 1.4%

35. A student measured the length and breadth of a rectangular lawn as 59.6 cm and 40.3 cm respectively instead of 60 cm and 40 cm. The percentage error in his calculation of the perimeter of the lawn is:
    [A] 1.4% [B] 0.1% [C] 0.2% [D] 0.7%

36. A boy estimated his transport fare for a journey as 1900 FRS instead of 2000 FRS. The percentage error in his estimate is:
    [A] 95% [B] 47.5% [C] 5.26% [D] 5%

37. To two significant figures, the percentage error in approximating 0.375 to 0.4 is:
    [A] 6.7% [B] 6.6% [C] 2.5% [D] 2.0%

Topic 2

SURDS, INDICES AND LOGARITHMS

Objectives

At the end of this topic, the learner should be able to:

1. Differentiate between rational and irrational numbers.
2. Simplify expressions involving surds.
3. Rationalize the denominators of surd expressions.
4. Apply laws of indices.
5. Solve simple exponential equations.
6. Apply properties of indices to find values.
7. State and apply properties of logarithms to find quantities.
8. Change logarithmic expressions from one base to another.

Module 15, Topic 2: Surds, Indices and Logarithms

This topic is an extension of form 3 work. The learner is advised to revise module 10 section 1.1 of topic 1 and topic 2.

1.14 Rational and Irrational Numbers

> **? Brainstorming Exercise**
>
> 1. What is a rational number?
> 2. State the symbol used to denote the set of rational numbers.
> 3. List some examples of elements of the set of rational numbers.
> 4. What is an irrational number?
> 5. State the symbol used to denote the set of rational numbers.
> 6. List some examples of elements of the set of rational numbers.

Recall that:
1. A **rational number** is a number which can be expressed as a quotient of two integers. The set of rational numbers is denoted by $\mathbb{Q}$. Examples of rational numbers are $\frac{3}{4}, -\frac{5}{8}, \frac{7}{2}, 1, 0, -6, 1\frac{3}{8}$.
2. Irrational numbers are numbers that cannot be expressed as the quotient or ratio of two integers. We denote irrational numbers by $\mathbb{Q}'$. Examples of irrational numbers include $\pi, \sqrt{2}, \sqrt{7}$.

SURDS

2.1 Concept of a surd

A surd is an expression, which involve roots. Examples of irrational numbers include $3 + \sqrt{2}, \sqrt{5} - 1, \sqrt[3]{4}$.

2.2 Laws of Surds

> **Investigative Activity**
>
> 1. Evaluate the following. (i) $\sqrt{9}$ (ii) $\sqrt{25}$
>   (iii) $\sqrt{9} \times \sqrt{25}$ (iv) $\sqrt{9 \times 25}$ (v) $\sqrt{\frac{9}{25}}$ (vi) $\frac{\sqrt{9}}{\sqrt{25}}$
> 2. Given that $a \neq 0$ and $b \neq 0$. What conclusions do you draw concerning $\sqrt{a \times b}$ and $\sqrt{\frac{a}{b}}$?

3. Evaluate the following. (i) $\left(\sqrt[3]{8}\right)^2$ (ii) $\sqrt[3]{8^2}$
4. What conclusions do you draw concerning $\left(\sqrt[3]{8}\right)^2$ and $\sqrt[3]{8^2}$?
5. Evaluate (i) $\sqrt[6]{64}$ (ii) $\sqrt[2\times3]{64}$ (iii) $\sqrt[3]{\sqrt{64}}$ (iv) $\sqrt{\sqrt[3]{64}}$
6. Given that $a \neq 0$. What conclusions do you draw concerning $\sqrt[mn]{a}$, $\sqrt[m]{\sqrt[n]{a}}$ and $\sqrt[n]{\sqrt[m]{a}}$?

From the above investigations, we see that:

1. $\sqrt[n]{ab} = \sqrt[n]{a} \times \sqrt[n]{b}$
2. $\sqrt[n]{\dfrac{a}{b}} = \dfrac{\sqrt[n]{a}}{\sqrt[n]{b}}$
3. $\sqrt[n]{a^m} = \left(\sqrt[n]{a}\right)^m$
4. $\sqrt[mn]{a} = \sqrt[m]{\sqrt[n]{a}} = \sqrt[n]{\sqrt[m]{a}}$

Note!!

$\sqrt[n]{a} + \sqrt[n]{b} \neq \sqrt[n]{a+b}$ and $\sqrt[n]{a} - \sqrt[n]{b} \neq \sqrt[n]{a-b}$

The above rules are very useful in manipulating surds and we need to be aware of them.

Example

Evaluate the following.

(a) $\sqrt{1296}$ (b) $\sqrt{\dfrac{342225}{38025}}$ (c) $\sqrt[6]{729}$ (d) $\sqrt[6]{4096}$

Solution

(a) $\sqrt{1296} = \sqrt{(16)(81)} \Rightarrow \sqrt{1296} = \left(\sqrt{16}\right)\left(\sqrt{81}\right) = 4 \times 9 = 36$

(b) $\sqrt{\dfrac{342225}{38025}} = \sqrt{\dfrac{25 \times 81 \times 169}{25 \times 9 \times 169}} = \sqrt{9} = 3$

(c) $\sqrt[6]{729} = \sqrt[6]{9 \times 9 \times 9} = \sqrt[6]{3^6} = 3$

(d) $\sqrt[6]{4096} = \sqrt[6]{64 \times 64} = \sqrt[6]{4^6} = 4$

Module 15, Topic 2: Surds, Indices and Logarithms

2.3   The Conjugate of a Surd

The **conjugate** of a surd is the surd resulting from reversing the sign of the radical part of the surd. The conjugate of $a + \sqrt{b}$ is $a - \sqrt{b}$.

To multiply a surd by its conjugate, we can take advantage of the expansion of the factors of the difference of two squares. Thus,

$$\left(a+\sqrt{b}\right)\left(a-\sqrt{b}\right) = a^2 - \left(\sqrt{b}\right)^2 = a^2 - b$$

 Example

Evaluate the following.

(a) $(3 + \sqrt{2})(3 - \sqrt{2})$ (b) $(5 - \sqrt{7})(5 + \sqrt{7})$

Solution
(a) $(3 + \sqrt{2})(3 - \sqrt{2}) = 3^2 - 2 = 7$
(b) $(5 - \sqrt{7})(5 + \sqrt{7}) = 5^2 - 7 = 25 - 7 = 18$

2.4   Rationalizing the Denominator

We call the process by which we remove the radical sign from the denominator of a surd expression in order to simplify or evaluate it as **rationalizing the denominator**. Usually when the denominator is of the form $a + \sqrt{b}$ or $a - \sqrt{b}$ we multiply both numerator and denominator by the conjugate.

 Example

1. Simplify   (a) $\dfrac{2}{\sqrt{3}}$ (b) $\dfrac{1}{\sqrt{5}}$

   Solution
   (a) $\dfrac{2}{\sqrt{3}} = \dfrac{2}{\sqrt{3}} \times \dfrac{-\sqrt{3}}{-\sqrt{3}} = \dfrac{2\sqrt{3}}{3}$ (b) $\dfrac{1}{\sqrt{5}} = \dfrac{1}{\sqrt{5}} \times \dfrac{-\sqrt{5}}{-\sqrt{5}} = \dfrac{\sqrt{5}}{5}$

2. Rationalize the denominator of $\dfrac{\sqrt{7}}{2\sqrt{7} - 5}$

Solution

$$\frac{\sqrt{7}}{2\sqrt{7}-5} = \frac{\sqrt{7}}{2\sqrt{7}-5}\left(\frac{-2\sqrt{7}-5}{-2\sqrt{7}-5}\right) = \frac{2(7)+5\sqrt{7}}{4(7)-25} = \frac{14+5\sqrt{7}}{3}$$

2.5 Surd Equations

Surd equations or radical equations are equations involving surds. We solve radical equations by applying the laws of surds.

Example

Solve the following equations

(a) $\sqrt{2x-1} = 3$

(b) $\sqrt{4x+1} = x-5$

Solution

(a) $\sqrt{2x-1} = 3$

Squaring both sides

$2x-1 = 9$

$2x = 10$

$x = 5$

(b) $\sqrt{4x+1} = x-5$

Squaring both sides

$4x+1 = (x-5)^2$

$4x+1 = x^2 - 10x + 25$

$x^2 - 14x + 24 = 0$

$(x-12)(x-2) = 0$

$x = 12$ or $x = 2$

Exercise 2:1

1. Evaluate the following.

   (a) $3\sqrt{4} + \sqrt{25} - 2\sqrt{9}$

   (b) $\sqrt{\frac{25}{9}} + \sqrt{\frac{64}{9}}$

   (c) $\sqrt[4]{\frac{81}{4}}$

   (d) $\frac{1}{3}\sqrt{36} + \frac{2}{5}\sqrt{100}$

2. Rationalize the denominator in each of the following.

   (a) $\frac{12}{\sqrt{6}}$

   (b) $\frac{\sqrt{3}}{\sqrt{10}}$

   (c) $\frac{24}{\sqrt{8}}$

   (d) $\frac{3+\sqrt{2}}{\sqrt{2}}$

(e) $\dfrac{3}{\sqrt{3}}$ (f) $\dfrac{c^3}{\sqrt{c^3}}$ (g) $\dfrac{7}{\sqrt{7}}$ (h) $\dfrac{20}{\sqrt{50}}$

3. Solve the following equations
   (a) $\sqrt{7x+5} = 3$ (b) $2\sqrt{x-8} = 3$
   (c) $\sqrt{8-4x} = x+1$ (d) $\sqrt{-x-5} = x+5$

4. Find the value of each of the following
   (a) $\sqrt{81} - \sqrt{25}$ (b) $2\sqrt{9} + 3\sqrt{4}$ (c) $\sqrt{\dfrac{4}{9}} + \sqrt{\dfrac{1}{9}}$
   (d) $\sqrt[8]{\dfrac{1}{16}} + \sqrt[4]{\dfrac{9}{16}}$ (e) $\sqrt{16} + \sqrt{1600} + \sqrt{160,000}$
   (f) $\left(-\sqrt{\dfrac{4}{9}}\right)\left(-\sqrt{\dfrac{81}{100}}\right)$

5. Simplify the following.
   (a) $\dfrac{1}{5-\sqrt{3}} + \dfrac{1}{5+\sqrt{3}}$ (b) $\sqrt[3]{216}$ (c) $\sqrt[3]{\dfrac{9}{27}}$ (d) $\sqrt[3]{64^2}$

6. Rationalize the denominator in each of the following.
   (a) $\dfrac{1}{2\sqrt{3}-1}$ (b) $\dfrac{1}{\sqrt{3}-\sqrt{2}}$

7. Physicians often use an index called BSA to calculate the approximate body surface area of an adult. According to this index, the BSA of an adult is given by $B = \sqrt{\dfrac{HW}{3600}}$ where B is the body surface area in square centimeters; H is the height in centimeters and W is the weight in kilograms. Find the BSA of an adult who is
   (a) 175 cm tall and weighs 72 kg.   (b) 160 cm tall and weighs 70 kg.

THEORY OF INDICES

2.6 Laws of Indices

We already treated the laws of indices in detail in section 2.2 of module 10, topic 2. Below is an outline of the laws as a reminder.

If a and b are non-zero real numbers and m and n are integers then:
1. **Exponent 1 Law:** $a^1 = a$.
2. **Exponent 0 Law:** $a^0 = 1$.
3. **Multiplication Law:** $a^m \times a^n = a^{m+n}$.

4. **Division Law:** $a^m \div a^n = a^{m-n}$.
5. **Negative Index Law:** $a^{-n} = \frac{1}{a^n}$.
6. **Product index law:** $(a^m)^n = (a^n)^m = a^{mn}$
7. **Fractional index laws:** $a^{\frac{m}{n}} = \left(\sqrt[n]{a}\right)^m = \sqrt[n]{a^m}$.
8. **Product of Numbers with the same Power:** $a^m b^m = (ab)^m$

2.7 Exponential or Index Equations

In module 10, topic 2, section 2.3, we introduced exponential equations. We shall embark on more challenging exponential equations.

 Example

1. Find the value of x for which $4^{x+1} - 9(2^x) = -2$.

   Solution

   $4^{x+1} - 9(2^x) = -2 \Rightarrow 4^x \times 4^1 - 9(2^x) = -2$
   $\Rightarrow 4(4^x) - 9(2^x) + 2 = 0 \Rightarrow 4(2^x)^2 - 9(2^x) + 2 = 0$
   Let $y = 2^x \Rightarrow 4y^2 - 9y + 2 = 0$
   $\qquad 4y^2 - 8y - y + 2 = 0$
   $\qquad 4y(y - 2) - 1(y - 2) = 0$
   $\qquad (4y - 1)(y - 2) = 0$
   $\qquad\qquad 4y = 1 \text{ or } y = 2$
   $\qquad\qquad y = \frac{1}{4} \text{ or } y = 2$

   Substituting for y;
   $\qquad 2^x = \frac{1}{4} \text{ or } 2^x = 2$
   $\qquad 2^x = 2^{-2} \text{ or } 2^x = 2^1$
   The values of x are $x = -2$ or $x = 1$

2. Solve the simultaneous equations $2^{x+y} = 8$; $3^{2x-y} = 27$

   Solution
   $2^{x+y} = 8 \Rightarrow 2^{x+y} = 2^3$ which leads to $x + y = 3$ .... ①
   $3^{2x-y} = 27 \Rightarrow 3^{2x-y} = 3^3$ which leads to $2x - y = 3$ .... ②
   ①+②: $3x = 6 \Rightarrow x = 2$ and $2 + y = 3 \Rightarrow y = 1$

Module 15, Topic 2: Surds, Indices and Logarithms

 Exercise 2:2

Simplify the following.

1. $\dfrac{81^{0.25}}{32^{0.2}}$

2. $\left(\dfrac{4}{25}\right)^{-\frac{1}{2}} \times 2^4 \div \left(\dfrac{15}{2}\right)^{-2}$

3. $5\dfrac{2}{5} \times \left(\dfrac{2}{3}\right)^2 \div \left(1\dfrac{1}{2}\right)^{-1}$

4. $\left(\dfrac{1}{343}\right)^{\frac{1}{3}} + 64^{\frac{1}{3}} - \left(\dfrac{4}{9}\right)^{-\frac{1}{2}}$

5. $27^{\frac{1}{3}} \times (64)^{\frac{1}{3}} - \left(\dfrac{4}{9}\right)^{-\frac{1}{2}}$

6. $27^{-\frac{1}{3}} \times 64^{-\frac{1}{2}} \times 4^{\frac{1}{2}}$

7. $125^{-\frac{1}{3}} \times 64^{-\frac{1}{3}} \times 81^0$

Solve the following equations.

8. $9 \times 3^{3+x} = 27^{-x}$

9. $3 \times 9^{1+x} = 27^{-x}$

10. $9^{2x+1} = \dfrac{81^{x-1}}{3^x}$

11. $\dfrac{9^{2x-1}}{3^{x+3}} = 1$

12. $9^{2x-1} \times 3^{x+1} = 27^{x+3}$

13. $27^{2x+1} = 3^{x-1}$

14. $3^{y+x} = 9^{y-x}$ and $2^{x-y} = 8^{x-3}$.

THEORY OF LOGARITHMS

 Review

1. Write the statement $x^y = z$ in logarithmic form
2. Write the statement $\log_x z = y$ in exponential form
3. State the values of (a) $\log_b b$ (b) $\log_b b^n$
4. Solve the equation $\log_3 81 = x + 1$.

2.8 Laws of Logarithms

The Addition Law of logarithms

$$\log_b xy = \log_b x + \log_b y$$

23

Proof

Let $\log_b x = t$, $\log_b y = u$ and $\log_b xy = v$

Then by definition of logarithms

$$x = b^t \quad \text{............................①}$$

$$y = b^u \quad \text{............................②}$$

and $xy = b^v \quad \text{............................③}$

Multiply equation ① by equation ②

$$xy = (b^t)(b^u)$$

Using the multiplication law of indices

$$xy = b^{t+u} \quad \text{.......................④}$$

Substituting equation ④ in equation ③

$$b^v = b^{t+u}$$

Comparing powers

$$v = t + u$$

$$\Rightarrow \log_b xy = \log_b x + \log_b y$$

The Subtraction Law of logarithms

$$\log_b \left(\frac{x}{y}\right) = \log_b x - \log_b y$$

Proof

Let $\log_b x = t$, $\log_b y = u$ and $\log_b \left(\frac{x}{y}\right) = w$

Then by definition of logarithms

$$x = b^t \quad \text{............................①}$$

$$y = b^u \quad \text{............................②}$$

and $\dfrac{x}{y} = b^w \quad \text{............................③}$

Dividing equation ① by equation ②

$$\frac{x}{y} = \frac{b^t}{b^u}$$

Using the division law of indices

$$\frac{x}{y} = b^{t-u} \quad \text{................④}$$

Substituting equation ④ in equation ③
$$b^w = b^{t-u}$$
Comparing powers,
$$w = t - u$$
$$\Rightarrow \log_b\left(\frac{x}{y}\right) = \log_b x - \log_b y$$

The Exponential Law of Logarithms

$$\log_b x^n = n \log_b x$$

Proof

Let $\log_b x = t$, and $\log_b x^n = r$

Then by definition of logarithms
$$x = b^t \quad \text{................①}$$
$$\text{and } x^n = b^r \quad \text{................②}$$

Substituting equation ① in equation ②
$$b^{nt} = b^r$$
Comparing powers, $r = nt$
Since $r = \log_b x^n$ and $t = \log_b x$
$\log_b x^n = n \log_b x$ as required.

The Logarithm of 1

$$\log_b 1 = 0$$

The logarithm of 1 to any base is zero.

Proof

Let $\log_b 1 = p \Rightarrow$ by definition of logarithms $b^p = 1$.
But $b^0 = 1 \Rightarrow b^p = b^0$
Equating exponents, $p = 0 \Rightarrow \log_b 1 = 0$.

Example

Without using tables or calculators, simplify the following.

1. $\dfrac{\log_{10} \sqrt[3]{13}}{\log_{10} 13}$
2. $\log_{10} 75 + 2\log_{10} 2 - \log_{10} 3$
3. $\log_{10}\left(\dfrac{15}{8}\right) - 2\log_{10}\left(\dfrac{5}{9}\right) + \log_{10}\left(\dfrac{400}{243}\right)$
4. $\dfrac{\log_{10} 6 - \log_{10} 3}{\log_{10} 8 - \log_{10} 4} \div \dfrac{\log_{10} 5}{\log_{10} 0.2}$
5. $\log_{10} \sqrt{35} + \log_{10} \sqrt{2} - \log_{10} \sqrt{7}$

Solutions

1. $\dfrac{\log_{10} \sqrt[3]{13}}{\log_{10} 13} = \dfrac{\log_{10}(13)^{\frac{1}{3}}}{\log_{10} 13} = \dfrac{\frac{1}{3}\log_{10} 13}{\log_{10} 13} = \dfrac{1}{3}$

2. $\log_{10} 75 + 2\log_{10} 2 - \log_{10} 3 = \log_{10}\left(\dfrac{75}{3}\right) + \log_{10} 2^2 = \log_{10}(25)(4) = \log_{10} 100 = 2$

3. $\log_{10}\left(\dfrac{15}{8}\right) - 2\log_{10}\left(\dfrac{5}{9}\right) + \log_{10}\left(\dfrac{400}{243}\right) = \log_{10}\left(\dfrac{15}{8}\right)\left(\dfrac{400}{243}\right) - \log_{10}\left(\dfrac{5}{9}\right)^2$

$= \log_{10}\left(\dfrac{15}{8}\right)\left(\dfrac{400}{243}\right)\left(\dfrac{81}{25}\right) = \log_{10} 10 = 1$

4. $\dfrac{\log_{10} 6 - \log_{10} 3}{\log_{10} 8 - \log_{10} 4} \div \dfrac{\log_{10} 5}{\log_{10} 0.2} = \dfrac{\log_{10}\left(\frac{6}{3}\right)}{\log_{10}\left(\frac{8}{4}\right)} \times \dfrac{\log_{10} 0.2}{\log_{10} 5}$

$= \dfrac{\log_{10} 2}{\log_{10} 2} \times \dfrac{\log_{10}\left(\frac{1}{5}\right)}{\log_{10} 5} = \dfrac{\log_{10} 5^{-1}}{\log_{10} 5} = \dfrac{-1\log_{10} 5}{\log_{10} 5} = -1$

5. $\log_{10} \sqrt{35} + \log_{10} \sqrt{2} - \log_{10} \sqrt{7} = \log_{10}\left(\dfrac{(35)(2)}{(7)}\right)^{\frac{1}{2}} = \dfrac{1}{2}\log_{10} 10 = \dfrac{1}{2}$

2.9 Change of Base Formula

$$\log_b x = \frac{\log_a x}{\log_a b}$$

Proof

Let $\log_b x = t$
Then by definition of logarithms
$$x = b^t$$
Taking the log to the base a of both sides
$$\log_a b^t = \log_a x$$
Applying the exponential law
$$t \log_a b = \log_a x$$
$$t = \frac{\log_a x}{\log_a b}$$
$$\Rightarrow \log_b x = \frac{\log_a x}{\log_a b}$$

We can use the change of base formula to change the logarithm of a number to one base b to the logarithm of the number to another base a

 Example

Given that $\log_{10} 3 = 0.4771$ and $\log_{10} 4 = 0.6021$. Find the value of $\log_4 3$ to three significant figures.

Solution

$$\log_4 3 = \frac{\log_{10} 3}{\log_{10} 4} = \frac{0.4771}{0.6021} = 0.792$$

Exercise 2:3

(1) Simplify the following without using tables or calculators.

(a) $\dfrac{\log_{10} 81}{\log_{10} \frac{1}{3}}$

(b) $\dfrac{\log_{10} 8}{\log_{10} 12 - \log_{10} 3}$

(c) $\log_{10} 10 - 2 \log_{10}\left(\dfrac{1}{5}\right) - \log_{10} 2.5$

(d) $\log_2 8 - \log_2 0.25 + 2 \log_2 \dfrac{1}{8}$

(e) $\log_{10} 64 + \log_{10} 25 - 2 \log_{10} 4$ (f) $\log_2 8 - \log_3\left(\dfrac{1}{27}\right)$

(2) Given that $\log_{10} 8 = 0.9030$, write down without the use of tables, the values of (a) $\log_{10} 64$ (b) $\log_{10} 2$ (c) $\log_{10} \sqrt{2}$

(3) Given that $\log 3 = 0.4771$ and $\log 2 = 0.3010$, find the values of the following without using tables or calculators.

(i) $\log_{10} \sqrt{6}$ (ii) $\log_{10} \sqrt[3]{0.3}$

(4) Given that $\log_{10} 5 = 0.699$ and $\log_{10} 3 = 0.477$, find $\log_{10} 45$ without using tables or calculators.

Multiple Choice Exercise 2

1. The result of simplifying $\dfrac{4^{-\frac{1}{2}} \times 16^{\frac{3}{4}}}{4^{\frac{1}{2}}}$ is:

   [A] $\dfrac{1}{4}$ [B] 0   [C] 1   [D] 2

2. On evaluation, the result of $\dfrac{27^{\frac{1}{3}}}{16^{-\frac{1}{4}}}$ is:

   [A] 6   [B] 2   [C] 4   [D] 3

3. $16^{\frac{5}{4}} \times 2^{-3} \times 3^0$ is equal to:

   [A] 20   [B] 2   [C] 4   [D] 10

4. The result of evaluating $0.027^{-\frac{1}{2}}$ is:

   [A] $3\dfrac{1}{3}$ [B] 3   [C] $\dfrac{3}{10}$ [D] $\dfrac{1}{3}$

Module 15, Topic 2: Surds, Indices and Logarithms

5. $\dfrac{8^{\frac{2}{3}} \times 27^{-\frac{1}{3}}}{64^{\frac{1}{3}}}$ simplifies to:

   [A] $\dfrac{1}{3}$ [B] $\dfrac{1}{9}$ [C] $\dfrac{16}{3}$ [D] $\dfrac{27}{8}$

6. After evaluating $5\dfrac{2}{3} \times \left(\dfrac{2}{3}\right)^2 \div \left(1\dfrac{1}{2}\right)^{-1}$, the result is:

   [A] $\dfrac{12}{5}$ [B] $\dfrac{8}{5}$ [C] $3\dfrac{3}{5}$ [D] $4\dfrac{1}{8}$

7. The result of simplifying $\dfrac{2^{\frac{1}{2}} \times 8^{\frac{1}{2}}}{4}$ is:

   [A] 1   [B] 2   [C] 4   [D] 16

8. $\left(\dfrac{1}{4}\right)^{-1\frac{1}{2}}$ is equal to:

   [A] 8   [B] 4   [C] $\dfrac{1}{4}$ [D] $\dfrac{1}{16}$

9. The result of evaluating $\left(\dfrac{16}{81}\right)^{-\frac{3}{4}} \times \left(\dfrac{100}{81}\right)$ is:

   [A] $\dfrac{80}{243}$ [B] $\dfrac{20}{27}$ [C] $\dfrac{25}{6}$ [D] $\dfrac{15}{4}$

10. When $10a^6$ is divided by $5a^3$ the result is:
    [A] $3a^3$ [B] $2a$ [C] a^3 [D] $2a^3$

11. If $2^x \times 3^2 = 144$, the value of x is:
    [A] 7   [B] 5   [C] 4   [D] 8

12. If $x^2 \times 3^2 \times 1^2 = 144$, the value of x is:
    [A] -4 [B] 2   [C] -2 [D] 16

13. If $3^x + 6 = 87$, the value of x is:
    [A] 1   [B] 2   [C] 3   [D] 4

14. If $3^{2x} = 27$ the value of x is:
    [A] 1   [B] 1.5   [C] 4.5   [D] 18

15. The value of t for which $\dfrac{64}{27} = \left(\dfrac{3}{4}\right)^{t-1}$ is:

    [A] -4 [B] 2   [C] 4   [D] -2

16. Given that $27^{(1+x)} = 9$. The value of x is:

[A] −3   [B] $-\frac{1}{3}$ [C] $\frac{1}{3}$ [D] 2

17. If $16(4)^{2x} = \left(\frac{1}{2}\right)^x$, the value of x is:

[A] −3   [B] $-\frac{4}{5}$ [C] $-\frac{4}{3}$ [D] $\frac{4}{3}$

18. $\left(\frac{1}{4}\right)^{2-y} = 1$, the value of y is:

[A] −2   [B] $-\frac{1}{2}$ [C] $\frac{1}{2}$ [D] 2

19. The solution of the equation $2\sqrt{x} = 4$ is:
   [A] −2   [B] 2   [C] 4   [D] 6

20. The value of x for which $2^{-6x} = 8^{(1-x)}$ is true is:

[A] $-\frac{7}{3}$ [B] $\frac{1}{3}$ [C] −1   [D] $\frac{7}{9}$

21. The value of n which satisfies the equation $\frac{3^{(1-n)}}{9^{-2n}} = \frac{1}{9}$ is:

[A] $-\frac{3}{2}$ [B] $\frac{1}{3}$ [C] −3   [D] 1

22. When simplified $\frac{1}{4}(2^n - 2^{n+2})$ becomes:
   [A] $2^n - 2^n$ [B] $2^{n-2}(1 - 2^n)$ [C] $2^{n+2} + 2$ [D] 2^n

23. When $56x^{-4} \div 14x^{-8}$ is simplified the result is:
   [A] $2x^{-12}$ [B] $4x^{-4}$ [C] $4x^{+4}$ [D] $4x^{-3}$

24. Given that $81 \times 2^{2n-2} = k$, the value of $\sqrt{k}$ which satisfies this equation is:
   [A] 4.5×2^{2n} [B] 4.5×2^{2n} [C] $9 \times 2^{n-1}$ [D] 9×2^n

25. If $\log_{10} 5.4440 = 0.7359$, then $\log_{10} 54440$ is:
   [A] 3.7359   [B] 4.7359   [C] 5.7359   [D] 6.7359

26. If $\log_{10} x = 2.8765$, the value of x is:
   [A] Greater than 100   [B] between 1 and 10
   [C] less than 10   [D] between 10 and 100

27. On simplification $\frac{\log \sqrt{8}}{\log 8}$ is equal to:

[A] $\frac{1}{3}$ [B] $\frac{1}{3}\log \sqrt{8}$ [C] $\frac{1}{3}\log \sqrt{2}$ [D] $\frac{1}{2}$

28. Simplifying $\frac{\log 27^{\frac{1}{3}}}{\log 81}$ gives:

Module 15, Topic 2: Surds, Indices and Logarithms

[A] $\frac{1}{4}$ [B] $\frac{3}{8}$ [C] $\frac{1}{2}$ [D] $\frac{3}{4}$

29. Simplifying $\log_7 8 - \log_7 2 + \log_7 4$, gives:
    [A] 0 [B] 2 [C] $2\log_2 7$ [D] $4\log_7 2$

30. The value of $\log_{10} 5 + \log_{10} 20$ is:
    [A] 2 [B] 3 [C] 4 [D] 5

31. If $\log_{10} 2 = 0.3010$ and $\log_{10} 2^y = 1.8062$. The value of y to the nearest whole number is:
    [A] 4 [B] 6 [C] 5 [D] 2

32. If $3\log_{10} a = \log_{10} 64$, the value of a is:
    [A] 4 [B] 6 [C] 8 [D] 16

33. Given that $\log p = 2\log x + 3\log q$. The correct expression of p in terms of x and q is:
    [A] $p = 6xq$ [B] $p = x^2 q^3$ [C] $p = x^2 + q^3$ [D] $p = 2x + 3q$

34. The solution of the equation $\log_8 x - 4\log_8 x = 2$ is:
    [A] $\frac{1}{4}$ [B] $\frac{1}{2}$ [C] 4 [D] 2

35. $7^{x-1} = \log_5 5$, then x is equal to:
    [A] 1 [B] 7 [C] -1 [D] -7

36. Given that $\frac{1}{3}\log_{10} p = 1$. The value of p is:
    [A] 3 [B] 10 [C] 100 [D] 1000

37. $\log_2 a = \log_8 4$ only if a is equal to:
    [A] $2^{\frac{1}{2}}$ [B] $2^{\frac{2}{3}}$ [C] $4^{\frac{2}{3}}$ [D] $4^{\frac{1}{3}}$

38. If $\log_a x = p$, then in terms of a and p, x is equal to:
    [A] a^p [B] $\frac{a}{p}$ [C] p^a [D] ap

39. The value of p for which $\frac{1}{2}\log_{10} p = 1$ is true is:
    [A] 10^{-1} [B] 10^3 [C] 10^2 [D] 10^1

40. Given that $\log_4 x = -3$. The value of x is:
    [A] $\frac{1}{81}$ [B] $\frac{1}{64}$ [C] 64 [D] 81

41. On simplification $\frac{4\sqrt{18}}{\sqrt{8}}$ becomes:
    [A] 2 [B] 3 [C] 6 [D] 12

42. The value of $\sqrt{96} + \sqrt{54} - \sqrt{24}$ is:
    [A] $\sqrt{6}$ [B] $2\sqrt{6}$ [C] $3\sqrt{6}$ [D] $5\sqrt{6}$

43. $\sqrt{32} - \sqrt{98} + 5\sqrt{2}$ is equal to:
    [A] $\frac{1}{2}\sqrt{2}$ [B] $2\sqrt{2}$ [C] $3\sqrt{2}$ [D] $4\sqrt{2}$

44. The value of $\sqrt{12} + 10\sqrt{3} - \frac{6}{\sqrt{3}}$ is:

[A] $7\sqrt{3}$ [B] $10\sqrt{3}$ [C] $14\sqrt{3}$ [D] $18\sqrt{3}$

45. Given that $\frac{1}{\sqrt{2}} = 0.7071$. Then, $\frac{3\sqrt{2}}{2}$ is greater than $\frac{1}{\sqrt{2}}$ by:
    [A] -3 [B] -1.4142 [C] 1.4142 [D] 3

46. Evaluating $\sqrt{20} \times (\sqrt{5})^3$ gives:
    [A] 10 [B] 20 [C] 25 [D] 50

47. If $k\sqrt{28} + \sqrt{63} - \sqrt{7} = 0$, the value of k is:
    [A] -2 [B] -1 [C] 1 [D] 2

48. If $\sqrt{128} + \sqrt{18} - \sqrt{k} = 7\sqrt{2}$, then k must be:
    [A] 8 [B] 16 [C] 32 [D] 48

49. When the denominator is rationalized, $\frac{10}{\sqrt{32}}$ becomes:
    [A] $\frac{5}{4}\sqrt{2}$ [B] $\frac{4}{5}\sqrt{2}$ [C] $\frac{5}{16}\sqrt{2}$ [D] $\frac{16}{5}\sqrt{2}$

50. The number $\frac{6}{\sqrt{2}}$ is equal to:
    [A] $4\sqrt{2}$ [B] $3\sqrt{2}$ [C] $2\sqrt{2}$ [D] 2

Topic 3

MATRICES

Objectives

At the end of this topic, the learner should be able to:

1. Determine the value of an umknown parameter in a 2 by 2 singular matrix.
2. Find the inverse of a 2 by 2 matrix by definition and by formula.
3. Find the multiplicative inverse of a 2 by 2 matrix.
4. Solve linear simultaneous equations using matrices.
5. Apply matrices to real life situations.

3.1 Review and Revision

Review

1. State the size of the matrix $\begin{pmatrix} 2 & 4 & 7 \\ 3 & 6 & 5 \end{pmatrix}$.
2. Find the unknowns in each of the following:

   (a) $\begin{pmatrix} 2 & 4 \\ 3 & z \end{pmatrix} + \begin{pmatrix} x & y \\ 3 & 4 \end{pmatrix} = \begin{pmatrix} 4 & 4 \\ w & 0 \end{pmatrix}$ (b) $\begin{pmatrix} x & y \\ z & w \end{pmatrix} - \begin{pmatrix} 2 & 1 \\ 5 & 3 \end{pmatrix} = \begin{pmatrix} 1 & 0 \\ 0 & 1 \end{pmatrix}$

3. Find **AB** given that $A = \begin{pmatrix} 0 & -1 \\ 3 & -2 \end{pmatrix}$ and $B = \begin{pmatrix} 2 & 4 & 3 \\ 1 & 5 & 0 \end{pmatrix}$.

4. Given that $A = \begin{pmatrix} 4 & 2 \\ -2 & 3 \end{pmatrix}$ and $= \begin{pmatrix} 2 & -1 \\ 4 & -2 \end{pmatrix}$.
   Find the det **A** and det **B** and state which of **A** and **B** is singular.

3.2 The Determinant of a 2×2 Singular Matrix

Recall that:

(i) the **determinant** of the 2×2 matrix $A = \begin{pmatrix} a & b \\ c & d \end{pmatrix}$, is denoted by det **A**

and is given by $\text{Det } A = ad - bc$.

(ii) If det **A** = 0, then a is singular.

Example

Find the value of k for which the matrix $\begin{pmatrix} k & -1 \\ 2 & k-3 \end{pmatrix}$ is singular.

Solution

If the matrix is singular then, $k(k-3) - 2(-1) = 0$.
$$\Rightarrow k^2 - 3k + 2 = 0$$
$$(k-2)(k-1) = 0$$
$$k = 2 \text{ or } k = 1$$

34

Exercise 3:1

1. The matrices $\mathbf{P} = \begin{pmatrix} m & 2 \\ -2 & 0 \end{pmatrix}$ and $\mathbf{Q} = \begin{pmatrix} 2 & -m \\ -2 & 1 \end{pmatrix}$ are such that $\mathbf{P} + \mathbf{Q}$ is singular. Find the value of m.

2. Given that the matrix $\begin{pmatrix} a & 1 \\ 4 & a \end{pmatrix}$ is singular, find the possible values of a.

3. Which of the following matrices is/are singular?
$\begin{pmatrix} -2 & 2 \\ 2 & 2 \end{pmatrix}, \begin{pmatrix} -2 & -2 \\ 2 & 2 \end{pmatrix}, \begin{pmatrix} -2 & -2 \\ 2 & -2 \end{pmatrix}, \begin{pmatrix} -2 & -2 \\ -2 & 2 \end{pmatrix}, \begin{pmatrix} 2 & -2 \\ -2 & -2 \end{pmatrix}.$

4. Find $\begin{pmatrix} 2 & -3 \\ -2 & 4 \end{pmatrix}\begin{pmatrix} -2 & 1 \\ -5 & 3 \end{pmatrix}$ and calculate its determinant.

5. Find the values of m for which the matrix $\begin{pmatrix} 6-m & 2 \\ 25 & 1-m \end{pmatrix}$ is singular.

6. Find the values of x for which the matrix $\begin{pmatrix} 2x & 2 \\ 3 & x-2 \end{pmatrix}$ is singular.

7. Solve the equation $\det\begin{pmatrix} x & 10 \\ x & x \end{pmatrix} = -21$.

3.3 The Adjoint of a 2×2 Matrix

The adjoint of the 2×2 matrix $\mathbf{A} = \begin{pmatrix} a & b \\ c & d \end{pmatrix}$ is denoted by adj $\mathbf{A}$ or Adj$\begin{pmatrix} a & b \\ c & d \end{pmatrix}$ and is obtained by interchanging the elements along the leading diagonal and changing the sign of the elements along the minor diagonal. Hence,

$$\text{Adj}\begin{pmatrix} a & b \\ c & d \end{pmatrix} = \text{Adj}\mathbf{A} = \begin{pmatrix} d & -b \\ -c & a \end{pmatrix}$$

Example

Find the adjoint of the following matrices.

(a) $\begin{pmatrix} 15 & 9 \\ 10 & 5 \end{pmatrix}$ (b) $\begin{pmatrix} 3 & 7 \\ -4 & 1 \end{pmatrix}$

Solution

(a) $\text{Adj}\begin{pmatrix} 15 & 9 \\ 10 & 5 \end{pmatrix} = \begin{pmatrix} 5 & -9 \\ -10 & 15 \end{pmatrix}$ (b) $\text{Adj}\begin{pmatrix} 1 & -7 \\ 4 & 3 \end{pmatrix} = \begin{pmatrix} 1 & -7 \\ 4 & 3 \end{pmatrix}$

 Exercise 3:2

Write down the adjoint of the following matrices.

(a) $\begin{pmatrix} 5 & -3 \\ -4 & 1 \end{pmatrix}$ (b) $\begin{pmatrix} 3 & 4 \\ 5 & 1 \end{pmatrix}$ (c) $\begin{pmatrix} 2 & 1 \\ -2 & 3 \end{pmatrix}$ (d) $\begin{pmatrix} 4 & 1 \\ 2 & 0 \end{pmatrix}$ (e) $\begin{pmatrix} 4 & 2 \\ 3 & 5 \end{pmatrix}$

(f) $\begin{pmatrix} -2 & -3 \\ -2 & 1 \end{pmatrix}$ (g) $\begin{pmatrix} 1 & 0 \\ 0 & 1 \end{pmatrix}$ (h) $\begin{pmatrix} \frac{1}{2} & 1 \\ 2 & 4 \end{pmatrix}$ (i) $\begin{pmatrix} 2 & 5 \\ 3 & 7 \end{pmatrix}$

3.4   The Inverse of a 2×2 Matrix

The inverse or reciprocal of a square matrix **A** is denoted by A^{-1}, read "A inverse". A^{-1} is the inverse of **A** if and only if

$$A^{-1}A = AA^{-1} = I$$

Where, **I** is a compatible identity matrix.

Example

1. Find the inverse of the matrix $\begin{pmatrix} 4 & 1 \\ 6 & 2 \end{pmatrix}$.

Solution
The definition $\mathbf{A}^{-1}\mathbf{A} = \mathbf{I}$ may be used. Thus,

Let $\mathbf{A} = \begin{pmatrix} 4 & 1 \\ 6 & 2 \end{pmatrix}$ and $\mathbf{A}^{-1} = \begin{pmatrix} a & b \\ c & d \end{pmatrix}$

Then, $\begin{pmatrix} a & b \\ c & d \end{pmatrix}\begin{pmatrix} 4 & 1 \\ 6 & 2 \end{pmatrix} = \begin{pmatrix} 1 & 0 \\ 0 & 1 \end{pmatrix}$.

$\Rightarrow \begin{pmatrix} 4a+6b & a+2b \\ 4c+6d & c+2d \end{pmatrix} = \begin{pmatrix} 1 & 0 \\ 0 & 1 \end{pmatrix}$

$$4a+6b = 1 \quad \ldots \ldots \ldots \text{①}$$
$$a+2b = 0 \quad \ldots \ldots \ldots \text{②}$$

① − 3② $\quad \Rightarrow a = 1$

Substitute ② in $\Rightarrow b = -\dfrac{1}{2}$

Similarly,
$$4c+6d = 0 \quad \ldots \ldots \ldots \text{③}$$
$$c+2d = 1 \quad \ldots \ldots \ldots \text{④}$$

③ − 3④ $\quad \Rightarrow c = -3$

Substitute ④ in $\Rightarrow -3+2d = 1 \Rightarrow d = 2$

Therefore the inverse of $\begin{pmatrix} 4 & 1 \\ 6 & 2 \end{pmatrix}$ is $\begin{pmatrix} 1 & -\dfrac{1}{2} \\ -3 & 2 \end{pmatrix}$

2. Find the inverse of $\mathbf{A} = \begin{pmatrix} a & b \\ c & d \end{pmatrix}$.

Solution

Let $\mathbf{A}^{-1} = \begin{pmatrix} w & x \\ y & z \end{pmatrix}$. Then by definition $\mathbf{A}^{-1}\mathbf{A} = \mathbf{I}$.

$$\Rightarrow \begin{pmatrix} w & x \\ y & z \end{pmatrix} \begin{pmatrix} a & b \\ c & d \end{pmatrix} = \begin{pmatrix} 1 & 0 \\ 0 & 1 \end{pmatrix}$$

Multiplying and equating corresponding entries,

$aw + cx = 1$ .... ①
$bw + dx = 0$ .... ②
$ay + cz = 0$ .... ③
$by + dz = 1$ .... ④

Solving for w, x, y and z in these equations,

$$w = \frac{d}{ad-bc}, \quad x = -\frac{b}{ad-bc}, \quad y = -\frac{c}{ad-bc}, \quad z = \frac{a}{ad-bc}$$

$$\mathbf{A}^{-1} = \begin{pmatrix} \frac{d}{ad-bc} & -\frac{b}{ad-bc} \\ -\frac{c}{ad-bc} & \frac{a}{ad-bc} \end{pmatrix} = \frac{1}{ad-bc} \begin{pmatrix} d & -b \\ -c & a \end{pmatrix}$$

Since the adjoint and determinant of the 2×2 matrix $\mathbf{A} = \begin{pmatrix} a & b \\ c & d \end{pmatrix}$ are given by $\text{Adj}\,\mathbf{A} = \begin{pmatrix} d & -c \\ -b & a \end{pmatrix}$ and $\det \mathbf{A} = ad - bc$ respectively,

$$\mathbf{A}^{-1} = \frac{1}{ad-bc} \begin{pmatrix} d & -b \\ -c & a \end{pmatrix} = \frac{1}{\text{Det}\,\mathbf{A}} \times \text{Adj}\,\mathbf{A}$$

We can use this formula to find the inverse of any 2×2 matrix, which has an inverse.

3. Find the inverse of $\mathbf{A} = \begin{pmatrix} 4 & 1 \\ 6 & 2 \end{pmatrix}$.

Solution

$$\mathbf{A}^{-1} = \frac{1}{ad-bc} \begin{pmatrix} d & -b \\ -c & a \end{pmatrix} = \frac{1}{\text{Det}\,\mathbf{A}} \times \text{Adj}\,\mathbf{A}$$

$$\mathbf{A}^{-1} = \frac{1}{4(2)-1(6)} \begin{pmatrix} 2 & -1 \\ -6 & 4 \end{pmatrix} = \frac{1}{2} \begin{pmatrix} 2 & -1 \\ -6 & 4 \end{pmatrix}$$

$$\Rightarrow A^{-1} = \begin{pmatrix} 1 & -\frac{1}{2} \\ -3 & 2 \end{pmatrix}$$

4. Find the inverse of $\begin{pmatrix} 4 & 6 \\ 5 & 8 \end{pmatrix}$.

Solution

$$\begin{pmatrix} 4 & 6 \\ 5 & 8 \end{pmatrix}^{-1} = \frac{1}{4(8)-6(5)} \begin{pmatrix} 8 & -6 \\ -5 & 4 \end{pmatrix} = \frac{1}{2} \begin{pmatrix} 8 & -6 \\ -5 & 4 \end{pmatrix}$$

$$\Rightarrow \begin{pmatrix} 4 & 6 \\ 5 & 8 \end{pmatrix}^{-1} = \begin{pmatrix} 4 & -3 \\ -\frac{5}{2} & 2 \end{pmatrix}$$

Exercise 3:3

1. Find the inverses of the following matrices.

   (a) $\begin{pmatrix} 5 & -3 \\ -4 & 1 \end{pmatrix}$ (b) $\begin{pmatrix} 3 & 4 \\ 5 & 1 \end{pmatrix}$ (c) $\begin{pmatrix} 2 & 1 \\ -2 & 3 \end{pmatrix}$

   (d) $\begin{pmatrix} 4 & 1 \\ 2 & 0 \end{pmatrix}$ (e) $\begin{pmatrix} 4 & 2 \\ 3 & 5 \end{pmatrix}$ (f) $\begin{pmatrix} -2 & -3 \\ -2 & 1 \end{pmatrix}$

   (g) $\begin{pmatrix} 1 & 0 \\ 0 & 1 \end{pmatrix}$ (h) $\begin{pmatrix} \frac{1}{2} & 1 \\ 2 & 4 \end{pmatrix}$ (i) $\begin{pmatrix} 2 & 5 \\ 3 & 7 \end{pmatrix}$

2. Given that $A = \begin{pmatrix} 3 & 1 \\ 2 & 1 \end{pmatrix}$, find A^{-1} and hence $A^{-1}A$.

3. If $B = \begin{pmatrix} 1 & 3 \\ 1 & 5 \end{pmatrix}$, find B^{-1} and hence $B^{-1}B$.

4. Find the inverses of the following matrices.

   (i) $A = \begin{pmatrix} 4 & -2 \\ 1 & 3 \end{pmatrix}$ (ii) $B = \begin{pmatrix} 1 & 0 \\ 0 & 1 \end{pmatrix}$ (iii) $C = \begin{pmatrix} 3 & 0 \\ 0 & 2 \end{pmatrix}$ (iv) $D = \begin{pmatrix} 2 & -2 \\ 3 & 1 \end{pmatrix}$

5. Show that $\begin{pmatrix} 1 & 0 \\ 0 & 1 \end{pmatrix}$ is the identity matrix under matrix multiplication.

6. Show that $\begin{pmatrix} -2 & 3 \\ 3 & -4 \end{pmatrix}$ is the inverse of $\begin{pmatrix} 4 & 3 \\ 3 & 2 \end{pmatrix}$ under matrix multiplication.

3.5 Simultaneous Equations – the Matrix Method

Consider,
$$\begin{pmatrix} 4 & -1 \\ 3 & 1 \end{pmatrix}\begin{pmatrix} x \\ y \end{pmatrix} = \begin{pmatrix} 3 \\ 4 \end{pmatrix} \dots \text{①}$$

Computing the product in the left hand side leads to
$$\begin{pmatrix} 4x - y \\ 3x + y \end{pmatrix} = \begin{pmatrix} 3 \\ 4 \end{pmatrix}$$

By equating corresponding entries, we obtain the following simultaneous equations.
$$4x - y = 3 \dots \text{②}$$
$$3x + y = 4 \dots \text{③}$$

Therefore, we can arrange the simultaneous equations in ② and ③ to obtain ①. Since multiplying a matrix by its inverse gives rise to the identity matrix and multiplying a matrix by the identity matrix leaves the matrix unchanged, we can use this idea to solve simultaneous equations.

Example

1. Using the matrix method, solve the simultaneous equations
$$4x - y = 3$$
$$3x + y = 4$$

Solution

$$\begin{matrix} 4x - y = 3 \\ 3x + y = 4 \end{matrix} \Rightarrow \begin{pmatrix} 4 & -1 \\ 3 & 1 \end{pmatrix}\begin{pmatrix} x \\ y \end{pmatrix} = \begin{pmatrix} 3 \\ 4 \end{pmatrix} \dots \text{①}$$

Let $A = \begin{pmatrix} 4 & -1 \\ 3 & 1 \end{pmatrix}$

Then, $A^{-1} = \frac{1}{\det B}(\text{Adj } A) = \frac{1}{4(1)-3(-1)}\begin{pmatrix} 1 & 1 \\ -3 & 4 \end{pmatrix}$.

$$\Rightarrow \mathbf{A}^{-1} = \frac{1}{7}\begin{pmatrix} 1 & 1 \\ -3 & 4 \end{pmatrix} = \begin{pmatrix} \frac{1}{7} & \frac{1}{7} \\ -\frac{3}{7} & \frac{4}{7} \end{pmatrix}$$

Pre-multiply both sides of equation ① $\mathbf{A}^{-1}$

$$\begin{pmatrix} x \\ y \end{pmatrix} = \begin{pmatrix} \frac{1}{7} & \frac{1}{7} \\ -\frac{3}{7} & \frac{4}{7} \end{pmatrix} \begin{pmatrix} 3 \\ 4 \end{pmatrix} \Rightarrow \begin{pmatrix} x \\ y \end{pmatrix} = \begin{pmatrix} 1 \\ 1 \end{pmatrix} \text{ i.e. } x = 1 \text{ and } y = 1.$$

2. Using the matrix method, solve the equations
$$4x + y = 3$$
$$6x + 2y = 5$$

Solution

$$\begin{pmatrix} 4 & 1 \\ 6 & 2 \end{pmatrix} \begin{pmatrix} x \\ y \end{pmatrix} = \begin{pmatrix} 3 \\ 5 \end{pmatrix} \quad \dots\dots\dots\dots ①$$

$$\begin{pmatrix} 4 & 1 \\ 6 & 2 \end{pmatrix}^{-1} = \begin{pmatrix} 1 & -\frac{1}{2} \\ -3 & 2 \end{pmatrix}$$

Pre-multiply both sides of equation ① by $\begin{pmatrix} 4 & 1 \\ 6 & 2 \end{pmatrix}^{-1}$.

$$\begin{pmatrix} x \\ y \end{pmatrix} = \begin{pmatrix} \frac{1}{2} \\ 1 \end{pmatrix} \text{ i.e } x = \frac{1}{2} \text{ and } y = 1$$

3. Solve the following simultaneous equations using the matrix method.
   (i) $2x + 3y = 2$ (ii) $3p + 4q = 10$
         $4x + 5y = 5$ $2p + 3p = 7$

Solution
(i) $2x + 3y = 2$
      $4x + 5y = 5$

$$\Rightarrow \begin{pmatrix} 2 & 3 \\ 4 & 5 \end{pmatrix} \begin{pmatrix} x \\ y \end{pmatrix} = \begin{pmatrix} 2 \\ 5 \end{pmatrix}$$

$$\begin{pmatrix} 2 & 3 \\ 4 & 5 \end{pmatrix}^{-1} = \begin{pmatrix} -\frac{5}{2} & \frac{3}{2} \\ 2 & -1 \end{pmatrix}$$

$$\therefore \begin{pmatrix} x \\ y \end{pmatrix} = \begin{pmatrix} -\frac{5}{2} & \frac{3}{2} \\ 2 & -1 \end{pmatrix} \begin{pmatrix} 2 \\ 5 \end{pmatrix} \Rightarrow x = \frac{5}{2} \text{ and } y = -1$$

(ii) $3p + 4q = 10$
$2p + 3p = 7$

$$\Rightarrow \begin{pmatrix} 3 & 4 \\ 2 & 3 \end{pmatrix} \begin{pmatrix} p \\ q \end{pmatrix} = \begin{pmatrix} 10 \\ 7 \end{pmatrix}$$

$$\begin{pmatrix} 3 & 4 \\ 2 & 3 \end{pmatrix}^{-1} = \begin{pmatrix} 3 & -4 \\ -2 & 3 \end{pmatrix}$$

$$\begin{pmatrix} p \\ q \end{pmatrix} = \begin{pmatrix} 3 & -4 \\ -2 & 3 \end{pmatrix} \begin{pmatrix} 10 \\ 7 \end{pmatrix} = \begin{pmatrix} 2 \\ 1 \end{pmatrix}$$

$\therefore p = 2$ and $q = 1$

Exercise 3:4

Solve the following simultaneous equations using the matrix method.

1. $p - 3q = 10$
   $3p - 2q = 16$

2. $s = 2t - 1$
   $2s = 3t + 2$

3. $5x - 2y = 14$
   $x + y = 7$

4. $4m + 4n = 3$
   $m + 2n = 1$

5. $2x + y = 7$
   $3x - 2y = 7$

6. $3u - 7v = 1$
   $2u + v = 12$

3.6   Information Matrices

We often represent network information on matrices called information matrices. The most popular information matrices related to networks are route matrices and incidence matrices.

Route and Incidence Matrices

Consider the following network made up of the regions w, x, y and z; the junctions A, B and C in the region and the routes 1, 2, 3, 4 and 5 linking the junctions.

Module 15, Topic 3: Matrices

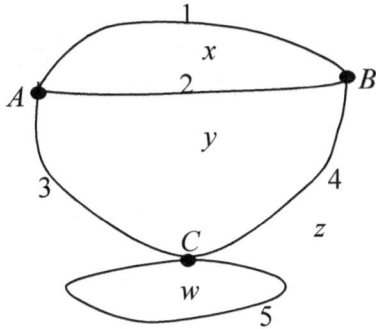

A **route matrix** (matrix i) is a matrix that represents the number of direct routes linking each node (junction, town or joint) in a network to another. For instance, the following is a route matrix representing the number of direct routes, which link the junctions A, B and C in the network in the figure above.

An **incidence matrix** on the other hand is a matrix that shows the relationship between each node and each region (matrix ii) or each node and each route (matrix iii) or each region and each route (matrix iv) in a network.

$$\begin{array}{c} \begin{array}{ccc} A & B & C \end{array} \\ \begin{array}{c} A \\ B \\ C \end{array}\!\!\left(\begin{array}{ccc} 0 & 2 & 1 \\ 2 & 0 & 1 \\ 1 & 1 & 1 \end{array}\right) \end{array}$$

Matrix i

$$\begin{array}{c} \begin{array}{cccc} w & x & y & z \end{array} \\ \begin{array}{c} A \\ B \\ C \end{array}\!\!\left(\begin{array}{cccc} 0 & 1 & 1 & 1 \\ 0 & 1 & 1 & 1 \\ 1 & 0 & 1 & 1 \end{array}\right) \end{array}$$

Matrix ii

$$\begin{array}{c} \begin{array}{ccccc} 1 & 2 & 3 & 4 & 5 \end{array} \\ \begin{array}{c} A \\ B \\ C \end{array}\!\!\left(\begin{array}{ccccc} 1 & 1 & 1 & 0 & 0 \\ 1 & 1 & 0 & 1 & 0 \\ 0 & 0 & 1 & 1 & 1 \end{array}\right) \end{array}$$

Matrix iii

$$\begin{array}{c} \begin{array}{cccc} w & x & y & z \end{array} \\ \begin{array}{c} 1 \\ 2 \\ 3 \\ 4 \\ 5 \end{array}\!\!\left(\begin{array}{cccc} 0 & 1 & 0 & 0 \\ 0 & 1 & 1 & 0 \\ 0 & 0 & 1 & 0 \\ 0 & 0 & 1 & 1 \\ 1 & 0 & 0 & 1 \end{array}\right) \end{array}$$

Matrix iv

 Exercise 3:5

1. Figure (a) shows the roads s, t, u, v, w, x, y and z linking four cities A, B, C and D. Build up
   (a) A route matrix to show the network.
   (b) An incidence matrix to show the network

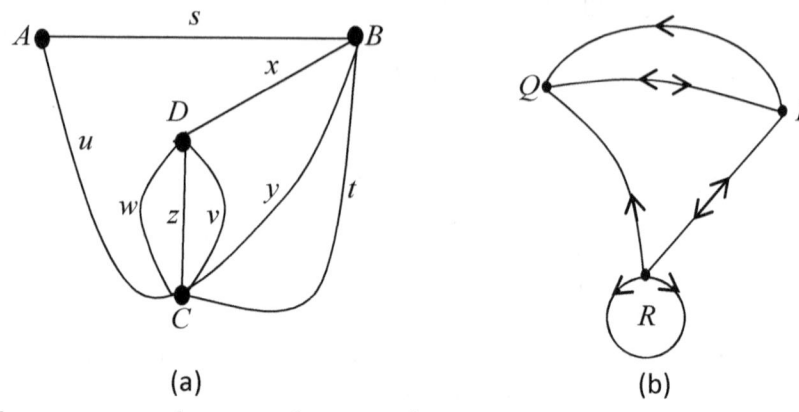

(a)   (b)

2. We can express the network connecting towns P, Q and R shown in Figure 43:3 as a route matrix, **M**, where the entries show the number of routes leading to the town. Complete the entries in the matrix, **M**.

$$\mathbf{M} = \text{from} \begin{array}{c} \\ P \\ Q \\ R \end{array} \begin{pmatrix} & 2 & \\ & 0 & \\ 1 & & 2 \end{pmatrix}$$

with columns labelled "To: $P\ Q\ R$".

 Integration Activity

1. A businessman has a store at Mankon and another at Nkwen. The following matrices show the income and expenditure in FCFA for three days on a certain week at each store.

	Income			Expenses	
	Mankon	Nkwen		Mankon	Nkwen
Monday	42500	85630	Monday	32000	58000
Tuesday	62850	41420	Tuesday	45000	33000
Wednesday	74850	59270	Wednesday	63000	68000

(a) Write a matrix that shows the daily profit for each store.
(b) In which store did he make the greatest net profit for the three days?
(c) Which store suffered the greatest lost and on which day?

2. You are a daily collector attached to a credit union. On a certain day, 5 hawkers save money to you in denominations of 25 FCFA, 50

FCFA and 100 FCFA as follows.

$$\begin{array}{c c} & \begin{array}{c c c} 100 & 50 & 25 \end{array} \\ \begin{array}{c} Yuh \\ Suh \\ Doh \\ Tah \\ Fru \end{array} & \begin{pmatrix} 2 & 7 & 1 \\ 0 & 10 & 3 \\ 1 & 6 & 8 \\ 3 & 3 & 0 \\ 5 & 2 & 4 \end{pmatrix} \end{array}$$

Write down two matrices and use them to determine how much money altogether that you collected from the hawkers on that day.

Multiple Choice Exercise 3

1. Given that $P = \begin{pmatrix} 3 & 2 \\ 4 & 5 \end{pmatrix}$ then Adj P is equal to:

   [A] $\begin{pmatrix} -3 & 4 \\ 2 & -5 \end{pmatrix}$ [B] $\begin{pmatrix} 5 & 4 \\ 2 & 3 \end{pmatrix}$ [C] $\begin{pmatrix} 5 & -2 \\ -4 & 3 \end{pmatrix}$ [D] $\begin{pmatrix} 5 & -4 \\ -2 & 3 \end{pmatrix}$

7. Given that **A** is a 3 by 2 matrix and **B** is a 2 by 4 matrix. The matrix product **AB** will have size:
   [A] 2 by 3   [B] 3 by 4   [C] 3 by 2   [D] 2 by 4

8. The values of m for which the matrix $\begin{pmatrix} 6-m & 2 \\ 25 & 1-m \end{pmatrix}$ is singular are:
   [A] -11 and 4 [B] -11 and -4 [C] 11 and 4 [D] 11 and -4

9. Given that $A = \begin{pmatrix} 2 & -3 \\ -2 & 4 \end{pmatrix}$ and $B = \begin{pmatrix} -2 & 1 \\ -5 & 3 \end{pmatrix}$. The matrix product **AB** is:

   [A] $\begin{pmatrix} -11 & 7 \\ 16 & 10 \end{pmatrix}$ [B] $\begin{pmatrix} 11 & -7 \\ 16 & 10 \end{pmatrix}$ [C] $\begin{pmatrix} 11 & -7 \\ -16 & 10 \end{pmatrix}$ [D] $\begin{pmatrix} 11 & 7 \\ -16 & 10 \end{pmatrix}$

10. Given that $A = \begin{pmatrix} 2 & -3 \\ -2 & 4 \end{pmatrix}$ and $B = \begin{pmatrix} -2 & 1 \\ -5 & 3 \end{pmatrix}$. The determinant of the product **AB** is:
    [A] 2   [B] 222   [C] -222 [D] -2

11. Given that $M = \begin{pmatrix} 4 & 2 \\ 6 & 1 \end{pmatrix}$, M^{-1} is equal to:

    [A] $\dfrac{1}{8}\begin{pmatrix} 1 & -2 \\ -6 & 4 \end{pmatrix}$ [B] $-\dfrac{1}{8}\begin{pmatrix} -1 & 2 \\ 6 & -4 \end{pmatrix}$

    [C] $-\dfrac{1}{8}\begin{pmatrix} 1 & 2 \\ 6 & -4 \end{pmatrix}$ [D] $\dfrac{1}{8}\begin{pmatrix} 1 & -2 \\ -6 & 4 \end{pmatrix}$

12. The singular matrix among the following matrices is:

[A] $\begin{pmatrix} -2 & 2 \\ 2 & 2 \end{pmatrix}$ [B] $\begin{pmatrix} -2 & -2 \\ 2 & 2 \end{pmatrix}$ [C] $\begin{pmatrix} -2 & -2 \\ 2 & -2 \end{pmatrix}$ [D] $\begin{pmatrix} -2 & -2 \\ -2 & 2 \end{pmatrix}$

13. Given that the matrix $\begin{pmatrix} a & 1 \\ 4 & a \end{pmatrix}$ is singular, the possible values of a are:

[A] 4 and -1 [B] 4 and 1   [C] -4 and 1   [D] -2 and 2

14. The inverse of the matrix $\begin{pmatrix} 5 & 2 \\ 7 & 3 \end{pmatrix}$ is:

[A] $\begin{pmatrix} -3 & -7 \\ -2 & -5 \end{pmatrix}$ [B] $\begin{pmatrix} -3 & 2 \\ 7 & -5 \end{pmatrix}$ [C] $\begin{pmatrix} 3 & -2 \\ -7 & 5 \end{pmatrix}$ [D] $\begin{pmatrix} 3 & -7 \\ -2 & 5 \end{pmatrix}$

15. The value of a for which the matrix $\begin{pmatrix} a & 3-a \\ 2 & 1 \end{pmatrix}$ is singular is:

[A] 2   [B] -2 [C] 3   [D] -3

16. Given that $\mathbf{PQ} = \mathbf{I}$, where $\mathbf{P} = \begin{pmatrix} 5 & -4 \\ -1 & 1 \end{pmatrix}$ and $\mathbf{I} = \begin{pmatrix} 1 & 0 \\ 0 & 1 \end{pmatrix}$. The matrix $\mathbf{Q}$ is:

[A] $\begin{pmatrix} 1 & 4 \\ -1 & 5 \end{pmatrix}$ [B] $\begin{pmatrix} 1 & -4 \\ -1 & 5 \end{pmatrix}$ [C] $\begin{pmatrix} 1 & -4 \\ 1 & 5 \end{pmatrix}$ [D] $\begin{pmatrix} 1 & 4 \\ 1 & 5 \end{pmatrix}$

17. The matrices $\mathbf{P} = \begin{pmatrix} m & 2 \\ 5 & 0 \end{pmatrix}$ and $\mathbf{Q} = \begin{pmatrix} 2 & -m \\ -2 & 1 \end{pmatrix}$ are such that $\mathbf{P} + \mathbf{Q}$ is singular. The value of m is:

[A] -1 [B] 1   [C] 4   [D] -4

18. The inverse of the 2×2 matrix $\begin{pmatrix} 6 & 10 \\ 2 & 4 \end{pmatrix}$ is:

[A] $\begin{pmatrix} 1 & -2\frac{1}{2} \\ -\frac{1}{2} & 1\frac{1}{2} \end{pmatrix}$

[B] $\begin{pmatrix} 4 & -10 \\ -2 & 6 \end{pmatrix}$

[C] $\begin{pmatrix} -2 & 6 \\ -10 & 4 \end{pmatrix}$

[D] $\begin{pmatrix} 1 & 2\frac{1}{2} \\ \frac{1}{2} & 1\frac{1}{2} \end{pmatrix}$

19. The transpose of the matrix $\begin{pmatrix} 7 & 3 \\ -1 & 5 \end{pmatrix}$ is:

[A] $\begin{pmatrix} 5 & -3 \\ 1 & 7 \end{pmatrix}$ [B] $\begin{pmatrix} 7 & -1 \\ 3 & 5 \end{pmatrix}$ [C] $\begin{pmatrix} -7 & -1 \\ 3 & -5 \end{pmatrix}$ [D] $\begin{pmatrix} -7 & 1 \\ -3 & -5 \end{pmatrix}$

20. Given that $\begin{pmatrix} 2 & 1 \\ 1 & -1 \end{pmatrix}\begin{pmatrix} 1 & -2 \\ 1 & 2 \end{pmatrix} = 3\mathbf{A}$. The matrix $\mathbf{A}$ is:

[A] $\begin{pmatrix} 3 & -2 \\ 0 & -4 \end{pmatrix}$ [B] $\begin{pmatrix} 1 & \frac{2}{3} \\ 0 & \frac{4}{3} \end{pmatrix}$ [C] $\begin{pmatrix} 3 & \frac{2}{3} \\ 0 & \frac{4}{3} \end{pmatrix}$ [D] $\begin{pmatrix} 3 & -\frac{2}{3} \\ 0 & -\frac{4}{3} \end{pmatrix}$

21. The determinant of the matrix $\mathbf{A} = \begin{pmatrix} x & 7 \\ 4 & 2 \end{pmatrix}$ is 6. The value of x is:

   [A] −17   [B] −34   [C] 17   [D] 34

22. Let $\mathbf{M} = \begin{pmatrix} 3 & 1 \\ 1 & 2 \end{pmatrix}$ and $\mathbf{N} = \begin{pmatrix} 1 & 4 \\ -3 & 2 \end{pmatrix}$. As a single matrix, $\mathbf{M} + 3\mathbf{N}$ is equal to:

   [A] $\begin{pmatrix} 10 & 7 \\ 1 & 4 \end{pmatrix}$ [B] $\begin{pmatrix} 10 & 7 \\ -1 & 4 \end{pmatrix}$ [C] $\begin{pmatrix} 6 & 13 \\ 8 & -8 \end{pmatrix}$ [D] $\begin{pmatrix} 6 & 13 \\ -8 & 8 \end{pmatrix}$

23. The inverse of the 2×2 matrix $\begin{pmatrix} 6 & 11 \\ 2 & 4 \end{pmatrix}$ is:

   [A] $\begin{pmatrix} 2 & -5\frac{1}{2} \\ -1 & 3 \end{pmatrix}$ [B] $\begin{pmatrix} 4 & 2 \\ -11 & 6 \end{pmatrix}$ [C] $\begin{pmatrix} 3 & -5\frac{1}{2} \\ -1 & 2 \end{pmatrix}$ [D] $\begin{pmatrix} 0 & 0 \\ 0 & 0 \end{pmatrix}$

24. The adjoint of the matrix $\begin{pmatrix} 7 & 3 \\ -1 & 5 \end{pmatrix}$ is:

   [A] $\begin{pmatrix} 5 & -3 \\ 1 & 7 \end{pmatrix}$ [B] $\begin{pmatrix} 7 & -1 \\ 3 & 5 \end{pmatrix}$ [C] $\begin{pmatrix} -7 & -1 \\ 3 & -5 \end{pmatrix}$ [D] $\begin{pmatrix} -7 & 1 \\ -3 & -5 \end{pmatrix}$

25. Given that $p = 2q − 1$ and $2p = 3q + 2$, then:

   [A] $\begin{pmatrix} 1 & 2 \\ 2 & 3 \end{pmatrix}\begin{pmatrix} p \\ q \end{pmatrix} = \begin{pmatrix} -1 \\ 2 \end{pmatrix}$ [B] $\begin{pmatrix} 2 & -1 \\ 3 & 2 \end{pmatrix}\begin{pmatrix} p \\ q \end{pmatrix} = \begin{pmatrix} 1 \\ 2 \end{pmatrix}$

   [C] $\begin{pmatrix} 1 & -1 \\ 2 & 2 \end{pmatrix}\begin{pmatrix} p \\ q \end{pmatrix} = \begin{pmatrix} 2 \\ 3 \end{pmatrix}$ [D] $\begin{pmatrix} 1 & -2 \\ 2 & -3 \end{pmatrix}\begin{pmatrix} p \\ q \end{pmatrix} = \begin{pmatrix} -1 \\ 2 \end{pmatrix}$

Module 16
Plane Geometry

Family of Situations
Module 16 is an extension of module 2, 6 and 11. At the end of the module; the student is expected to have acquired many more competencies within the **families of situations** *'Representation and transformation of Plane Shapes within the Environment'*.

Categories of Action
The categories of action for module 16 include:
1. Perception of the physical environment,
2. Production of plane shapes, transformation of the physical environment
3. Determination of measures and position within the physical environment.

Credit
The module is expected to be covered within 11 weeks teaching 4 periods of 50 minutes per week (or within 44 periods).

Topic 4

VECTORS IN 2-DIMENSIONS

Objectives

At the end of this topic, the learner should be able to:

1. Convert coordinates into components of a vector and vice versa.
2. Name some vector quantities.
3. Represent vectors.
4. Express vectors in terms of other vectors.
5. Find the direction of a vector.
6. Calculate displacement vectors in terms of i and j.
7. Find scalar products.
8. Find the angle between two vectors.
9. Use the midpoint theorem to solve problems on vector geometry.
10. Carry out calculations involving division of a vector in a given ratio.

4.1 Vectors as Coordinates in Two Dimensions

In module 11, topic 5 of book 3 we saw that a vector from the origin to the point $A(4,6)$ can be represented by the component form vector $4\mathbf{i} + 6\mathbf{j}$ or the column vector $\binom{4}{6}$. This vector can also be represented by the directed line segment **OA** shown in the graph below. Conversely the vectors **ST**, **PQ** and **YZ** shown in the graph can be denoted by $\mathbf{ST} = \binom{7}{0}$ or $\mathbf{ST} = 7\mathbf{i}$; $\mathbf{PQ} = \binom{-4}{4}$ or $\mathbf{PQ} = -4\mathbf{i} + 4\mathbf{j}$; $\mathbf{YZ} = \binom{-2}{-5}$ or $\mathbf{YZ} = -2\mathbf{i} - 5\mathbf{j}$.

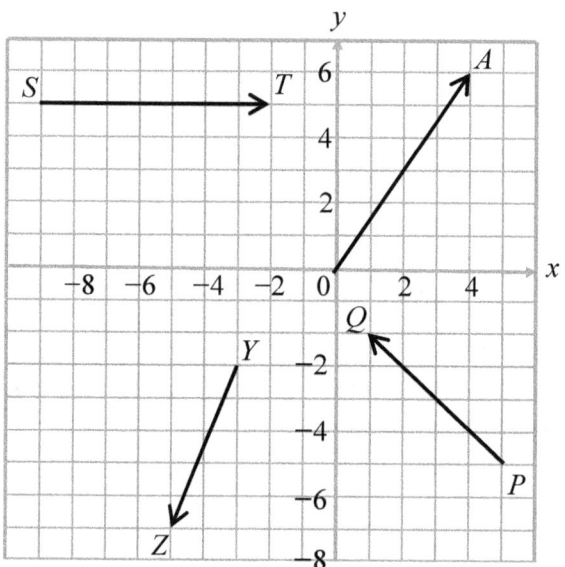

4.2 Parallel and Perpendicular Vectors

Parallel vectors are vectors which have the same direction. If **a** and **b** are parallel vectors then we can express **a** as a scalar multiple of **b**. i.e. $\mathbf{a} = n\mathbf{b}, n \in \mathbb{R}$. A special case of parallel vectors are collinear vectors.
Collinear vectors are vectors that lie on the same straight line.

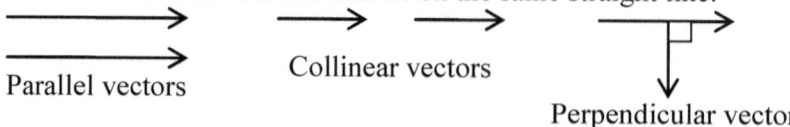

Perpendicular vectors are vectors which are at right angle to each other. If two vectors are perpendicular, we say they are **orthogonal**. Since the x and y-axes are perpendicular, **i** and **j** are orthogonal vectors. We call unit vectors such as **i** and **j**,

Module 16, Topic 4: Vectors in 2-Dimensions

which are perpendicular, **orthonormal vectors**. Thus, orthonormal means the vectors are:

(i) Unit vectors   (ii) Perpendicular (orthogonal).

? Brainstorming Exercise

1. A fruit is hanging on the tree in still air. Describe, name and draw all the vector quantities acting on the fruit.
2. The fruit suddenly cuts and is falling in still air. Describe, name and draw all the vector quantities acting on the fruit.

The **force of gravity** is a vector quantity and always acts from the centre of mass of every object towards the centre of the earth. For a fruit hanging on a tree in still air, the collinear vectors acting on the fruit are the force of gravity and the **up thrust** exerted by the branch upon the fruit.

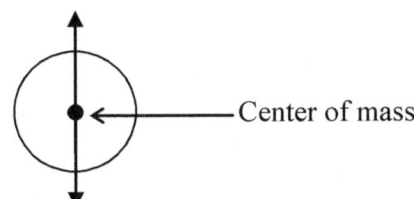

If the fruit suddenly cuts off it will fall to the surface below it. Since the distance covered by the fruit is in a specific direction(downwards), this distance is called the **displacement** and is a vector quantity. The speed at which the fruit is falling is equally directed towards the centre of the earth and is also a vector quantity. This speed is called the **velocity**. Another vector acting upon the fruit is the **acceleration due to gravity**. Your physics teacher may tell you more about these vectors and is directed towards the centre of the earth.

Exercise 4:1

1. Represent the following on a Cartesian plane.

   $OA = \begin{pmatrix} 5 \\ 2 \end{pmatrix}$, $OB = \begin{pmatrix} -5 \\ 3 \end{pmatrix}$, $OC = \begin{pmatrix} -4 \\ 1 \end{pmatrix}$, $OD = \begin{pmatrix} 6 \\ -2 \end{pmatrix}$

2. Write down the vectors in the diagram below as column vectors.

Competency Based Mathematics for Secondary Schools. Book 4

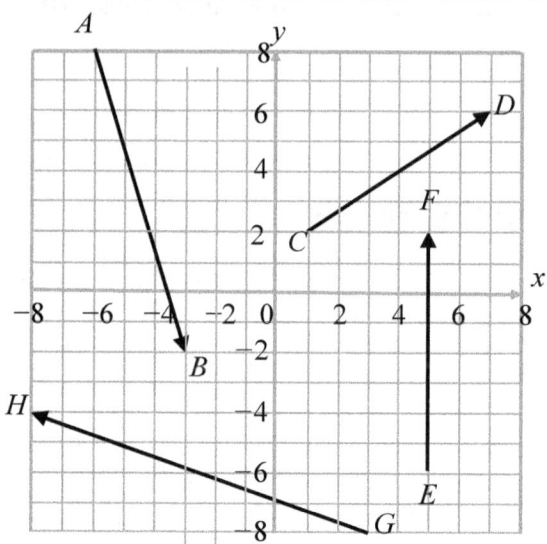

3. Given that A, B, C, D and E are the points (2,5), (−1,4), (0,3), (−5, −6) and (7, −2) respectively. Represent the following vectors on a Cartesian plane.

$$AV = \begin{pmatrix} 2 \\ 4 \end{pmatrix}, BW = \begin{pmatrix} -3 \\ 5 \end{pmatrix}, CX = \begin{pmatrix} -2 \\ -3 \end{pmatrix}, DY = \begin{pmatrix} 0 \\ 6 \end{pmatrix}, EZ = \begin{pmatrix} -5 \\ 0 \end{pmatrix}$$

4. Given that A, B, C, D, E are the points (2, 5), (−1, 4), (0, 3), (−5, −6), and (7, −2) respectively. Represent the vectors **AB**, **CE** and **BD** on a Cartesian plane.

4.3 Position Vectors

If $P(x, y)$ is any point in the x-y plane, then we denote the position vector of the point P by **r** given by

$$\mathbf{r} = \begin{pmatrix} x \\ y \end{pmatrix} \text{ or } \mathbf{r} = x\mathbf{i} + y\mathbf{j}$$

A position vector is another example of a fixed vector discussed earlier and is never a free vector. This is because the tail of a position vector is always at the origin. This means that tail must be at the origin.

Example

1. Draw the position vectors of the points $A(3,1)$ and $B(-5,4)$ on a Cartesian plane.

Solution

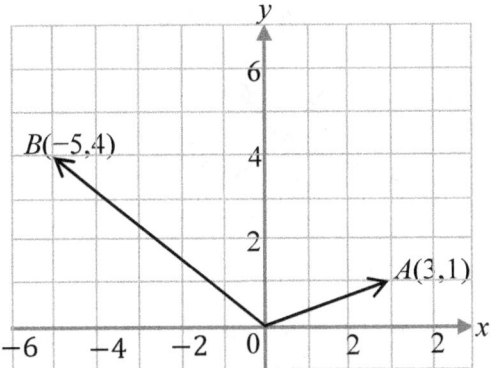

2. Write the position vector of the point $H(-7, 5)$ in
   (a) Column vector form   (b) Component form.

   Solution

   (a) In column vector form, $\mathbf{r} = \begin{pmatrix} -7 \\ 5 \end{pmatrix}$

   (b) In component form, $\mathbf{r} = -7\mathbf{i} + 5\mathbf{j}$.

 Exercise 4:2

1. Write the position vectors of the points labeled A to F in the diagram below in:
   (a) Component form   (b) Column vector form.

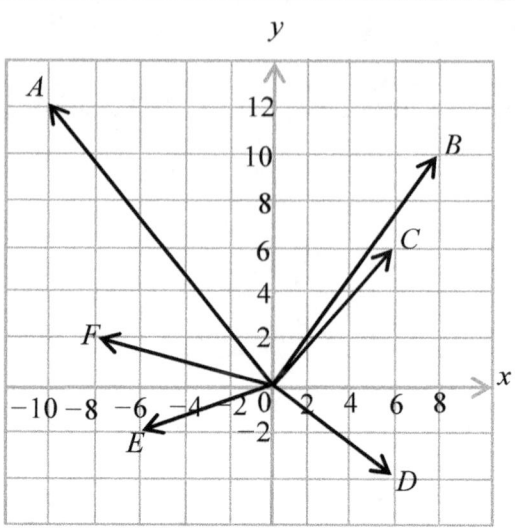

2. (i) Write the position vector of the points $A(-1,5)$, $B(4,-7)$, $C(-9, 3)$ and $D(-3, 6)$
   (a) Component form   (b) Column vector form.
   (ii) Represent these position vectors on the Cartesian plane.
3. In the figure below, which of the vectors are position vectors relative to the origin?

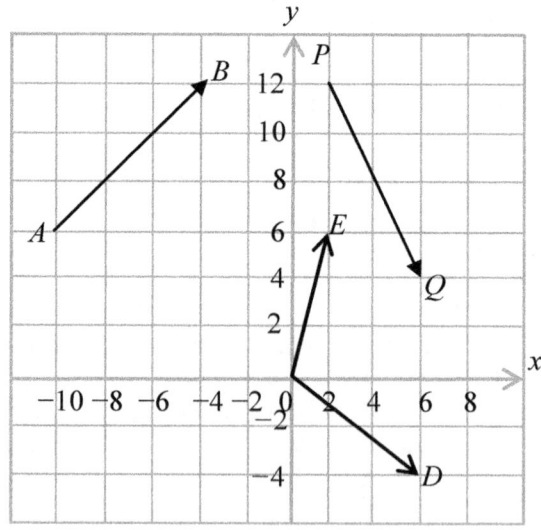

VECTOR GEOMETRY

4.4 The Position Vector of the Midpoint

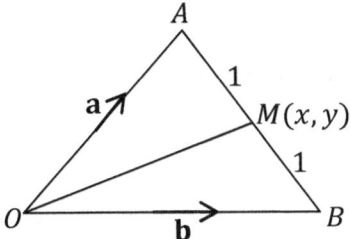

Let A and B be two points with position vectors **a** and **b** respectively and suppose M is the midpoint of AB. Then M divides AB in the ratio $1:1$.
Then the position vector **OM** of M is **OM** = **OA** + **AM** = **OA** + $\frac{1}{2}$**AB**.
Also **AB** = **AO** + **OB** = $-$ **OA** + **OB** = $-$ **a** + **b**

$\Rightarrow$ **OM** = **a** + $\frac{1}{2}(-\mathbf{a} + \mathbf{b})$

$$\mathbf{OM} = \frac{1}{2}(\mathbf{a} + \mathbf{b})$$

4.5 Proportional Division of a vector

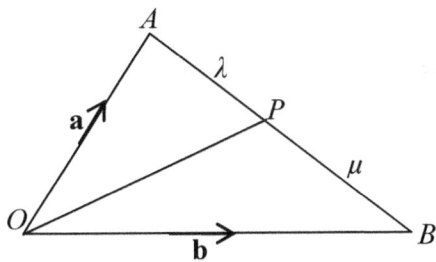

Let A and B be two points with position vectors **a** and **b** respectively and suppose P is a point dividing the line AB in the ratio $\lambda : \mu$.
Then the position vector **OP** of P is **OP** = **OA** + **AP**.

But $\dfrac{AP}{AB} = \dfrac{\lambda}{\lambda + \mu} \Rightarrow \mathbf{AP} = \dfrac{\lambda}{\lambda + \mu}\mathbf{AB}$

$\Rightarrow \mathbf{OP} = \mathbf{OA} \dfrac{\lambda}{\lambda + \mu} \mathbf{AB}$

Also **AB** = **AO** + **OB** = $-$ **OA** + **OB** = $-$ **a** + **b**

$$\Rightarrow OP = a + \frac{\lambda}{\lambda + \mu}(-a + b)$$

i.e. $OP = a + \frac{\lambda}{\lambda + \mu}(b - a)$

We call this theorem the **section theorem**.

Example

1. Given that **OP** = **p**, **OQ** = **q**, **OR** = **r** and that R divides PQ in the ratio 2:3. Express **r** in terms of **p** and **q**.

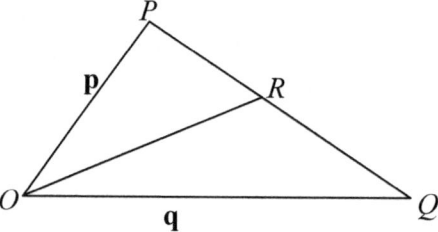

Solution
By section theorem,
$$r = p + \frac{2}{2+3}(q - p) = p + \frac{2}{5}q - \frac{2}{5}p \Rightarrow r = \frac{3}{5}p + \frac{2}{5}q$$

2. In the parallelogram $OACB$, the point X divides **OC** in the ratio 2:1. Given that **OA** = **a** and **OB** = **b**, express the vector **XB** in terms of **a** and **b**.

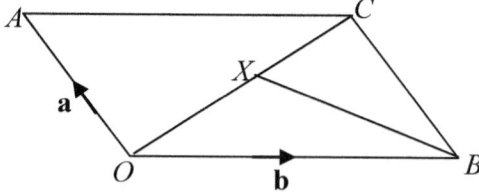

Solution
By the section theorem,
$$BX = BC + \frac{1}{2+1}(-OB - BC)$$
OA = **BC** = **a**   [opposite sides of a parallelogram]
$$BX = a + \frac{1}{2+1}(-b - a) = \frac{2}{3}a - \frac{1}{3}b$$

$$XB = -BX \Rightarrow XB = -\frac{2}{3}a + \frac{1}{3}b$$

3. In the figure below $OABC$ is a trapezium with $OA = a$, $AB = 2b$, $OC = 3b$. The point P is the midpoint of BC, the point X is the midpoint of OP and the point Y is the midpoint of AC. Express OX and PY in terms of a and b.

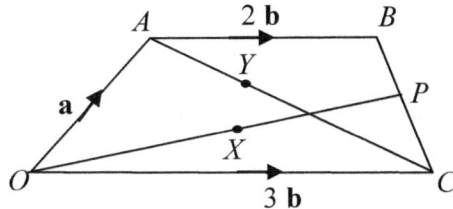

Solution
By the section theorem,

$$OY = a\frac{1}{2}(3b - a) = \frac{1}{2}a + \frac{3}{2}b$$

$$OX = \frac{1}{2}OP = \frac{1}{2}(OA + AB + BP)$$

$$\Rightarrow OX = \frac{1}{2}\left(a + 2b + \frac{1}{2}BC\right)$$

$$BC = -AB + (-OA) + OC$$
$$= -2b - a + 3b$$
$$= b - a$$

$$OX = \frac{1}{2}\left(a + 2b + \frac{1}{2}(b - a)\right)$$
$$= \frac{1}{2}\left(\frac{1}{2}a + \frac{5}{2}b\right)$$
$$= \frac{1}{4}(a + 5b)$$

$$PY = PC + CO + AY$$
$$= -\frac{1}{2}BC - 2b + \frac{1}{2}AC$$
$$= -\frac{1}{2}(b - a) - 2b + \frac{1}{2}(-a + 3b)$$
$$= -\frac{1}{2}b + \frac{1}{2}a - 2b - \frac{1}{2}a + \frac{3}{2}b$$
$$= -b$$

Exercise 4:3

1. In the quadrilateral $OABC$, D is the mid-point of BC and G is the point on AD such that $AG:GD = 2:1$. Given that **OA** = **a**, **OB** = **b** and **OC** = **c**, express **OD** and **OG** in terms of **a**, **b** and **c**.

2. In the figure below, $OPQR$ is a parallelogram, **TR** = 2**OT** and **RM** = **MQ**. If **OP** = **p** and **OR** = **r**, express the vectors **TM** and **PM** in terms of **p** and **r**.

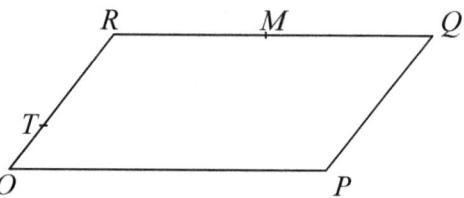

3. Using the figure below,
   (a) express **AC** in terms of **a** and **d** only
   (b) find expressions for **OB** in terms of
      (i) **a** and **b**   (ii) **c** and **d**
   Hence, express **c** in terms of **a** and **d** given that $\mathbf{b} = \frac{3}{2}\mathbf{d}$.

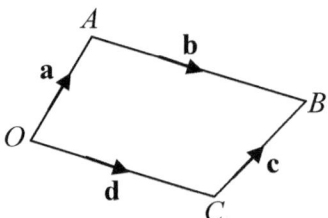

4. The figure below shows a triangle OBA. Given that **OA** = **a**, **OB** = **b** and $AM:MB = 2:1$. Find in terms of **a** and **b**, the vectors
   (a) **AB**   (b) **OM**

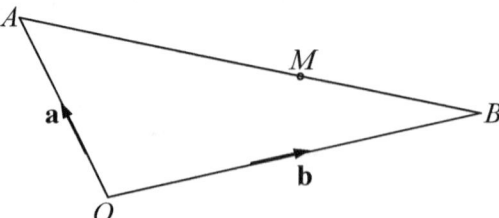

5. In the following figure shows a parallelogram, divided into smaller congruent parallelograms. Given that **OA** = **a** and **OB** = **b**, express in terms of **a** and **b** the displacements **PQ**, **QR** and **RP**.

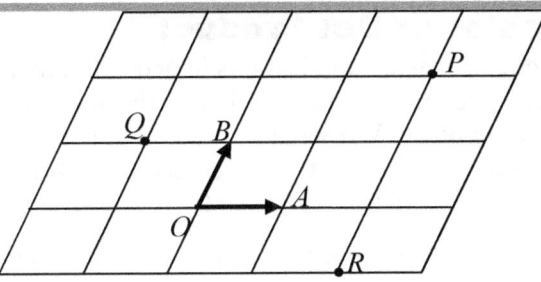

6. In the triangle OAB below, the point C divides AB in the ratio 3:1. Given that **OA** = **a** and **OB** = **b**, express **OC** in terms of **a** and **b**.

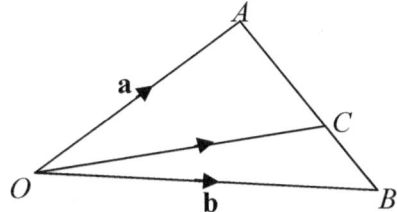

7. A and B are two points whose position vectors are $\mathbf{a} = -3\mathbf{i} + 6\mathbf{j}$ and $\mathbf{b} = 9\mathbf{i} + 3\mathbf{j}$ respectively. Points L and M divide **OA** and **OB** internally in the ratio $m:n$. i.e. $OL:LA = OM:MB = m:n$
   (a) Find in terms of **i** and **j**
      (i) The vector **AB**.
      (ii) The position vectors of the points L and M.
      (iii) The vector **LM**.
   (b) Hence or otherwise, show that **AB** is parallel to **LM**.
   (c) Given that $m:n = 2:1$, find:
      (i) The ratio of the length of LM to the length of AB.
      (ii) The area of the quadrilateral $ALMB$.

8. In the figure below, $OPQR$ is a trapezium with OP and RQ parallel. Given that **RQ** = 2**a**, **OR** = **b**, $\mathbf{RQ} = \dfrac{2}{3}$ **OP** and that M is the midpoint of RQ.
   Find the vector (a) **OM** in terms of **a** and **b**   (b) **QP** in terms of **a** and **b**

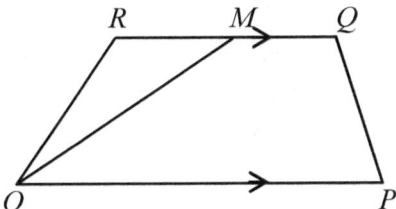

4.6 The Scalar or Dot Product

The scalar or dot product has many applications in geometry, dynamics and many other branches of science. We use it in finding the angle between two vectors, acting in the same plane. We denote the dot or scalar product of two vectors **a** and **b** acting on the same plane by **a** · **b** read 'a dot b' and define it as

$$\mathbf{a} \cdot \mathbf{b} = |\mathbf{a}||\mathbf{b}|\cos\theta$$

Where $0° \leq \theta \leq 180°$ is the angle between the two vectors **a** and **b**.
From the definition, it follows that

(a) If **a** is perpendicular to **b**, $\theta = 90°$, $\cos 90° = 0 \Rightarrow \mathbf{a} \cdot \mathbf{b} = 0$

(b) If **a** is parallel to **b**, $\theta = 0°$, $\cos 0 = 1 \Rightarrow \mathbf{a} \cdot \mathbf{b} = |\mathbf{a}||\mathbf{b}|$.

From (a) and (b) above, we can deduce that
$\mathbf{i} \cdot \mathbf{i} = 1$, $\mathbf{j} \cdot \mathbf{j} = 1$ and $\mathbf{i} \cdot \mathbf{j} = \mathbf{j} \cdot \mathbf{i} = 0$.
Therefore, if $\mathbf{a} = x_1\mathbf{i} + y_1\mathbf{j}$ and $\mathbf{b} = x_2\mathbf{i} + y_2\mathbf{j}$,
Then, $\mathbf{a} \cdot \mathbf{b} = (x_1\mathbf{i} + y_1\mathbf{j}) \cdot (x_2\mathbf{i} + y_2\mathbf{j})$
$= x_1\mathbf{i} \cdot (x_2\mathbf{i} + y_2\mathbf{j}) + y_1\mathbf{j} \cdot (x_2\mathbf{i} + y_2\mathbf{j})$
$\Rightarrow \mathbf{a} \cdot \mathbf{b} = x_1 x_2 + y_1 y_2$

 Example

1. The magnitudes of two vectors **a** and **b** are 5 and 8 and the angle between them is 60°. Find **a** · **b**.

   Solution

   $\mathbf{a}.\mathbf{b} = |\mathbf{a}||\mathbf{b}|\cos\theta = (5)(8)\cos 60° = (5)(8)\left(\dfrac{1}{2}\right) = 20$

2. Find the scalar product of the following pairs of vectors and state which of the pairs are parallel, perpendicular or not parallel or perpendicular.
   (a) 3i+4j and 2i−j   (b) 2i+3j and 3i−2j   (c) 10i+25j and 6i+15j

   Solution
   (a) $(3\mathbf{i} + 4\mathbf{j}) \cdot (2\mathbf{i} - \mathbf{j}) = 3(2) + 4(-1) = 2$

$$|3\mathbf{i}+4\mathbf{j}||2\mathbf{i}-\mathbf{j}| = \left(\sqrt{3^2+4^2}\right)\left(\sqrt{2^2+(-1)^2}\right)$$
$$= \left(\sqrt{25}\right)\left(\sqrt{5}\right) = 5\left(\sqrt{5}\right)$$

Therefore $(3\mathbf{i}+4\mathbf{j})$ and $(2\mathbf{i}-\mathbf{j})$ are neither perpendicular nor parallel.

(b) $(2\mathbf{i}+3\mathbf{j})\cdot(3\mathbf{i}-2\mathbf{j}) = 2(3)+3(-2) = 0$
$\therefore (2\mathbf{i}+3\mathbf{j}) \perp (3\mathbf{i}-2\mathbf{j})$

(c) $(10\mathbf{i}+25\mathbf{j})\cdot(6\mathbf{i}+15\mathbf{j}) = 10(6)+25(15) = 435$
$$|10\mathbf{i}+25\mathbf{j}||6\mathbf{i}+15\mathbf{j}| = \left(\sqrt{10^2+25^2}\right)\left(\sqrt{6^2+15^2}\right)$$
$$= \left(\sqrt{725}\right)\left(\sqrt{261}\right) = \sqrt{189225}$$
$$\Rightarrow |10\mathbf{i}+25\mathbf{j}||6\mathbf{i}+15\mathbf{j}| = 435$$

Also, notice that $10\mathbf{i}+25\mathbf{j} = \frac{5}{3}(6\mathbf{i}+15\mathbf{j})$

Therefore, $10\mathbf{i}+25\mathbf{j} \parallel 6\mathbf{i}+15\mathbf{j}$.

4.7 Angle between two vectors

The angle θ between two vectors is the angle of rotation between the directions of the two vectors. In other words, it is the angle θ between the two divergent rays or the two convergent rays representing the vectors. In the following figures, the angle α is the wrong angle.

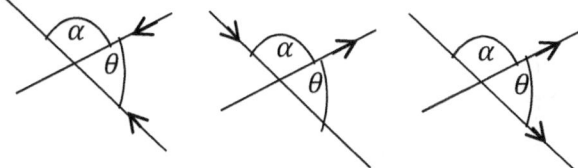

Angle between Two Vectors by the Scalar Product Method

Since $\mathbf{a}\cdot\mathbf{b}=|\mathbf{a}||\mathbf{b}|\cos\theta$, we can use the scalar product to find the angle between any two vectors on the same plane. Hence,

$$\cos\theta = \frac{\mathbf{a}\cdot\mathbf{b}}{|\mathbf{a}||\mathbf{b}|} \Leftrightarrow \theta = \cos^{-1}\left(\frac{\mathbf{a}\cdot\mathbf{b}}{|\mathbf{a}||\mathbf{b}|}\right)$$

This method is best because it does not require any intuitive judgment.

Example

Use the scalar product method to find the angle between the vectors 6i + 8j and 4i −2j.

Solution

$$\theta = \cos^{-1}\left(\frac{\mathbf{a} \cdot \mathbf{b}}{|\mathbf{a}||\mathbf{b}|}\right) = \cos^{-1}\left(\frac{(6\mathbf{i}+8\mathbf{j}).(4\mathbf{i}-2\mathbf{j})}{|6\mathbf{i}+8\mathbf{j}||4\mathbf{i}-2\mathbf{j}|}\right)$$

$$\Rightarrow \theta = \cos^{-1}\left(\frac{(24-16)}{\left(\sqrt{6^2+8^2}\right)\left(\sqrt{4^2+(-2)^2}\right)}\right)$$

$$\theta = \cos^{-1}(0.1789) = 79.7°$$

Finding the Angle between Two Vectors by the Graphical Method

 Example

Using the graphical method find the angle θ between the vectors 6i + 8j and 4i −2j.

Solution

Plot the vectors as position vectors on graph or square paper as below.

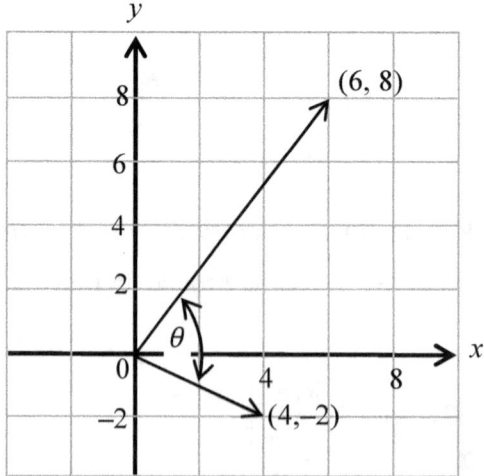

Measure the angle θ using a protractor. Depending on the accuracy of the graph, we should find that the angle is about 80°.

Module 16, Topic 4: Vectors in 2-Dimensions

Formula Method

The angle θ between two vectors $\mathbf{a} = x_1\mathbf{i} + y_1\mathbf{j}$ and $\mathbf{b} = x_2\mathbf{i} + y_2\mathbf{j}$ is given by the formula

$$\theta = \left|\tan^{-1}\left(\frac{y_1}{x_1}\right) - \tan^{-1}\left(\frac{y_2}{x_2}\right)\right|$$

In using this method, we should first draw a sketch of the vectors. Even after doing this, we must take serious care to ensure that the angle is not the wrong angle. Unless the student has a very good sense of judgment, this method is least recommended.

Example

Calculate the angle between the vectors 6i+8j and 4i –2j.

Solution

$$\theta = \left|\tan^{-1}\frac{y_1}{x_1} - \tan^{-1}\frac{y_2}{x_2}\right|$$

$$= \left|\tan^{-1}\left(\frac{8}{6}\right) - \tan^{-1}\left(\frac{-2}{4}\right)\right|$$

$$= 53.13° - (-26.57°)$$

$$= 79.7°$$

Exercise 4:5

1. Given that **u** = 3i+4j and **v** = 4i–3j, find
   (a) The modulus of **u**.   (b) The angle between **u** and **v**.
2. Given two vectors **u** = 2i+3j and **v** = 3i–2j, find
   (a) The modulus of **u**.   (b) The angle between **u** and **v**.
3. Given that |**x**| = 6, |**y**| = 6 and the angle between **x** and **y** is 45°. Calculate **x · y**.
4. Given that |**a**| = 9, |**b**| = 3, find **a · b** if:
   (a) **a** is perpendicular to **b**.   (b) **a** is parallel to **b**.
5. The magnitudes of two vectors **a** and **b** are 6 and 2 and their scalar product is –6. Calculate the angle between these vectors.
6. Find the scalar product of 3i +4j and 4i –3j and state whether these

vectors are parallel or perpendicular.
7. Find the angle between the vectors **u** = 3**i**+4**j** and **v** = –**i** + 8**j**.
8. Find the angle between the vectors **u** and **v**, where **u** = 3**i** + 4**j** and **v** = –6**i** + 8**j**.

Multiple Choice Exercise 4

1. Given that A, B and C are collinear and **OA** = **i** + **j**, **OB** = 2**i** –**j** and **OC** = 3**i** + a**j**. The value of a is:
   [A] 1   [B] –1   [C] –3   [D] –2

2. Given that $\mathbf{u} = \begin{pmatrix} 2 \\ 3 \end{pmatrix}$ and $\mathbf{v} = \begin{pmatrix} 0 \\ 1 \end{pmatrix}$, the numbers a and b such that
   $a\mathbf{u} + b\mathbf{v} = \begin{pmatrix} 4 \\ 5 \end{pmatrix}$
   [A] $a = -1, b = 2$ [B] $a = -2, b = -1$
   [C] $a = 2, b = -1$ [D] $a = 1, b = 2$

3. In the quadrilateral $OABC$, D is the mid-point of BC. Given that, **OA** = **a**, **OB** = **b** and **OC** = **c**. In terms of **a**, **b** and **c** **OD** is:
   [A] $\frac{1}{2}(\mathbf{b}-\mathbf{c})$ [B] $\frac{1}{2}(\mathbf{b}+\mathbf{c})$ [C] $\frac{1}{2}(-\mathbf{b}+\mathbf{c})$ [D] $-\frac{1}{2}(\mathbf{b}-\mathbf{c})$

4. In the quadrilateral $OABC$, D is the mid-point of BC and G is the point on AD such that $AG:GD = 2:1$. Given that, **OA** = **a**, **OB** = **b** and **OC** = **c**. **OG** in terms of **a**, **b** and **c** is:
   [A] $\frac{1}{3}(\mathbf{a}+\mathbf{b}-\mathbf{c})$ [B] $\frac{1}{3}(\mathbf{a}-\mathbf{b}+\mathbf{c})$ [C] $\frac{1}{3}(\mathbf{a}-\mathbf{b}-\mathbf{c})$ [D] $\frac{1}{3}(\mathbf{a}+\mathbf{b}+\mathbf{c})$

5. 5-6 In the figure below, $OPQR$ is a parallelogram $TR = 2OT$ and $RM = MQ$. Given that $\overline{OP} = \mathbf{p}$ and $\overline{OR} = \mathbf{r}$.

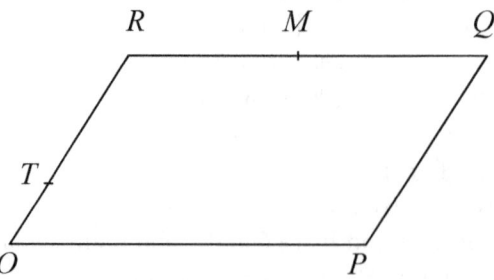

6. The vector **TM** in terms of **p** and **r** is:

   [A] $\frac{2}{3}r - \frac{1}{2}p$ [B] $-\frac{2}{3}r + \frac{1}{2}p$ [C] $-\frac{2}{3}r - \frac{1}{2}p$ [D] $\frac{2}{3}r + \frac{1}{2}p$

7. The vector **PM** in terms of **p** and **r** is:

   [A] $\frac{1}{2}p + r$ [B] $r - \frac{1}{2}p$ [C] $\frac{1}{2}r - p$ [D] $p - \frac{1}{2}r$

8. Let $a = \begin{pmatrix} 2 \\ 5 \end{pmatrix}$, $b = \begin{pmatrix} 4 \\ 4 \end{pmatrix}$ and $c = \begin{pmatrix} 1 \\ 3 \end{pmatrix}$. The relationship between **a**, **b**, and **c**, in the form $b = ua + vc$, where u and v are integers is:

   [A] $b = 3a - 2c$ [B] $b = 3a + 2c$
   [C] $b = -2a + 3c$ [D] $b = -3a + 2c$

9. The magnitude of $u = 2i + 3j$ is:

   [A]   5   [B]   13   [C] $\sqrt{13}$ [D] $\sqrt{5}$

10. 18. The angle between the vectors $u = 2i + 3j$ and $v = 3i - 2j$ is:
    [A] 30° [B] 45° [C] 60° [D] 90°

11. 11-12 In the figure below, $OABC$ is a trapezium with **OA** = **a** and **OB** = **b**. The point P is the midpoint of BC, the point X is the midpoint of OP and the point Y is the midpoint of AC.

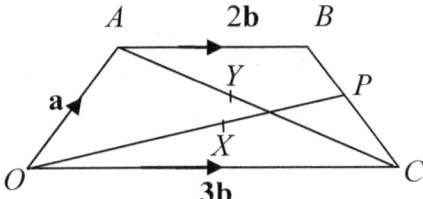

12. In terms of **a** and **b**, **OX** is equal to:

    [A] $\frac{1}{2}a + \frac{5}{2}b$ [B] $\frac{1}{4}(a + 5b)$ [C] $\frac{1}{2}a - \frac{5}{2}b$ [D] $\frac{1}{4}(a - 5b)$

13. In terms of **a** and **b**, **OY** is equal to:

    [A] $\frac{1}{4}(a - 3b)$ [B] $\frac{1}{4}(a + 3b)$ [C] $\frac{1}{2}a - \frac{3}{2}b$ [D] $\frac{1}{2}(a + 3b)$

14. Given that $a = 2i + 5j$, $b = 4i + 9j$, the vector **c** such that $b = 3a - 2c$ is:
    [A] $i - 3j$ [B] $-i + 3j$ [C] $i + 3j$ [D] $2i + 6j$

15. In the following figure, **AC** in terms of **a** and **d** only is:
    [A] $-a + d$ [B] $-a - d$ [C] $a - d$ [D] $a + d$

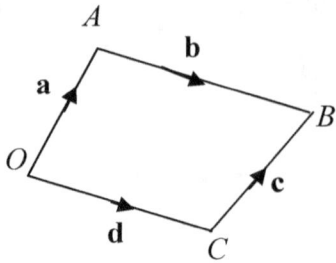

16. In the figure above, the right expression of **OB** in terms of **a** and **b** is:
    [A] −**a** + **b** [B] −**a** − **b** [C] **a** − **b** [D] **a** + **b**
17. In the figure above, the right expression of **OB** in terms of **c** and **d** is:
    [A] **c** + **d** [B] −**c** − **d** [C] −**c** + **d** [D] **c** − **d**
18. In the figure above, given that $\mathbf{b} = \frac{3}{2}\mathbf{d}$. In terms of **a** and **d**, **c** is:
    [A] $-\mathbf{a} + \frac{1}{2}\mathbf{d}$ [B] $-\mathbf{a} - \frac{1}{2}\mathbf{d}$ [C] $\mathbf{a} + \frac{1}{2}\mathbf{d}$ [D] $\mathbf{a} - \frac{1}{2}\mathbf{d}$
19. Given that the position vectors of A, B, C and D are respectively **a**, **b**, **c** and **d** where, $2\mathbf{c} = \mathbf{a}$ and $4\mathbf{d} = \mathbf{a} + \mathbf{b}$. We can write **CD** in terms of **AB** as:
    [A] 4**CD** = **AB**   [B] 4**CD** = −**AB**
    [C] **CD** = 4**AB**   [D] **CD** = −4**AB**
20. 27-28 In the following figure $OPQR$ is a trapezium with OP and RQ parallel. Given that **RQ** = 2**a**, **OR** = **b**, $\mathbf{RQ} = \frac{2}{3}\mathbf{OP}$ and that M is the midpoint of RQ

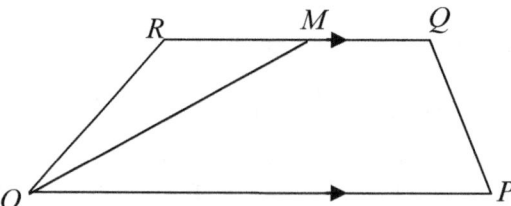

21. The vector **OM** in terms of **a** and **b** is:
    [A] −**a** + **b** [B] −**a** − **b** [C] **a** − **b** [D] **a** + **b**
22. The vector **QP** in terms of **a** and **b** is:
    [A] −**a** + **b** [B] −**a** − **b** [C] **a** − **b** [D] **a** + **b**
23. The following figure shows a triangle OBA. Given that **OA** = **a**, **OB** = **b** and $AM:MB = 2:1$. In terms of **a** and **b**, **AB** is equal to:
    [A] −**a** + **b** [B] −**a** − **b** [C] **a** − **b** [D] **a** + **b**

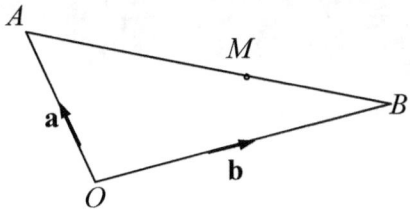

24. The figure above shows a triangle OAB such that $\mathbf{OA} = \mathbf{a}$, $\mathbf{OB} = \mathbf{b}$ and $AM : MB = 2 : 1$. In terms of $\mathbf{a}$ and $\mathbf{b}$, $\mathbf{OM}$ is equal to:

[A] $-\dfrac{1}{3}(\mathbf{a}+2\mathbf{b})$ [B] $\dfrac{1}{3}(\mathbf{a}+2\mathbf{b})$ [C] $\dfrac{1}{3}(\mathbf{a}-2\mathbf{b})$ [D] $\dfrac{1}{3}(-\mathbf{a}+2\mathbf{b})$

Topic 5

SIMPLE TRANSFORMATIONS

Objectives
At the end of this topic, the learner should be able to:

1. Relate object and image for a transformation.
2. Find the image of plane figures geometrically.
3. Find the image of plane figures using the matrix operator.
4. Identify and state the properties of isometries.
5. Determine the matrix operator for an isometry.
6. Establish the relationship between the area scale factor and the determinant of a matrix operator.
7. Find the image of a point by a singular matrix and the equation of a straight line containing all these images.
8. Perform successive transformations.

Module 16, Topic 5: Simple Transformations

TRANSFORMATIONS WITHOUT MATRICES

A **transformation** is a change in position of a point, a line an object.

5.1 Translation, Reflection, Rotation

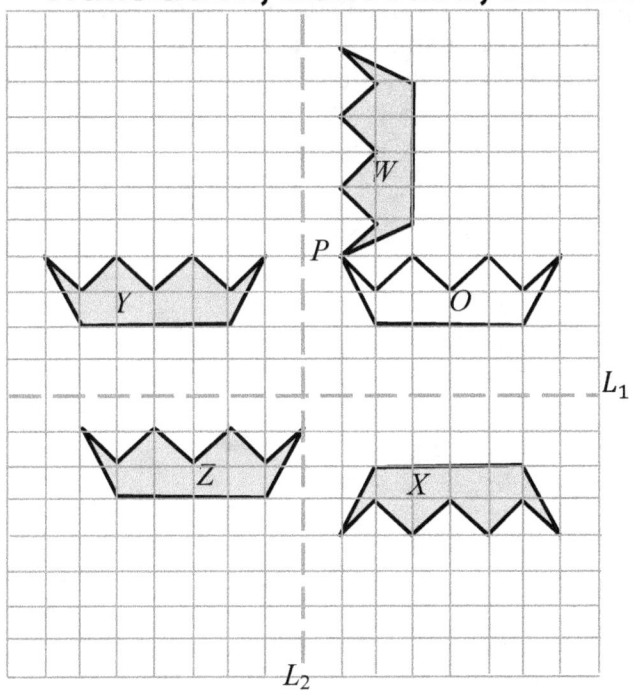

1. In the above diagram, W is a rotation of the plane figure O through an angle of $90°$ in an anticlockwise sense about the point P as axis of rotation. Note that conventionally we take clockwise rotations as negative.
2. X is a reflection of the plane figure O in the dotted line L_1 as mirror line. In a reflection, every point on the object is the same perpendicular distance from the mirror line as the corresponding point on the image.
3. Y is a reflection of the plane figure O in the dotted line L_2 as mirror line.
4. Z is a translation of the plane figure O, 7 units to the left and 5 units down the page. We can describe this translation by the column vector $\begin{pmatrix} -7 \\ -5 \end{pmatrix}$ called a **translation vector**.

In a translation, every point moves the same distance in the same direction, without any rotation.

NOTE!
In each of the above transformations, the shape and size of the figure remains unchanged (shape and size are invariant). An **isometry** is a transformation in which shape and size are invariant.

5.2 Enlargements

In the following figure, $A'B'C'$ is an enlargement of ABC, with enlargement factor 3 and centre O. If corresponding sides are compared, it will be noticed that each side of $\triangle ABC$ is tripled in $\triangle A'B'C'$.

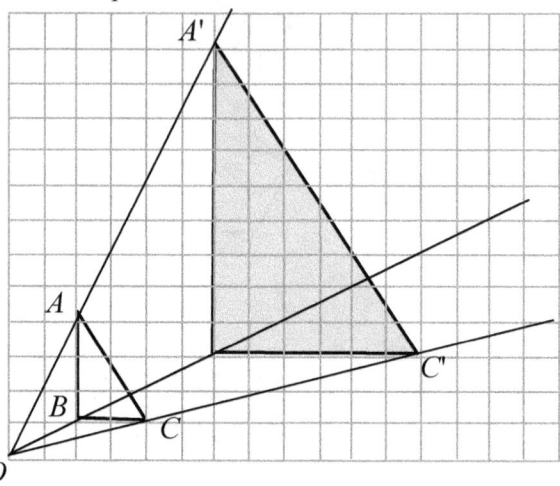

Remarks
1. If during an enlargement, each side of a figure increases k times, we say the enlargement has a scale factor k.
2. By counting squares, we can see that the area of the image $\triangle A'B'C'$ is 9 (or 3^2) times the area of $\triangle ABC$. Thus, the area of an enlargement scale factor k is k^2 times that of the original object.

$$k = \frac{\text{image length}}{\text{object length}}$$
$$\Rightarrow \text{image length} = k \times \text{object length}$$
$$k^2 = \frac{\text{image area}}{\text{object area}}$$
$$\Rightarrow \text{image area} = k^2 \times \text{object area}$$

3. If the object is reduced k times, the scale factor will be $\frac{1}{k}$ (i.e. a fraction). We then say the enlargement has a scale factor $\frac{1}{k}$ and understand that this description means a reduction.

Notice that in an enlargement object and image are similar.

Example

Find the area of the image of an irregular figure of area 252 cm² enlarged by scale factor $\frac{5}{6}$.

Solution

Image area = $k^2 \times$ object area = $\left(\frac{5}{6}\right)^2 \times 252 = 175$ cm²

5.3 Shear

In the following figure, the unit square maps into the parallelogram shaded. We call this type of transformation a shear. In particular, this shear is parallel to the x-axis and points on this axis are invariant. On the other hand, points on BC move through the greatest distance of 2 units to the positive x direction. In such a case, we describe the transformation as **a shear parallel to the x-axis with shear factor 2**.

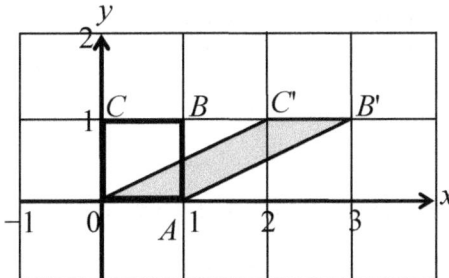

The figure below represents a shear parallel to the x- axis with shear factor –2.

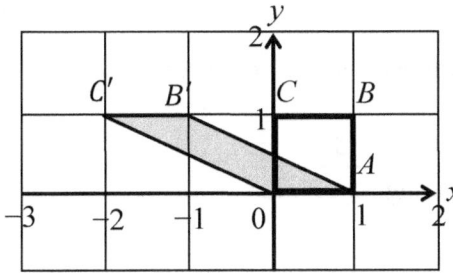

Example

In figure (a) and (b) below, the unit square $OABC$ is transformed into the parallelograms $OAB'C'$ and $OAB''C''$. Describe these transformations completely.

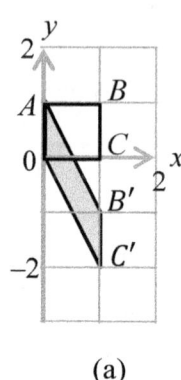

(a)

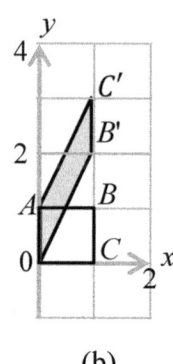

(b)

Solution
(a) A shear parallel to the y-axis with shear factor -2.
(b) A shear parallel to the y-axis with shear factor 2.

5.4 Stretch

In the figure below, $O'A'B'C'$ is a **stretch** of $OABC$ towards the right, **stretch factor** 3.

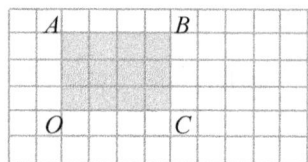 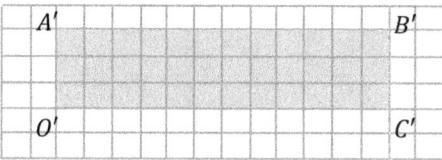

The stretch factor is important because it gives us some valuable information. For a stretch with stretch factor k

1. $k = \dfrac{\text{length of stretched side}}{\text{length of corresponding object side}}$

2. $k = \dfrac{\text{image area}}{\text{object area}}$

3. When $k < -1$, the stretch is in the negative direction and when $k > 1$, the stretch is in the positive direction.
4. When $-1 < k < 1$, the object is compressed in one direction only.

5.5 Invariant Points and Invariant Lines

In any transformation, points and lines which remain unchanged are called invariant points or lines. In the shears and stretches above the points along *OA* in all the transformations are invariant hence the line *OA* is an invariant line. In a rotation, the center of rotation is invariant. In a reflection, points on the mirror line are invariant. Sometimes in an enlargement, the center of the enlargement, if any is invariant as in the enlargement above

Exercise : 5:1

1. Write down the image of the point (–2, 3) if it is reflected in the *x*-axis followed by a reflection in the *y*-axis.
2. The points in the table below undergo a reflection in the lines shown. Fill in the images of the resulting points.

	(2,3)	(3,0)	(0,–5)	(3,3)	(–4,–6)
x-axis					
y-axis					
$y = x$					
$y = -x$					

3. Write down the image of the point (4, 2) if it is reflected in the line $y = x$ followed by a reflection in the *x*-axis.
4. The triangle *ABC* with vertices $A(-3,4)$, $B(5,3)$ and $C(-4,-2)$ is reflected in the *x*-axis. Make a sketch of the triangle *ABC* and its image $A'B'C'$.
5. Determine the images of each of the following points reflected in the line $y = x \tan 60°$.
   (i) (2, 3)   (ii) (3, 0)   (iii) (0,–5)   (iv) (3, 3)   (v) (–4, –6)
6. Find the image of the point (3, –3) after a rotation of 180° anticlockwise about the point (–2, 2).
7. Given that, a translation *T* takes the origin to the point (2, 3); find the image of the following points under *T*.
   (a) (1, 0)   (b) (0, 1)   (c) (–3,–1)   (d) (–1, 2)   (e) (–2, –2)
   (f) (–2, –3)   (g) (*a, b*)   (h) (1, 1)   (i) (–11,15)   (j) (–4, 3)
8. The triangle *ABC* with vertices $A(1,-2)$, $B(5,3)$ and $C(-3,4)$ is translated by *T* which takes the origin to the point (2,3). Make a sketch of the triangle *ABC* and its image $A'B'C'$. Given that T_1 is the translation that takes the origin to the point (–2,–3), what will be the result of *T* followed by T_1? Name this type of transformation.
9. A translation takes the point (2, 3) to $(3,-4)$. What is the image of the

following points under the translation?

(a) (0, 0) (b) $(1,1)$ (c) (1, 0) (d) (–1, 2) (e) $(4,-2)$ (f) (–2, –3)

10. An equilateral triangle ABC of side 1.5 cm maps onto $A_1B_1C_1$ by an anticlockwise rotation of 120° about A. $A_1B_1C_1$ then maps onto $A_2B_2C_2$ by an anticlockwise rotation of 120° about B. Finally, $A_2B_2C_2$ maps onto $A_3B_3C_3$ by an anticlockwise rotation of 120° about C. Using a plane paper accurately construct the complete figure and state a single geometrical transformation, which maps ABC onto $A_3B_3C_3$.
11. A translation takes the origin to the point (2,4). Find the image of the equation $y = x^2 + 4x$ under the transformation.
12. Given that **v** is the vector from the origin to the point (3, 2), state the image of the point (–4, 2) under the translation by the vector **v**.
13. P and Q are transformations defined as follows:
    P: 'Reflect in the y-axis'
    Q: 'Translate +3 units parallel to the x-axis.
    If PQ means P followed by Q, find
    (a) $PQ(2,4)$ (b) $QP(2,4)$
    (c) (x, y) if $QP(x, y) = (4,3)$ (d) (x, y) if $PQ(x, y) = (4,3)$
14. On graph paper, taking 1 cm to represent 1 unit, draw the x- and y-axes from –6 to +12 for each axis.
    (a) Draw the triangle T with vertices $P(2,1)$, $Q(2,3)$ and $R(5,1)$, constructing on the graph and writing down the coordinates of the vertices.
    (b) Translate by vector $\binom{-6}{5}$ the triangle T and label it T_1.
    (c) Reflect T in the y-axis and label it T_2.
    (d) Rotate T through 90° clockwise and label it T_3.
    (e) Enlarge T about the origin by scale factor of 2 and label it T_4.

TRANSFORMATIONS USING MATRICES

5.6 Transformation Matrix (Matrix Operator)

We can easily perform transformations using matrices. In the following pages, we will consider only transformations in the x-y plane. Consider the point $P(x, y)$. We earlier learnt that we can represent this point by the column matrix or vector $\mathbf{P} = \begin{pmatrix} x \\ y \end{pmatrix}$.

Module 16, Topic 5: Simple Transformations

Pre-multiply $\mathbf{P} = \begin{pmatrix} x \\ y \end{pmatrix}$ by the matrix $\mathbf{A} = \begin{pmatrix} a & b \\ c & d \end{pmatrix}$.

$$\mathbf{AP} = \begin{pmatrix} a & b \\ c & d \end{pmatrix}\begin{pmatrix} x \\ y \end{pmatrix} = \begin{pmatrix} ax+by \\ cx+dy \end{pmatrix} = \begin{pmatrix} X \\ Y \end{pmatrix}$$

Thus $\begin{pmatrix} X \\ Y \end{pmatrix}$ is the position vector of a new point.

Therefore, the matrix $\mathbf{A}$ has transformed $P(x,y)$ to the new point $P_1(X, Y)$. We call the matrix $\mathbf{A}$ a **transformation matrix** or a **matrix operator**. We call $P_1(X, Y)$ the **image** of $P(x,y)$. This transformation is denoted by $(x, y) \mapsto (X, Y)$ and read (x, y) maps to (X, Y).

Example

1. Find the image of the point $(2,-3)$ under the transformation defined by the matrix $\begin{pmatrix} 2 & 0 \\ 0 & 2 \end{pmatrix}$.

   Solution

   $$\mathbf{AP} = \begin{pmatrix} 2 & 0 \\ 0 & 2 \end{pmatrix}\begin{pmatrix} 2 \\ -3 \end{pmatrix} = \begin{pmatrix} 4 \\ -6 \end{pmatrix}$$

2. Plot the position vectors of the points $A(2,0)$ and $B(-2,4)$. Find the images A' and B' of A and B under the matrix operator $\mathbf{M} = \begin{pmatrix} 1 & -1 \\ 1 & 1 \end{pmatrix}$ and plot them on the same graph as A and B.

   Solution

   $$\begin{pmatrix} 1 & -1 \\ 1 & 1 \end{pmatrix}\begin{pmatrix} 2 \\ 0 \end{pmatrix} = \begin{pmatrix} 2 \\ 2 \end{pmatrix}, \quad \begin{pmatrix} 1 & -1 \\ 1 & 1 \end{pmatrix}\begin{pmatrix} -2 \\ 4 \end{pmatrix} = \begin{pmatrix} -6 \\ 2 \end{pmatrix}$$

   $$\therefore (2,0) \mapsto (2,2) \text{ and } (-2,4) \mapsto (-6,2)$$

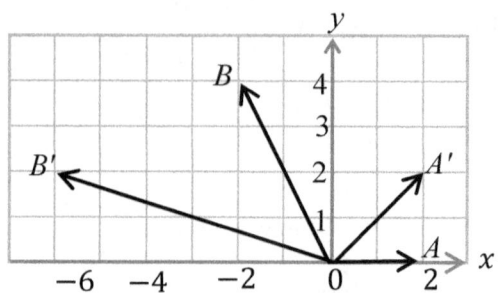

3. Show that the origin (0, 0) is invariant (i.e. remains unchanged) under any transformation defined by a 2×2 matrix.

   Solution

   Let the transformation matrix be $\begin{pmatrix} a & b \\ c & d \end{pmatrix}$.

   Then, $\begin{pmatrix} a & b \\ c & d \end{pmatrix}\begin{pmatrix} 0 \\ 0 \end{pmatrix} = \begin{pmatrix} 0 \\ 0 \end{pmatrix} \implies (0, 0) \mapsto (0, 0)$

   Therefore, origin (0, 0) is invariant.

 Exercise 5:2

1. Find the image of the point (−5, 3) under the transformation defined by each of the following matrices.

   (a) $\begin{pmatrix} 1 & 0 \\ 0 & -1 \end{pmatrix}$ (b) $\begin{pmatrix} 2 & 0 \\ 0 & -2 \end{pmatrix}$ (c) $\begin{pmatrix} \frac{1}{2} & 0 \\ 0 & \frac{1}{2} \end{pmatrix}$ (d) $\begin{pmatrix} 0 & -1 \\ 1 & 0 \end{pmatrix}$

2. Plot the object and image of the following set of points under the matrix indicated.

   (a) $A(-3,4)$, $B(-1,4)$, $C(-1,3)$ under $\begin{pmatrix} 0 & -1 \\ -1 & 0 \end{pmatrix}$.

   (b) $P(2,1)$, $Q(2,2)$, $R(0,2)$ under $\begin{pmatrix} 2 & 0 \\ 0 & -2 \end{pmatrix}$.

5.7 Transformations Involving Many Points

If the points whose images are required are many, we can write them in a single matrix where each column represents the position vector of a single point. We can then compute the product at once.

Example

Find the image of the square with vertices (0, 0), (1, 0), (1, 1) and (0, 1) under the transformation represented by the matrix $\begin{pmatrix} 1 & 2 \\ 0 & 1 \end{pmatrix}$.

Solution

As a single matrix the vertices of the square are $\begin{pmatrix} 0 & 1 & 1 & 0 \\ 0 & 0 & 1 & 1 \end{pmatrix}$.

Pre-multiply by $\begin{pmatrix} 1 & 2 \\ 0 & 1 \end{pmatrix}$.

$\begin{pmatrix} 1 & 2 \\ 0 & 1 \end{pmatrix}\begin{pmatrix} 0 & 1 & 1 & 0 \\ 0 & 0 & 1 & 1 \end{pmatrix} = \begin{pmatrix} 0 & 1 & 3 & 2 \\ 0 & 0 & 1 & 1 \end{pmatrix}$

Exercise 5:3

1. The transformation matrix $\begin{pmatrix} 1 & 0 \\ 1 & 1 \end{pmatrix}$ is applied to the quadrilateral whose vertices are (0, 0), (2, 0), (2, 3), (0, 3). Determine the vertices of the resulting quadrilateral.

2. Find and draw the image of the square with vertices (0, 0), (1, 1), (0, 2), (−1, 1) under the transformation represented by the matrix $\begin{pmatrix} 4 & 3 \\ -3 & -2 \end{pmatrix}$.

3. Plot the rectangle P (0, 0), Q (0, 2), R (3, 2), S (3, 0). Determine and draw the image of PQRS under the following transformation matrices.

   (a) $\begin{pmatrix} 3 & 0 \\ 0 & 2 \end{pmatrix}$ (b) $\begin{pmatrix} 1 & 2 \\ 0 & 1 \end{pmatrix}$ (c) $\begin{pmatrix} 2 & 0 \\ 0 & 2 \end{pmatrix}$

5.8 Describing Transformation

Given any transformation matrix, it is possible to determine and describe completely the nature of the transformation. To do this, consider the first and second columns of the matrix operator to be the images of the base vectors $\mathbf{i} = \begin{pmatrix} 1 \\ 0 \end{pmatrix}$ and $\mathbf{j} = \begin{pmatrix} 0 \\ 1 \end{pmatrix}$ respectively. We can pre-multiply the vertices of a given figure by the matrix operator to obtain the vertices of the image.

Example

Describe the transformation represented by the matrix $\begin{pmatrix} -1 & 0 \\ 0 & 1 \end{pmatrix}$.

Solution

$\begin{pmatrix} 1 \\ 0 \end{pmatrix} \mapsto \begin{pmatrix} -1 \\ 0 \end{pmatrix}$ and $\begin{pmatrix} 0 \\ 1 \end{pmatrix} \mapsto \begin{pmatrix} 0 \\ 1 \end{pmatrix}$

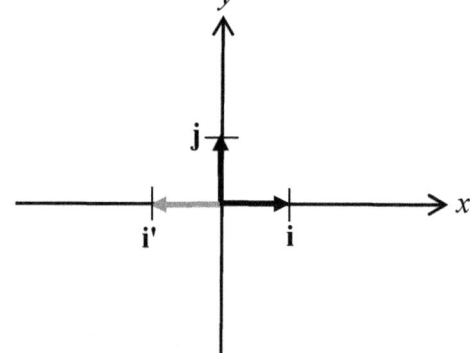

This shows that points on the positive x-axis are reflected on the negative x-axis and points on the y-axis are invariant. Therefore, the transformation is a reflection in the y-axis (i.e. the line $x = 0$)

For more obscure cases, we can pre-multiply the vertices of a regular figure such as the rectangle with vertices $O(0,0)$, $A(2,0)$, $B(2,1)$ and $C(0,1)$ by the transformation matrix to obtain the vertices of the image. We then sketch the two figures on the same Cartesian plane and compare them.

Module 16, Topic 5: Simple Transformations

 Example

1. Using the rectangle with vertices $O(0,0)$, $A(3,0)$, $B(3,2)$ and $C(0,2)$ show that the matrix $\begin{pmatrix} 1 & 0 \\ 0 & -1 \end{pmatrix}$ is a matrix of reflection in the line $y = 0$ (i.e. the x-axis).

Solution
We can write the single matrix, representing the vertices of the rectangle as $R = \begin{pmatrix} 0 & 3 & 3 & 0 \\ 0 & 0 & 2 & 2 \end{pmatrix}$.

Pre-multiplying R by $\begin{pmatrix} 1 & 0 \\ 0 & -1 \end{pmatrix}$ gives

$$\begin{pmatrix} 1 & 0 \\ 0 & -1 \end{pmatrix}\begin{pmatrix} 0 & 3 & 3 & 0 \\ 0 & 0 & 2 & 2 \end{pmatrix} = \begin{pmatrix} 0 & 3 & 3 & 0 \\ 0 & 0 & -2 & -2 \end{pmatrix}$$

Sketching R (not shaded) and its image (shaded) on the same Cartesian plane gives:

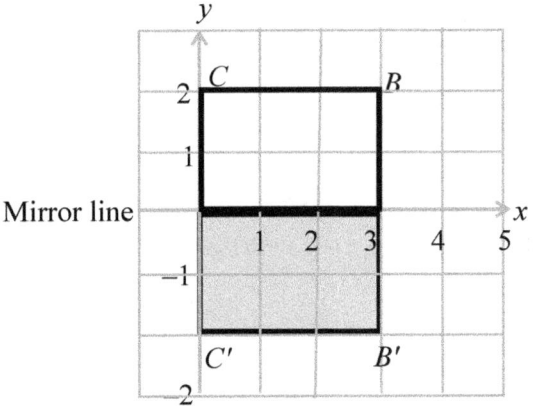

From the graph, $\begin{pmatrix} 1 & 0 \\ 0 & -1 \end{pmatrix}$ is clearly a matrix of reflection in the line $y = 0$ or the x-axis.

2. The matrix $\begin{pmatrix} 0 & 1 \\ -1 & 0 \end{pmatrix}$ transforms the triangle with vertices $A(1,1)$,

$B(3,1)$, $C(3,2)$. Find and describe the image of the transformation.

Solution

We can write the single matrix, representing the vertices of the triangle as $T = \begin{pmatrix} 1 & 3 & 3 \\ 1 & 1 & 2 \end{pmatrix}$.

Pre-multiplying T by $\begin{pmatrix} 0 & 1 \\ -1 & 0 \end{pmatrix}$ gives

$$\begin{pmatrix} 0 & 1 \\ -1 & 0 \end{pmatrix}\begin{pmatrix} 1 & 3 & 3 \\ 1 & 1 & 2 \end{pmatrix} = \begin{pmatrix} 1 & 1 & 2 \\ -1 & -3 & -3 \end{pmatrix}$$

The figure below shows the object and image under this transformation.

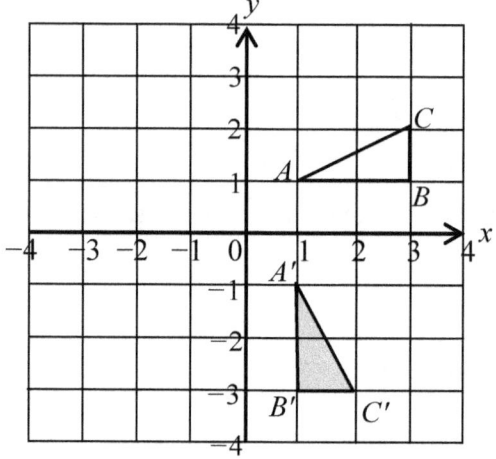

Therefore, from the graph, $\begin{pmatrix} 0 & 1 \\ -1 & 0 \end{pmatrix}$ is a matrix of rotation through 270° about (0,0) as center.

Remarks!
1. Rotations in the anti-clockwise sense are positive and rotations in the clockwise sense are negative.
2. A rotation through 270° is the same as a rotation through −90°.
3. A rotation through 180° is the same as a rotation through −180°
4. A rotation through 360° is an invariant transformation. Therefore, its transformation matrix is the identity matrix $\mathbf{I} = \begin{pmatrix} 1 & 0 \\ 0 & 1 \end{pmatrix}$.

Module 16, Topic 5: Simple Transformations

 Example

1. Find the image of the rectangle with vertices $O(0,0)$, $A(3,0)$, $B(3,2)$ and $C(0,2)$ under the transformation with matrix $\begin{pmatrix} 4 & 0 \\ 0 & 4 \end{pmatrix}$. Hence, describe the transformation completely.

   Solution

   Pre-multiplying the vertices by $\begin{pmatrix} 4 & 0 \\ 0 & 4 \end{pmatrix}$ gives

   $$\begin{pmatrix} 4 & 0 \\ 0 & 4 \end{pmatrix}\begin{pmatrix} 0 & 3 & 3 & 0 \\ 0 & 0 & 2 & 2 \end{pmatrix} = \begin{pmatrix} 0 & 12 & 12 & 0 \\ 0 & 0 & 8 & 8 \end{pmatrix}$$

   Figure 46:5 shows the rectangle $OABC$ and its image $OA'B'C'$. We call such a transformation an enlargement. Careful observation will reveal that each side of the image is 4 times longer. In this case, we say the enlargement has scale factor 4.

   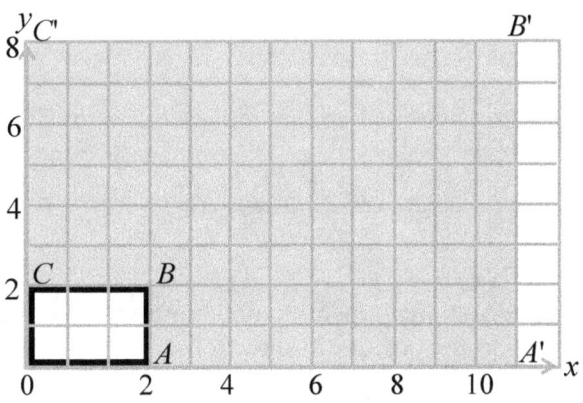

2. Use the rectangle with vertices $O(0,0)$, $A(3,0)$, $B(3,2)$ and $C(0,2)$ to determine and describe each of the following transformations:

   (i) $\begin{pmatrix} 1 & 2 \\ 0 & 1 \end{pmatrix}$ (ii) $\begin{pmatrix} 2 & 0 \\ 0 & 1 \end{pmatrix}$

   Solution

   (i) Pre-multiplying the vertices by $\begin{pmatrix} 1 & 2 \\ 0 & 1 \end{pmatrix}$ gives

$$\begin{pmatrix} 1 & 2 \\ 0 & 1 \end{pmatrix} \begin{pmatrix} 0 & 3 & 3 & 0 \\ 0 & 0 & 2 & 2 \end{pmatrix} = \begin{pmatrix} 0 & 3 & 7 & 4 \\ 0 & 0 & 2 & 2 \end{pmatrix}$$

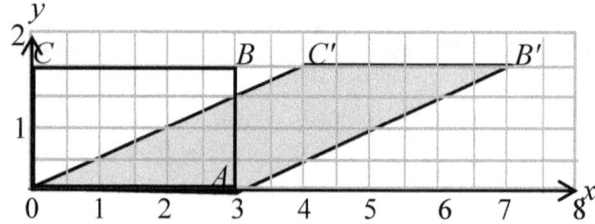

From the graph, we can see that, the transformation represents a shear with scale factor 2, parallel to the positive x- direction.

(ii) Pre-multiplying the vertices by $\begin{pmatrix} 2 & 0 \\ 0 & 1 \end{pmatrix}$ gives

$$\begin{pmatrix} 2 & 0 \\ 0 & 1 \end{pmatrix} \begin{pmatrix} 0 & 3 & 3 & 0 \\ 0 & 0 & 2 & 2 \end{pmatrix} = \begin{pmatrix} 0 & 6 & 6 & 0 \\ 0 & 0 & 2 & 2 \end{pmatrix}$$

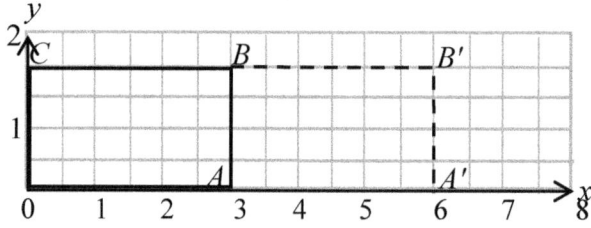

From the graph, we can see that, the transformation represents a stretch with scale factor 2, parallel to the positive x- direction.

Remarks

1. The transformation matrix representing a shear parallel to the x-axis is always of the form $\begin{pmatrix} 1 & k \\ 0 & 1 \end{pmatrix}$. When $k < 0$, the shear is in the negative x- direction and when $k > 0$, the shear is in the positive x- direction. We can show that the transformation matrix representing a shear parallel to the y-axis is of the form $\begin{pmatrix} 1 & 0 \\ k & 1 \end{pmatrix}$.

2. The transformation matrix representing a stretch parallel to the x-axis is

always of the form $\begin{pmatrix} k & 0 \\ 0 & 1 \end{pmatrix}$. We can show that the transformation matrix representing a stretch parallel to the y-axis is of the form $\begin{pmatrix} 1 & 0 \\ 0 & k \end{pmatrix}$.

Exercise 5:4

1. Use base vectors to determine and describe the transformations whose matrices are as follows:

$A = \begin{pmatrix} 1 & 0 \\ 0 & 1 \end{pmatrix}$, $B = \begin{pmatrix} 0 & -1 \\ 1 & 0 \end{pmatrix}$, $C = \begin{pmatrix} 0 & 1 \\ -1 & 0 \end{pmatrix}$, $D = \begin{pmatrix} -1 & 0 \\ 0 & -1 \end{pmatrix}$, $E = \begin{pmatrix} 0 & 1 \\ 1 & 0 \end{pmatrix}$,

$F = \begin{pmatrix} -1 & 0 \\ 0 & 1 \end{pmatrix}$, $G = \begin{pmatrix} 0 & -1 \\ -1 & 0 \end{pmatrix}$, $H = \begin{pmatrix} 1 & 0 \\ 0 & -1 \end{pmatrix}$, $P = \begin{pmatrix} 2 & 0 \\ 0 & 2 \end{pmatrix}$,

$Q = \begin{pmatrix} \frac{1}{2} & 0 \\ 0 & \frac{1}{2} \end{pmatrix}$, $R = \begin{pmatrix} -\frac{1}{2} & 0 \\ 0 & -\frac{1}{2} \end{pmatrix}$, $S = \begin{pmatrix} -2 & 0 \\ 0 & -2 \end{pmatrix}$, $T = \begin{pmatrix} 1 & 0 \\ 0 & 2 \end{pmatrix}$,

$U = \begin{pmatrix} 1 & 0 \\ 2 & 1 \end{pmatrix}$, $V = \begin{pmatrix} 1 & 1 \\ 0 & 1 \end{pmatrix}$, $W = \begin{pmatrix} 3 & 0 \\ 0 & 1 \end{pmatrix}$, $Y = \begin{pmatrix} 1 & 0 \\ 0 & 3 \end{pmatrix}$.

2. Using the matrix in question (1) draw, the object and the image of the rectangle which the matrix $\begin{pmatrix} 0 & 2 & 2 & 0 \\ 0 & 0 & 1 & 1 \end{pmatrix}$ represent and describe the transformation in each case.

5.9 Finding Transformation Matrices

There are two ways of finding transformation matrices.

1. Find the images of the base vectors $\begin{pmatrix} 1 \\ 0 \end{pmatrix}$ and $\begin{pmatrix} 0 \\ 1 \end{pmatrix}$ under the transformation.

   Thus if under the given transformation

   $\begin{pmatrix} 1 \\ 0 \end{pmatrix} \mapsto \begin{pmatrix} a \\ c \end{pmatrix}$ and $\begin{pmatrix} 0 \\ 1 \end{pmatrix} \mapsto \begin{pmatrix} b \\ d \end{pmatrix}$ then the transformation matrix is $\begin{pmatrix} a & b \\ c & d \end{pmatrix}$.

Example

1. Find the matrix for a reflection in the x-axis.

   Solution

   $\begin{pmatrix}1\\0\end{pmatrix} \mapsto \begin{pmatrix}1\\0\end{pmatrix}$ and $\begin{pmatrix}0\\1\end{pmatrix} \mapsto \begin{pmatrix}0\\-1\end{pmatrix}$.

   Therefore, the transformation matrix is $\begin{pmatrix}1 & 0\\0 & -1\end{pmatrix}$.

2. What is the transformation matrix for a stretch by a factor 5 parallel to Ox?

   Solution

   $\begin{pmatrix}1\\0\end{pmatrix} \mapsto \begin{pmatrix}5\\0\end{pmatrix}$ and $\begin{pmatrix}0\\1\end{pmatrix} \mapsto \begin{pmatrix}0\\1\end{pmatrix}$

   Therefore, the transformation matrix is $\begin{pmatrix}5 & 0\\0 & 1\end{pmatrix}$.

   If we know two object points, which are out of the invariant line, and different from the origin, and two corresponding image points, then we pre-multiply the object points by $\begin{pmatrix}a & b\\c & d\end{pmatrix}$ and equate to the image points. Then we solve for a, b, c and d.

3. Under a certain transformation, $\begin{pmatrix}3\\-2\end{pmatrix} \mapsto \begin{pmatrix}9\\8\end{pmatrix}$ and $\begin{pmatrix}5\\1\end{pmatrix} \mapsto \begin{pmatrix}-3\\4\end{pmatrix}$. Find the transformation matrix.

   Solution

   Let the transformation matrix be $\begin{pmatrix}a & b\\c & d\end{pmatrix}$.

   Then, $\begin{pmatrix}a & b\\c & d\end{pmatrix}\begin{pmatrix}3\\-2\end{pmatrix} = \begin{pmatrix}9\\8\end{pmatrix}$.

   $\Rightarrow 3a - 2b = 9$ .... ①
   $3c - 2d = 8$ .... ②

Module 16, Topic 5: Simple Transformations

$$\begin{pmatrix} a & b \\ c & d \end{pmatrix} \begin{pmatrix} 5 \\ 1 \end{pmatrix} = \begin{pmatrix} -3 \\ 4 \end{pmatrix}$$

$\Rightarrow 5a + b = -3$ ....③
$ 5c + d = 4$ ....④

$2 \times ③ + ①: \quad 13a = 3 \Rightarrow a = \dfrac{3}{13}$

Substitute in ①: $3\left(\dfrac{3}{13}\right) - 2b = 8 \Rightarrow b = -\dfrac{54}{13}$

$2 \times ④ + ②: \quad 13c = 16 \Rightarrow c = \dfrac{16}{13}$

Substitute in ②: $3\left(\dfrac{3}{13}\right) - 2d = 8 \Rightarrow d = -\dfrac{28}{13}$

Therefore, the transformation matrix is $\begin{pmatrix} \dfrac{3}{13} & -\dfrac{54}{13} \\ \dfrac{16}{13} & -\dfrac{28}{13} \end{pmatrix}$.

Exercise 5:5

Determine the transformation matrix, which maps the unit square $\begin{pmatrix} 0 & 1 & 1 & 0 \\ 0 & 0 & 1 & 1 \end{pmatrix}$ into each of the following figures. Illustrate each mapping with a sketch and describe the transformation in each case.

1. $\begin{pmatrix} 0 & 3 & 5 & 2 \\ 0 & 1 & 2 & 1 \end{pmatrix}$
2. $\begin{pmatrix} 0 & 3 & 4 & 1 \\ 0 & 0 & 1 & 1 \end{pmatrix}$
3. $\begin{pmatrix} 0 & 2 & 2 & 0 \\ 0 & 0 & 1 & 1 \end{pmatrix}$
4. $\begin{pmatrix} 0 & 2 & 2 & 0 \\ 0 & 0 & 2 & 2 \end{pmatrix}$
5. $\begin{pmatrix} 0 & 4 & 5 & 1 \\ 0 & 0 & 2 & 1 \end{pmatrix}$
6. $\begin{pmatrix} 0 & 1 & 1 & 0 \\ 0 & 0 & -1 & -1 \end{pmatrix}$
7. $\begin{pmatrix} 0 & 3 & 3 & 0 \\ 0 & 0 & -3 & -3 \end{pmatrix}$
8. $\begin{pmatrix} 0 & 0 & -1 & -1 \\ 0 & 1 & 1 & 0 \end{pmatrix}$
9. $\begin{pmatrix} 0 & 3 & 3 & 0 \\ 0 & 0 & 3 & 3 \end{pmatrix}$
10. $\begin{pmatrix} 0 & 1 & -1 & -2 \\ 0 & 0 & 1 & 1 \end{pmatrix}$
11. $\begin{pmatrix} 0 & 3 & 4 & 1 \\ 0 & 1 & 3 & 2 \end{pmatrix}$
12. $\begin{pmatrix} 0 & 0 & 1 & 1 \\ 0 & \dfrac{1}{2} & 1\dfrac{1}{2} & 1 \end{pmatrix}$

5.10 Inverse Transformations

Suppose **A** is a transformation matrix. Provided the determinant of **A** is not zero i.e. $|A| \neq 0$, we can find the inverse of **A** (i.e. A^{-1}). A^{-1} performs the reverse of what **A** does. For instance, if **A** is a rotation of 90° anticlockwise, then A^{-1} will be a rotation of 90° clockwise.

Therefore, if A maps $\begin{pmatrix} x \\ y \end{pmatrix}$ to $\begin{pmatrix} X \\ Y \end{pmatrix}$

Then A^{-1} maps $\begin{pmatrix} X \\ Y \end{pmatrix}$ to $\begin{pmatrix} x \\ y \end{pmatrix}$

Example

1. The matrix $\begin{pmatrix} 0 & -1 \\ 1 & 1 \end{pmatrix}$ transforms the point (2,–3) onto the point (3,2).

   Find the matrix, which transforms the point (3,2) onto the point (2,–3).

   Solution

   Let $\begin{pmatrix} 0 & -1 \\ 1 & 1 \end{pmatrix} = A$. Then the required matrix is

   $$A^{-1} = \frac{1}{\det A}(\text{Adj}A) = \frac{1}{1}\begin{pmatrix} 0 & 1 \\ -1 & 1 \end{pmatrix} = \begin{pmatrix} 0 & 1 \\ -1 & 1 \end{pmatrix}$$

2. The matrix $M = \begin{pmatrix} 1 & 0 \\ -2 & 1 \end{pmatrix}$ maps a point (x, y) to $(6,-9)$. Find the values of x and y.

   Solution

   $$\begin{pmatrix} 1 & 0 \\ -2 & 1 \end{pmatrix}\begin{pmatrix} x \\ y \end{pmatrix} = \begin{pmatrix} 6 \\ -9 \end{pmatrix}$$

   $$\begin{pmatrix} 1 & 0 \\ -2 & 1 \end{pmatrix}^{-1} = \frac{1}{1}\begin{pmatrix} 1 & 0 \\ 2 & 1 \end{pmatrix} = \begin{pmatrix} 1 & 0 \\ 2 & 1 \end{pmatrix}$$

$$\Rightarrow \begin{pmatrix} x \\ y \end{pmatrix} = \begin{pmatrix} 1 & 0 \\ 2 & 1 \end{pmatrix} \begin{pmatrix} 6 \\ -9 \end{pmatrix}$$

$$\Rightarrow \begin{pmatrix} x \\ y \end{pmatrix} = \begin{pmatrix} 6 \\ 3 \end{pmatrix}$$

Therefore, the values of x and y are $x = 6$ and $y = 3$.

5.11 Transformations by Singular Matrices

The matrix $\begin{pmatrix} 4 & 2 \\ 2 & 1 \end{pmatrix}$ transforms the rectangle $\begin{pmatrix} 0 & 3 & 3 & 0 \\ 0 & 0 & 2 & 2 \end{pmatrix}$ as follows:

$$\begin{pmatrix} 4 & 2 \\ 2 & 1 \end{pmatrix} \begin{pmatrix} 0 & 3 & 3 & 0 \\ 0 & 0 & 2 & 2 \end{pmatrix} = \begin{pmatrix} 0 & 12 & 16 & 4 \\ 0 & 6 & 8 & 2 \end{pmatrix}$$

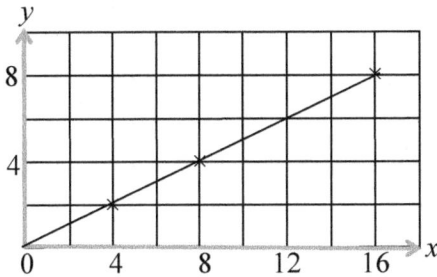

Verify that $\det \begin{pmatrix} 4 & 2 \\ 2 & 1 \end{pmatrix} = 0$.

Hence $\begin{pmatrix} 4 & 2 \\ 2 & 1 \end{pmatrix}$ is a non-zero singular matrix. Therefore, non-zero singular matrices map all points to a straight line. A **dilation** is a transformation, which maps geometrical shapes to a straight line.

5.12 Transformations involving Change of Area

For any transformation, the ratio of the image area to the object area is equal to the absolute value of the determinant of the transformation matrix **A**.

$$\text{image area} = |\det \mathbf{A}| \times \text{object area}$$

Example

1. A rectangle with vertices (4, 4), (4,–1), (10,–1), (10, 4) is transformed by the matrix $\begin{pmatrix} 3 & 3 \\ 2 & 3 \end{pmatrix}$. Find the area of its image.

   Solution

   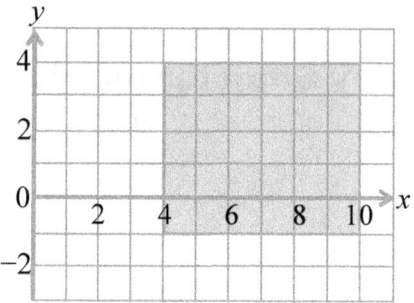

   Image area = |det **A**|×object area
   Object area = 5×6 = 30 un²

   $\Rightarrow$ Image area = $\left|\det\begin{pmatrix} 3 & 3 \\ 2 & 3 \end{pmatrix}\right| \times 30$

   $= |9 - 6|30 = 90$ un²

2. Find the ratio of the image area to the object area under a transformation defined by the matrix $\begin{pmatrix} 4 & 2 \\ 1 & 2 \end{pmatrix}$.

   Solution

   Let **A** = $\begin{pmatrix} 4 & 2 \\ 1 & 2 \end{pmatrix}$

   Image area = |det **A**|×object area
   Image area: object area = |det **A**|:1
       Det **A** = 8–2 = 6
   ∴Image area: object area = 6:1

3. The base of a triangle is 6 cm and its height is 8 cm. Find the area of its image under a transformation defined by the matrix operator $\begin{pmatrix} 16 & 4 \\ 4 & 2 \end{pmatrix}$.

Module 16, Topic 5: Simple Transformations

Solution

Let $\mathbf{A} = \begin{pmatrix} 16 & 4 \\ 4 & 2 \end{pmatrix}$

Image area = $|\det \mathbf{A}| \times$ object area

$|\det \mathbf{A}| = |16(2) - (4)(4)| = |32 - 16| = 16$

∴ Image area = $16 \times 24 = 384$ cm^2

5.13 Composite or Successive Transformations

Suppose **A** and **B** are two matrices representing two transformations for which the origin is invariant, then we can perform a transformation **A** followed by **B** by the matrix **BA**.

Example

The triangle T_1 with vertices $A_1(1,1)$, $B_1(2,3)$ and $C_1(4,3)$, is mapped into triangle T_2 with vertices $A_2B_2C_2$ by the transformation P whose matrix is

$\mathbf{P} = \begin{pmatrix} 0 & 1 \\ -1 & 0 \end{pmatrix}$.

The transformation Q whose matrix is $\mathbf{Q} = \begin{pmatrix} 1 & 2 \\ 0 & 1 \end{pmatrix}$ further maps T_2 to T_3 whose vertices are $A_3B_3C_3$. Find the image of T_3 without finding the image of T_2.

Solution

$\mathbf{QP} = \begin{pmatrix} 1 & 2 \\ 0 & 1 \end{pmatrix}\begin{pmatrix} 0 & 1 \\ -1 & 0 \end{pmatrix} = \begin{pmatrix} -2 & 1 \\ -1 & 0 \end{pmatrix}$

∴ $T_3 = \begin{pmatrix} -2 & 1 \\ -1 & 0 \end{pmatrix}\begin{pmatrix} 1 & 2 & 4 \\ 1 & 3 & 3 \end{pmatrix}$

5.14 Translations and Translation Vectors

A translation is a transformation in which every object point moves the same distance in the same direction. Translations are the only transformations treated in this book, which we cannot represent using a 2×2 matrix. However, we represent translations using translation vectors.

The translation vector $\mathbf{t} = \begin{pmatrix} p \\ q \end{pmatrix}$ transforms any point $A(a, b)$ by moving it p units in the Ox direction and q units in the Oy direction. In this light, the image A' of the point $A(a,b)$ under a translation represent with translation vector $\mathbf{t} = \begin{pmatrix} p \\ q \end{pmatrix}$ is

$$\mathbf{OA'} = \mathbf{OA} + \mathbf{t} = \begin{pmatrix} a \\ b \end{pmatrix} + \begin{pmatrix} p \\ q \end{pmatrix} = \begin{pmatrix} a+p \\ b+q \end{pmatrix}$$

Example

Determine the image of the triangle ABC whose vertices are $A(1, 1)$, $B(-2, 1)$ and $C(3, -4)$ under the translation whose vector is $\begin{pmatrix} 4 \\ -7 \end{pmatrix}$.

Solution

$$\begin{pmatrix} 1 \\ 1 \end{pmatrix} \rightarrow \begin{pmatrix} 1 \\ 1 \end{pmatrix} + \begin{pmatrix} 4 \\ -7 \end{pmatrix} = \begin{pmatrix} 5 \\ -6 \end{pmatrix}$$

$$\begin{pmatrix} -2 \\ 1 \end{pmatrix} \rightarrow \begin{pmatrix} -2 \\ 1 \end{pmatrix} + \begin{pmatrix} 4 \\ -7 \end{pmatrix} = \begin{pmatrix} 2 \\ -6 \end{pmatrix}$$

$$\begin{pmatrix} 3 \\ -4 \end{pmatrix} \rightarrow \begin{pmatrix} 3 \\ -4 \end{pmatrix} + \begin{pmatrix} 4 \\ -7 \end{pmatrix} = \begin{pmatrix} 7 \\ -11 \end{pmatrix}$$

Therefore, the image of the triangle ABC is $A'(5, -6), B'(2, -6)$ and $C'(7, -11)$

Module 16, Topic 5: Simple Transformations

SUMMARY

It is very imperative to take note of the following transformation matrices, which are very common at this level.

	Operator	Geometrical Effect
1	$\begin{pmatrix} 1 & 0 \\ 0 & -1 \end{pmatrix}$	A reflection in the x-axis
2	$\begin{pmatrix} -1 & 0 \\ 0 & 1 \end{pmatrix}$	A reflection in the y-axis
3	$\begin{pmatrix} 0 & 1 \\ 1 & 0 \end{pmatrix}$	A reflection in the line $y = x$
4	$\begin{pmatrix} 0 & -1 \\ -1 & 0 \end{pmatrix}$	A reflection in line $y = -x$
5	$\begin{pmatrix} 0 & -1 \\ 1 & 0 \end{pmatrix}$	An anticlockwise rotation through 90° about the origin (0,0)
6	$\begin{pmatrix} -1 & 0 \\ 0 & -1 \end{pmatrix}$	An anticlockwise rotation through 180° about the origin (0,0)
7	$\begin{pmatrix} 0 & 1 \\ -1 & 0 \end{pmatrix}$	An anticlockwise rotation through 270° about the origin (0,0)
8	$\begin{pmatrix} 1 & 0 \\ 0 & 1 \end{pmatrix}$	The identity transformation leaves all points invariant
9	$\begin{pmatrix} k & 0 \\ 0 & 1 \end{pmatrix}$	A stretch in the direction Ox, stretch factor k
10	$\begin{pmatrix} 1 & 0 \\ 0 & k \end{pmatrix}$	A stretch in the direction Oy, stretch factor k
11	$\begin{pmatrix} k & 0 \\ 0 & k \end{pmatrix}$	An enlargement centre (0,0), enlargement factor k
12	$\begin{pmatrix} 1 & k \\ 0 & 1 \end{pmatrix}$	A shear in the direction Ox, shear factor k
13	$\begin{pmatrix} 1 & 0 \\ k & 1 \end{pmatrix}$	A shear in the direction Oy, shear factor k

Exercise 5:6

1. (a) Draw the quadrilateral with vertices $A(0,0)$, $B(3,4)$, $C(4,0)$, $D(3,1)$. Find the image of $ABCD$ under the transformation represented by the matrix $\begin{pmatrix} -2 & 0 \\ 0 & -2 \end{pmatrix}$. Find the ratio $\left(\dfrac{\text{area of image}}{\text{area of object}}\right)$.

   (b) Repeat (a) using the quadrilateral $P(-4,1)$, $Q(-2,4)$, $R(0,1)$, $S(-1,0)$.

2. (a) Draw the triangle $A(1,1)$, $B(4,1)$, $C(1,3)$ and find the image of ABC under the transformation with matrix $\begin{pmatrix} 2 & 0 \\ 0 & 2 \end{pmatrix}$.

   (b) Without calculating the actual areas of the triangles, determine the ratio $\left(\dfrac{\text{area of image}}{\text{area of object}}\right)$.

3. An irregular figure O whose area is 16 cm² is transformed to the image I by the matrix $\begin{pmatrix} 2 & 1 \\ 1 & 2 \end{pmatrix}$. Find the area of I.

4. The quadrilateral with vertices $A(0, 0)$, $B(3, 4)$, $C(7,4)$, $D(4, 0)$ undergoes a transformation represented by the matrix $\mathbf{Q} = \begin{pmatrix} 2 & 3 \\ 6 & 9 \end{pmatrix}$. Find the area of the image $A'B'C'D'$ of $ABCD$.

5. (a) Taking 1 cm for 1 unit on each axis, ranging from –6 to +6, draw the figure $OABC$ with $A(3, 2)$, $B(5, 2)$ and $C(2, 0)$. Given the transformation matrix $\mathbf{T} = \begin{pmatrix} 0 & -1 \\ 1 & 0 \end{pmatrix}$.

   (b) Plot the image $OA'B'C'$ of $OABC$ under the transformation T.
   (c) Determine T^2 and plot the image $OA"B"C"$ of $OABC$ under T^2.
   (d) Describe the geometrical effects on $OABC$ of T and T^2.

6. A transformation is represented by the 2×2 matrix $\mathbf{M} = \begin{pmatrix} 0 & -1 \\ -1 & 0 \end{pmatrix}$.

   Describe this transformation completely.
   Use this matrix to find the image of a quadrilateral $ABCD$ where $A(3, 1)$, $B(0, 3)$, $C(-2, 1)$ and $D(1, -1)$. Give the coordinates of the vertices of the image of $ABCD$.

7. The triangle ABC with vertices $A(1, 0)$, $B(2, 1)$ and $C(4, 0)$ is mapped

onto triangle $A'B'C'$ by the transformation matrix $T = \begin{pmatrix} 0 & 1 \\ -1 & 0 \end{pmatrix}$.

(a) Find the coordinates of the images A', B' and C'.
(b) Using the scale of 2 cm to 1 unit on both axes plot the triangle ABC and its image $A'B'C'$ on graph paper.
(c) Describe the transformation T geometrically.
(d) Determine, the inverse of the transformation matrix. Hence, or otherwise, describe completely the transformation represented by this inverse matrix.

8. The matrix $M = \begin{pmatrix} 0 & -1 \\ 1 & 0 \end{pmatrix}$ maps the point A on to the point B $(4, -6)$.

(a) Find the coordinates of A
(b) Describe completely the transformation whose matrix is M.

9. Using graph paper and with a scale of 1 cm for 1 unit on both axes, plot the points $A\,(0, 2)$, $B\,(1, 2)$, $C(3,1)$ and $D\,(2, 1)$. Name the type of quadrilateral $ABCD$.

Transform $ABCD$ using the matrix $M = \begin{pmatrix} -2 & 0 \\ 0 & -2 \end{pmatrix}$.

Plot the image $A'B'C'D'$. Describe completely the transformation M. By construction, find the image $A''B''C''D''$ of $A'B'C'D'$ under the transformation M. Write on your graph the coordinates of the vertices. Find the ratio of the area of $ABCD$ to that of $A''B''C''D''$.

Multiple Choice Exercise 5

1. In the figure below, triangle $A'B'C'$ is an enlargement of triangle ABC. The scale factor of the enlargement is:

   [A] $\dfrac{3}{2}$ [B] 2   [C] $\dfrac{5}{2}$ $\dfrac{5}{2}$ [D] 3

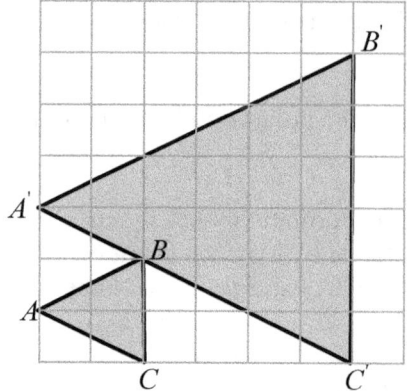

2. In the figure below, not drawn to scale, triangle $OA'B'$ is an enlargement of triangle OAB.

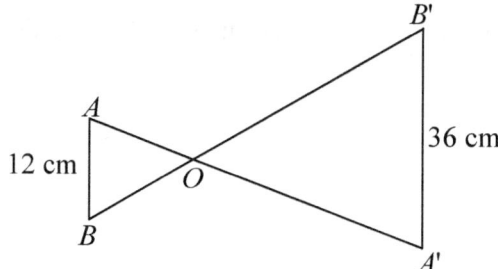

3. The scale factor of the enlargement is:

   [A] $-\dfrac{1}{3}$ [B] -3 [C] $\dfrac{1}{3}$ [D] 3

3. The point (4,3) is reflected in the x-axis followed by a reflection in the y-axis. The final image is:

   [A] (4,–3)   [B] (–4,3)   [C] (–4,–3)   [D] (–3,4)

4. The image of the point (–2, 6) under the transformation $T: (x,y) \mapsto (2x+y, -x+y)$ is:

   [A] (–2,6)   [B] (2,8)   [C] (2,–6)   [D] (2,–8)

5. The image of the point (–3,4) under the transformation $T: (x,y) \mapsto (2x, 2y)$ is:

   [A] (–6,–8)   [B] (6,–8)   [C] (6,8)   [D] (–6,8)

6. The statement that correctly describes the transformation

$T: (x, y) \mapsto (2x, 2y)$ is:
[A] An enlargement scale factor 2 centre (0, 0)
[B] An enlargement scale factor 2 centre (0, 2)
[C] An enlargement scale factor 2 centre (2, 2)
[D] A translation 2 units to the right

7. The square with vertices $W(0,0)$, $X(1,0)$, $Y(1,1)$, $Z(0,1)$ is transformed into a square with vertices $W(0,0)$, $X(1,0)$, $Y(3,1)$, $Z(2,1)$. The statements that correctly describes the transformation T is:
[A] A shear, shear factor 2 where points on the y-axis are invariant.
[B] A stretch, stretch factor 2 where points on the y-axis are invariant.
[C] A shear, shear factor 2 where points on the x-axis are invariant.
[D] A stretch, stretch factor 2 where points on the x-axis are invariant.

8. A triangle whose vertices are $A(1,2)$, $B(3,4)$ and $C(6,2)$ is transformed to a triangle whose vertices are $A'(1,-2)$, $B'(3,-4)$ and $C'(6,-2)$. The statement, which correctly describes the transformation, is:
[A] T is a reflection in the line $y = -1$ [B] T is a reflection in the line $y = x$
[C] T is a reflection in the line $y = -x$ [D] T is a reflection in the line $y = 0$

9. The figure below shows a triangle ABC.

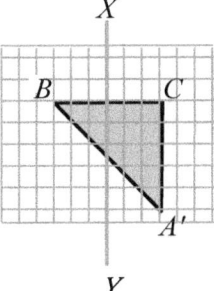

4. The graph below which shows the reflection of triangle ABC in the line XY is:

[A] 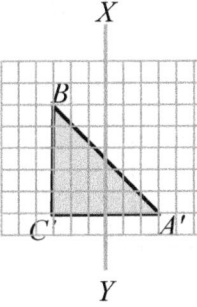   [B]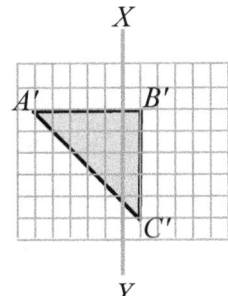

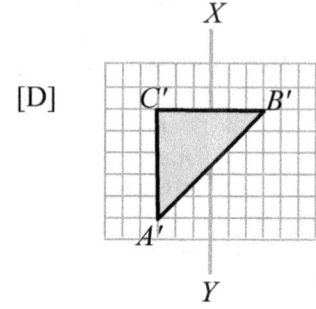

10. In figure below, the diagram that represents a shear, shear factor −3 parallel to the y-axis is:

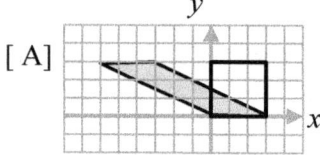
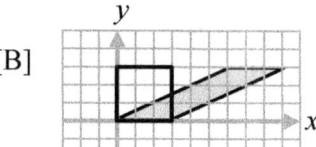
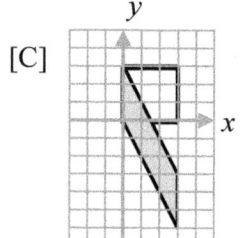
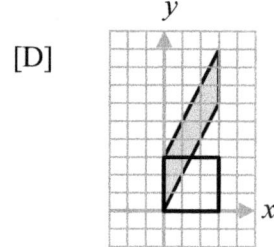

11. In the figure below, the invariant line is:
    [A] AD   [B] BC   [C] CD   [D] AB

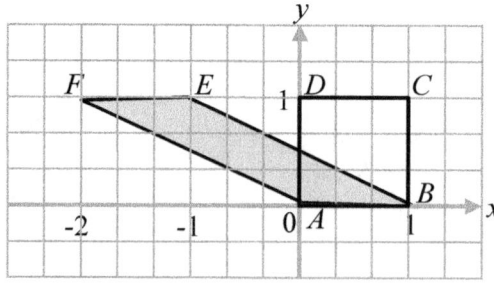

12. A non-zero singular matrix:
    [A] has no inverse.
    [B] does not map any plane figure to a straight line.
    [C] does not have a determinant equal to zero.
    [D] can be a unit matrix.
13. A non-zero matrix whose determinant is zero:
    [A] has an inverse.   [B] maps any plane figure to a straight line.

[C] is not a singular matrix.   [D] is an identity matrix.

14. The transformation matrix $\begin{pmatrix} 6 & 3 \\ 6 & 4 \end{pmatrix}$ maps the point $(x,1)$ onto the point (3, 4). The value of x is:
   [A] −3   [B] 4   [C] 0   [D] −6

15. The transformation matrix $\begin{pmatrix} a & 2a \\ b & 4b \end{pmatrix}$ maps the point (1, 2) to (5, 18). The values of a and b are:
   [A] $a = 2$ and $b = 1$ [B] $a = 1$ and $b = 2$
   [C] $a = 4$ and $b = 8$ [D] $a = 8$ and $b = 4$

16. The matrix, which represents a reflection in the line, $y = 0$ is:
   [A] $\begin{pmatrix} -1 & 0 \\ 0 & 1 \end{pmatrix}$ [B] $\begin{pmatrix} 0 & 1 \\ -1 & 0 \end{pmatrix}$ [C] $\begin{pmatrix} 0 & -1 \\ 1 & 0 \end{pmatrix}$ [D] $\begin{pmatrix} 1 & 0 \\ 0 & -1 \end{pmatrix}$

17. The transformation with matrix $\begin{pmatrix} -1 & 0 \\ 0 & 1 \end{pmatrix}$ is:
   [A] A reflection in the line $x = 0$.   [B] A reflection in the line $y = 0$.
   [C] A rotation through 180° center (0, 0)   [D] A shear with shear factor −1

18. The description of the transformation represented by the matrix $\begin{pmatrix} 1 & 0 \\ 2 & 1 \end{pmatrix}$ is:
   [A] A stretch with stretch factor 2 and points on the x-axis invariant.
   [B] A stretch with stretch factor 2 and points on the y-axis invariant.
   [C] A shear with shear factor 2 and points on the x-axis invariant.
   [D] A shear with shear factor 2 and points on the y-axis invariant.

19. The transformation matrix $\begin{pmatrix} 2 & 4 \\ 0 & 2 \end{pmatrix}$ maps the point P onto P'. Given that P' is the point (9, 4), then P must be the point:
   [A] $\left(\dfrac{1}{2}, 0\right)$ [B] (0, 2)   [C] $\left(2, \dfrac{1}{2}\right)$ [D] $\left(\dfrac{1}{2}, 2\right)$

20. We can represent a rotation through 90° center (0, 0) by the matrix:
   [A] $\begin{pmatrix} -1 & 0 \\ 0 & 1 \end{pmatrix}$ [B] $\begin{pmatrix} 0 & 1 \\ -1 & 0 \end{pmatrix}$ [C] $\begin{pmatrix} 0 & -1 \\ 1 & 0 \end{pmatrix}$ [D] $\begin{pmatrix} 1 & 0 \\ 0 & -1 \end{pmatrix}$

21. The transformation matrix $\begin{pmatrix} 2 & 4 \\ 0 & 2 \end{pmatrix}$ maps the point A onto B. Given that, A is the point (3, 2), then B is the point:
   [A] (14, 4)   [B] (4, 14)   [C] (6, 8)   [D] (0, 4)

22. A linear transformation $\mathbf{M}: \mathbb{R}^2 \to \mathbb{R}^2$ is defined by $\mathbf{P} = \mathbf{Mq}$, where $\mathbf{M}$ is a

2 x 2 matrix and **p**, **q** are 2×1 column vectors. Given that $\mathbf{p} = \begin{pmatrix} 3 \\ 7 \end{pmatrix}$ when $\mathbf{q} = \begin{pmatrix} 1 \\ 0 \end{pmatrix}$ and $\mathbf{p} = \begin{pmatrix} 6 \\ -1 \end{pmatrix}$ when $\mathbf{q} = \begin{pmatrix} 2 \\ -3 \end{pmatrix}$, **M** = :

[A] $\begin{pmatrix} 3 & 0 \\ 7 & 5 \end{pmatrix}$ [B] $\begin{pmatrix} 5 & 7 \\ 0 & 3 \end{pmatrix}$ [C] $\begin{pmatrix} 0 & 5 \\ 3 & 7 \end{pmatrix}$ [D] $\begin{pmatrix} 7 & 3 \\ 5 & 0 \end{pmatrix}$

23. The triangle with vertices $A(-1,-3)$, $B(2,1)$ and $C(-2,2)$ is transformed by the matrix $\begin{pmatrix} x & 0 \\ 0 & y \end{pmatrix}$ into the triangle with vertices $A'(-2,-3)$, $B'(4,1)$, $C'(-4,2)$.
The values of x and y is respectively:
[A] 2 and 0 [B] 0 and 1 [C] 1 and 2 [D] 2 and 1

24. The 2×2 matrix representing defined by $T:(x,y) \mapsto (2x+y, -x+y)$ is:

[A] $\begin{pmatrix} 2 & 1 \\ -1 & 1 \end{pmatrix}$ [B] $\begin{pmatrix} 1 & 1 \\ -1 & 2 \end{pmatrix}$ [C] $\begin{pmatrix} 1 & -1 \\ 1 & 2 \end{pmatrix}$ [D] $\begin{pmatrix} 1 & 1 \\ 2 & -1 \end{pmatrix}$

25. The quadrilateral $A(2,0)$, $B(2,2)$, $C(-2,-2)$, $D(-2,0)$ is transformed by the matrix $\mathbf{P} = \begin{pmatrix} 2 & -1 \\ 1 & 0 \end{pmatrix}$. The image $A'B'C'D'$ of $ABCD$ under **P** is:

[A] $A'(2,2)$, $B'(6,4)$, $C'(10,-6)$, $D'(6,-4)$
[B] $A'(4,2)$, $B'(-6,-2)$, $C'(2,2)$, $D'(-4,-2)$
[C] $A'(4,2)$, $B'(2,2)$, $C'(-2,-2)$, $D'(-4,-2)$
[D] $A'(4,2)$, $B'(2,2)$, $C'(-4,-2)$, $D'(-6,-2)$

26. The 2×2 matrix representing the transformation defined by $T:(x, y) \mapsto (2x, 2y)$ is:

[A] $\begin{pmatrix} 2 & 2 \\ 1 & 1 \end{pmatrix}$ [B] $\begin{pmatrix} 0 & 2 \\ 2 & 0 \end{pmatrix}$ [C] $\begin{pmatrix} 2 & 2 \\ 2 & 2 \end{pmatrix}$ [D] $\begin{pmatrix} 2 & 0 \\ 0 & 2 \end{pmatrix}$

27. The points $A(2,0)$, $B(2,2)$, $C(-2,-2)$ and $D(-2,0)$ are transformed by the matrix $\mathbf{P} = \begin{pmatrix} 2 & -1 \\ 1 & 0 \end{pmatrix}$. The invariant line is:

[A] the line $x = 0$ [B] the line $y = 0$ [C] the line $y = x$ [D] the line $y = 1$

28. We can describe the transformation defined by $T:(x, y) \mapsto (2x, 2y)$ as:
[A] A translation 2 units along Ox.   [B] A shear scale factor 2
[C] An enlargement scale factor 2   [D] A stretch scale factor 2.

Topic 6
CONSTRUCTIONS AND LOCI

Objectives

At the end of this topic, the learner should be able to:

1. Construct special angles. (30°, 45°, 60° and 90°)
2. Construct an angle bisector.
3. Construct the mediator of a line segment.
4. Divide a line segment into a given number of congruent segments.
5. Construct the circum-circle, in-circle of a given triangle.
6. Construct a line parallel to a given line and passing through a given point.
7. Construct a line perpendicular to a given line, and passing through a given point.
8. Construct simple locus of a point described under a given condition.

CONSTRUCTIONS

Normally we do constructions with pencil, ruler and a pair of compasses only, and unless otherwise stated we should not use any other device. After a construction, do not erase the construction lines.

6.1 Constructing a Line of Given Length

 Example

Construct a line AB of length 6 cm.

Procedure
1. Draw a line of any length longer than 6 cm and mark a point A on it.
2. With open compass measure 6 cm from a ruler, (We do this by placing the pin end of the ruler at the zero mark and opening it up until the tip of the pencil is exactly on the 6 cm mark).
3. With the centre A trace an arc at B.

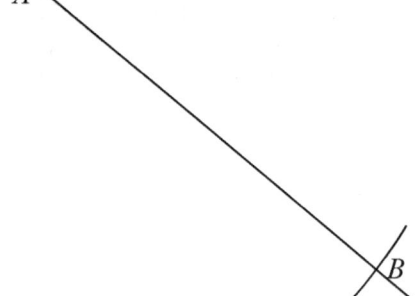

6.2 Constructing a Triangle ABC with Sides of Given Length

 Example

Construct a triangle ABC with sides of length $AB = 9$ cm, $BC = 6$ cm and $AC = 4$ cm.

Procedure
(i) Following the steps in 1, construct the line *AB* of length 9 cm.
(ii) With compass, measure 6 cm and with centre *B*, draw an arc on one side of the line.
(iii) With compass, measure 4 cm and with centre *A*, draw another arc to cut the first. Mark their point of intersection *C*.
(iv) Using ruler and pencil join *AC* and *BC*
 Note! Construction lines do not end at the vertices *A*, *B*, and *C*.

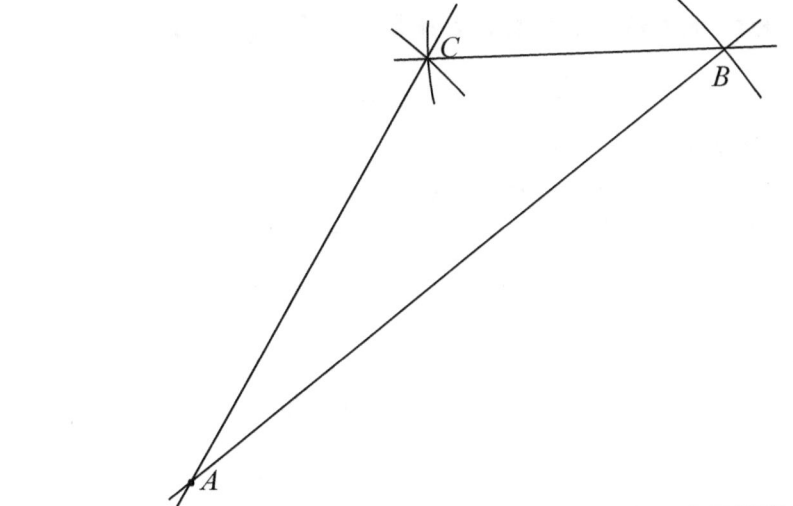

6.3 Height, Median and Perpendicular Bisector

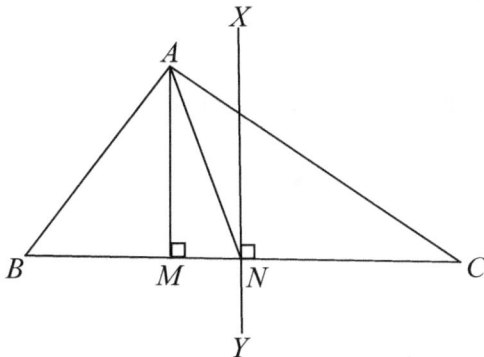

The **altitude** (or **height**) of a triangle is the perpendicular distance from one side of the triangle to the opposite vertex.

The **perpendicular bisector** or **mediator** of a side of a triangle is a line that is perpendicular to the side and passes through its midpoint.

The **median** of a triangle is a line from the midpoint of a side of the triangle to the opposite vertex.

In the figure above, AM is the altitude (or height), AN is the median and XY is the perpendicular bisector (or mediator) of the triangle ABC with respect to the side BC.

6.4 Constructing a Perpendicular Bisector (or Mediator) of a Given Line Segment

 Example

Construct the perpendicular bisector of the line segment AB.

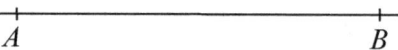

Procedure

This construction was earlier done in section 6.3 of book 2 and you may refer to it if necessary.

6.5 Constructing a Perpendicular to a Given Line Segment, from a Given Point

 Example

Construct the perpendicular to the line segment AB passing through the point P.

$P \bullet$

Procedure
(i) With centre P draw two arcs of equal radii to cut AB at two points C and D.

(ii) With centres *C* and *D* draw two arcs of equal radii to intersect on the opposite side of *P*. Name this point of intersection *Q*.
(iii) Now join *PQ*. *PQ* is perpendicular to *AB*.

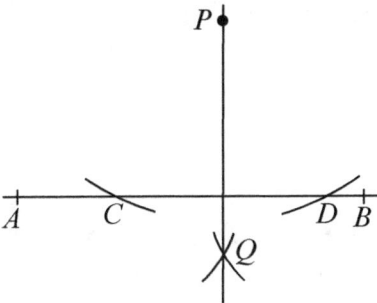

Note!!
The construction is the same event if *P* lies on *AB*.

6.6   Bisecting a Given Angle

 Example

Construct the bisector of the angle *PAR* below.

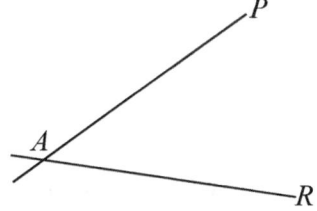

Procedure
(i) With centre *A* draw two arcs of equal radii to cut the adjacent sides to the given angle at two points *C* and *B*.
(ii) With centres *C* and *B* draw two arcs of equal radii to intersect at *D*.
(iii) Join *A* and *D* with ruler and pencil. *AD* is the bisector of the angle *PAR*.

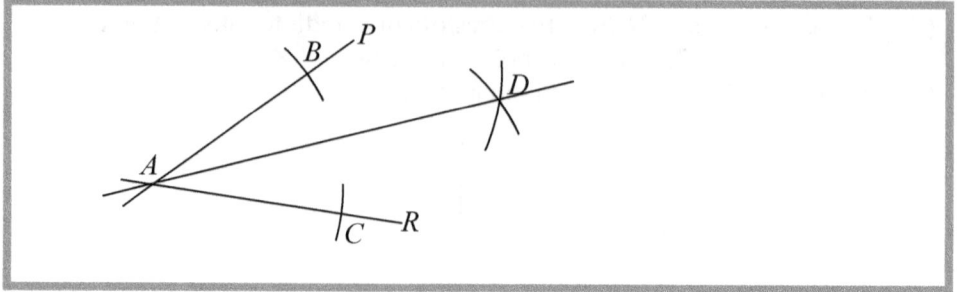

6.7 Construction of Some Special Angles

Example

1. Construct an angle of 90° at the point A.

   Procedure
   (i) With centre A draw 2 arcs C and D of equal radii to cut the line passing through A.

   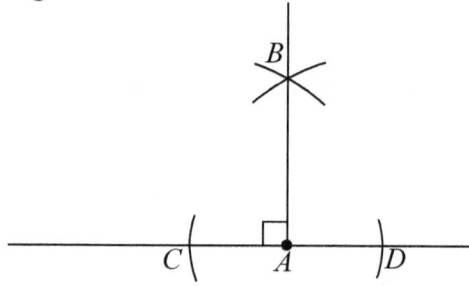

   (ii) With centres C and D draw two arcs of equal radii to intersect at B on one side of the line segment.
   (iii) Now join AB. The angle AB makes with the line segment will be 90°.

2. Construct an angle of 45°

   Procedure
   To construct an angle of 45° simply construct an angle of 90° and bisect it.

3. Construct an angle of 60°.

Procedure

Since all the three angles of an equilateral triangle are each 60°, we can use this idea to construct an angle of 60°.

(i) With centre A draw two arcs of equal radii one to cut the line segment AB at B and the other on one side of AB.
(ii) With centre B and the same radius, draw another arc to intersect the first at C.
(iii) Join AC. The angle BAC is 60°.

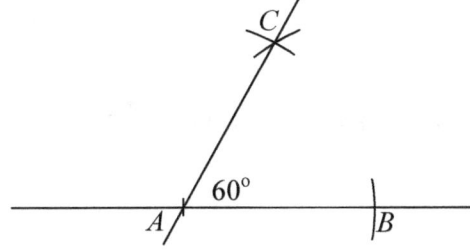

Note!
After constructing an angle of 60°, we obtain the 120° angle constructed as a bonus.

4. Construct an angle of 30°.
   To do this simply construct an angle of 60° and bisect it.

6.8 Trisecting a Right Angle

 Example

Trisect a right angle ABC which is already constructed.

Procedure

(i) With centre B, draw a large arc to cut BC at C and AB at A.
(ii) With centre C, draw an arc of the same radius to cut the large arc at D.
(iii) With centre, A draw an arc of the same radius to cut the large arc at E.
(iv) Join BD and BE.

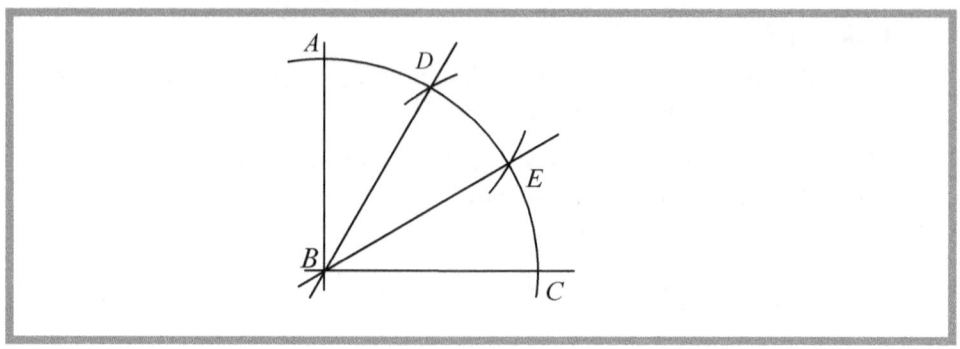

6.9 Constructing an Inscribed Circle of a Given Triangle

 Example

Construct the inscribed circle of triangle ABC.

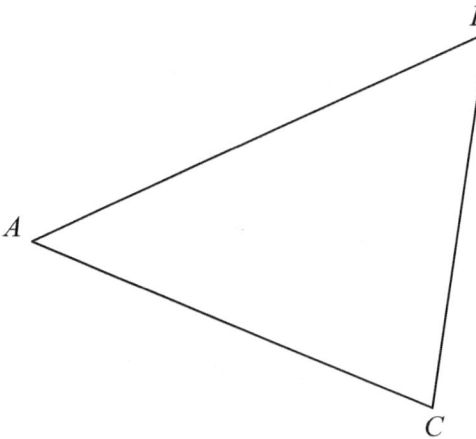

Procedure

(i) Construct the bisectors of each of the angles of the triangle ABC.
(ii) Where these bisectors intersect is the centre O of the circle.
(iii) Construct a perpendicular from O to any of the sides of the triangle. This helps to situate the radius of the circle.

Module 16, Topic 6: Constructions and Loci

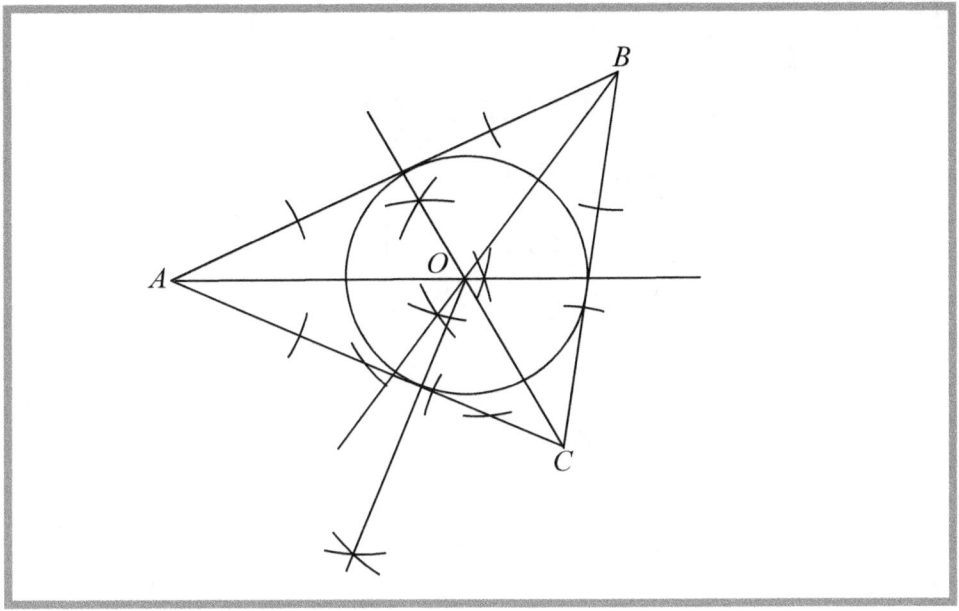

6.10 Constructing a Circumscribed Circle of a Given Triangle

 Example

Construct the circumscribed circle of triangle *ABC* below.

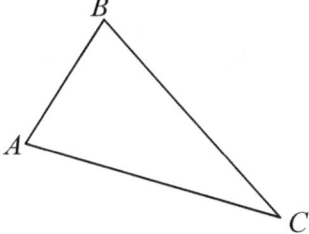

Procedure

(i) Construct the perpendicular bisectors of each of the sides of the triangle *ABC*.
(ii) Where these bisectors intersect is the centre *O* of the circle.

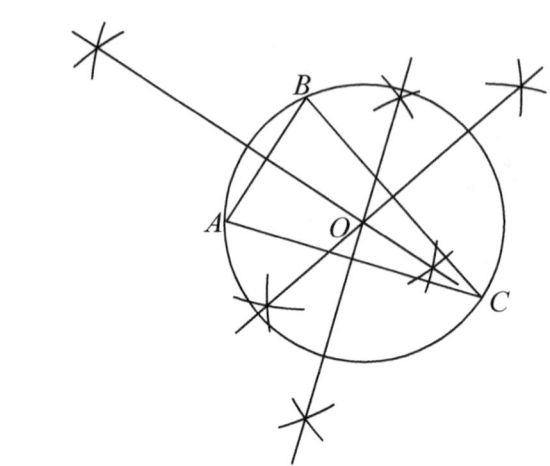

Note!!
(a) Any point P on the arc BAC is such that $\angle PCA = \angle PBA$.
(b) Any point Q on the arc CBA is such that angle QCB = angle QAB.
(c) Any point R on the arc ACB is such that angle RAC = angle RBC.

6.11 Dividing a Given Line Segment into Congruent Segments

Example

Divide the line segment AB below into five congruent segments.

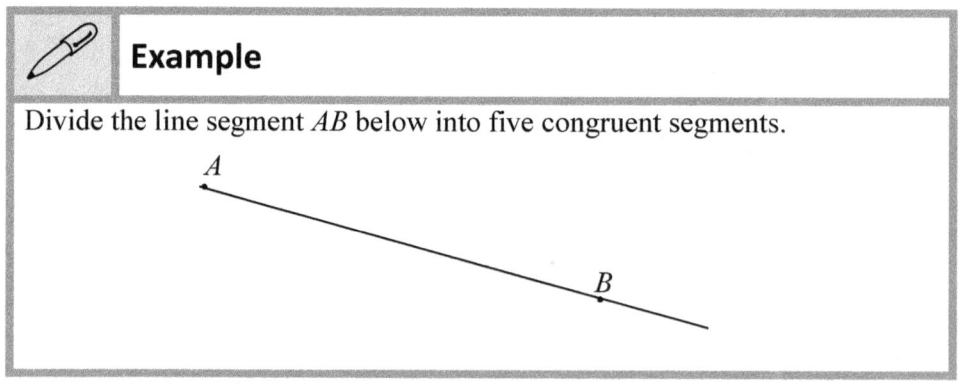

Module 16, Topic 6: Constructions and Loci

Procedure

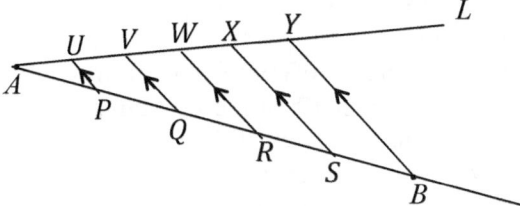

1. Draw any other line AL through A.
2. With a pair of compass, mark off any five equal distances AU, UV, VW, WX, XY on AL.
3. Join BY with a straight line using a ruler.
4. Draw PU, QV, RW, SX parallel to BY to meet AB at P, Q, R and S.

The points P, Q, R and S divide AB into five equal parts as required.

6.12 Construction of Polygons

Example

Construct a regular hexagon with sides 3 cm.

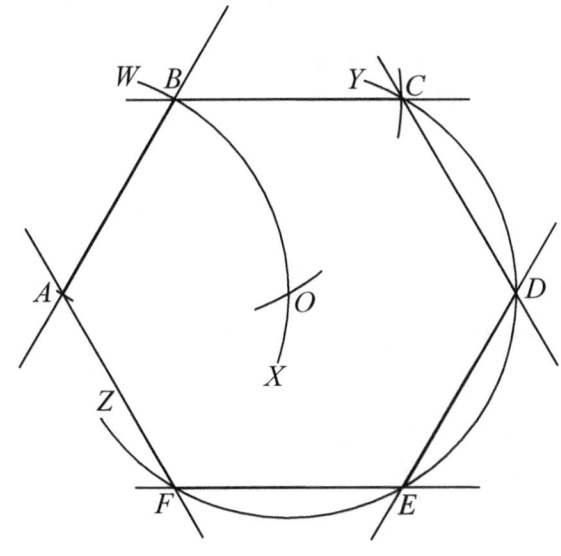

Procedure
1. Draw a straight line more than 3 cm.
2. Mark a point *A* on the line.
3. With a pair of compasses measure 3 cm from your ruler.
4. With centre *A*, draw a large arc *WX* of radius 3 cm to cut the line at *B*.
5. With centre *B*, draw another arc to cut *WX* at *O*.
7. With centre *B*, draw an arc to cut *YZ* at *C*.
8. With centre *C*, draw an arc to cut *YZ* at *D*.
9. With centre *D*, draw an arc to cut *YZ* at *E*.
10. With centre *E*, draw an arc to cut *YZ* at *F*.
11. Now connect the points by drawing the lines *BC*, *CD*, *DE*, *EF* and *FA*.

Exercise 6:1

In this exercise use only a ruler, a pair of compasses and a pencil, and show all construction lines.

1. (i) Draw a line *OA* of length 8 cm.
   (ii) Construct a line *OB* of length 6 cm perpendicular to *OA*.
   (iii) Measure *AB*.
   (iv) Bisect *AB* (by construction only) and mark the midpoint *M* of *AB*.
   (v) Measure *MO*.
   (vi) Construct the circle through *O*, *A* and *B*.

2. Draw a line segment, *PQ*, 7 cm long in the middle of a new page.
   (b) Bisect line *PQ* and label the mid-point, *X*.
   (c) Draw a circle such that *PQ* is the diameter
   (d) Locate a point *A* on the circumference of the circle such that *AP* = 4.5 cm.
   (e) Draw and measure line *AQ*.
   (f) Write down the value of angle *PAQ*.

3. (a) Draw a line *PQ* of length 8 cm.
   (b) Construct a line *OQ* of length 6 cm perpendicular to *PQ*.
   (c) Draw and measure the line *OP*.
   (d) Construct the perpendicular bisector of *OP* and label its foot *M*.
   (e) Construct the circle, which circumscribe triangle *OPQ*.

4. (a) Draw a line *AB* of length 8 cm.
   (b) Construct a line *AD* of length 6 cm such that angle *BAD* = 60°.
   (c) Construct the parallelogram *ABCD*.
   (d) Produce *DA* to *X* and bisect angle *BAX*.
   (e) Bisect the line segment *AB*.
       Given that the two bisecting lines meet at *O*,

Module 16, Topic 6: Constructions and Loci

   (f) Measure the distance of O from the line AB and from the point A.
   (g) Deduce the angle between the two bisecting lines.
5. (a) Draw a line ST, 5 cm long with T at the centre of a new page.
   (b) On the base ST, construct triangle RST such that $\angle S = 60°$ and $\angle T = 90°$ with the point R above the base ST.
   (c) Measure the length of SR.
   (d) The sides RS and RT are produced downwards to the points M and N such that RSM = 13 cm and RTN = 15 cm.
   (e) Construct the bisectors of angles TSM and STN and where these lines intersect, mark the point O.
   (f) From O construct a perpendicular to the line TN to meet TN at P.
   (g) With centre O and radius OP, construct a circle.
6. Given the triangle ABC below.

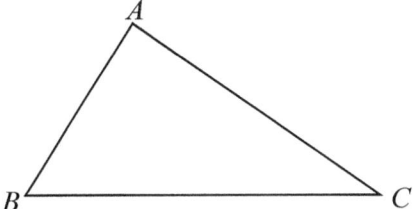

   (a) Using a ruler, pencil and compass only, copy triangle ABC.
   (b) (i) Construct the median passing through the point A.
       (ii) Construct the perpendicular bisector of the side BC of the triangle.
       (iii) Construct the perpendicular to the side BC, passing through the point A.

LOCI

A locus is the path of a moving point subject to some restrictions.

6.13 Loci in 2-Dimensions

The following are some examples of loci in 2-dimensions.

(1) The locus of a point which is moving such that its distance from a fixed point is always constant (the same) is a **circle**.

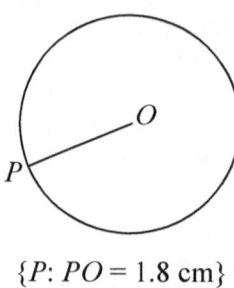

{P: PO = 1.8 cm}

(2) The locus of a point, which is moving such that its distance from two fixed points is always equal, is the **mediator** or **perpendicular bisector** of the line segment joining these two points.

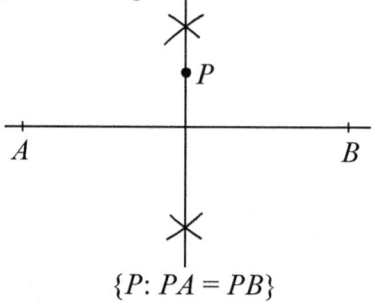

{P: PA = PB}

(3) The locus of a point, which is equidistant from two intersecting lines, is the bisector of the angle between the lines.

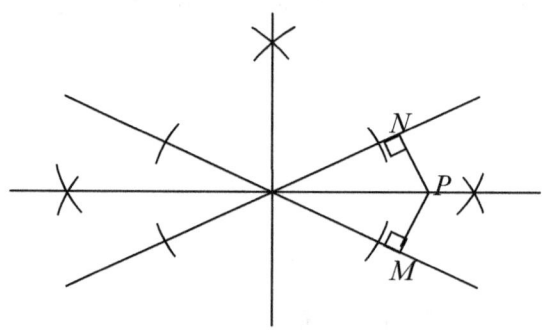

{P: PM = PN}

(4) The locus of a point, which is moving such that its distance d from a fixed line L_1 is always constant, is a line L_2 parallel to L_1.

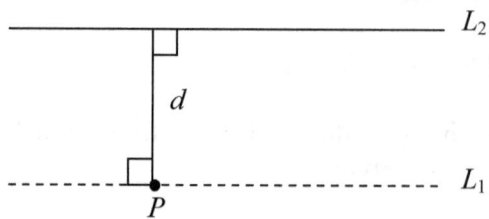

Module 16, Topic 6: Constructions and Loci

Exercise 6:2

1. State the locus of a point, which is equidistant from two fixed points.
2. *ABCD* is a quadrilateral with unequal sides. Show how to find a point *P* such that $PA = PB$ and $PC = PD$.
3. *YZ* is a straight line 14 cm long. A point *P* moves so that $PY^2 - PZ^2 = 56$ cm^2. If *X* is the foot of the perpendicular from *P* to *YZ*,
   (a) Calculate the length of *YX*.
   (b) State the locus of *P*, and show it clearly on a diagram (not necessarily drawn accurately or to scale)
4. Draw a rectangle *ABCD* with $AB = 10$ cm and $AD = 6$ cm. Draw the following loci in each case giving only that part of the locus which is inside the rectangle. The locus of the point, which is:
   (a) 2 cm from *A*.
   (b) equidistant from B and D.
   (c) 2 cm from *BD*.
   (d) equidistant from *BD* and *DC*.
5. Construct in a single diagram:
   (a) A triangle *LMN* with sides $LM = 7$ cm, $LN = 8$ cm and $MN = 6$ cm,
   (b) The locus of points, which are 5 cm from *L*.
   (c) The locus of points, which are equidistant from *MN* and *LN*. Given that,
   $\mathscr{E} = \{P : P \text{ lies inside } \triangle LMN\}$
   $X = \{P : LP < 5 \text{ cm}\}$ and
   $Y = \{P : P \text{ is nearer to } MN \text{ than } LN\}$
   (d) Indicate the region $X \cap Y$ by suitable shading of your diagram.

Multiple Choice Exercise 6

1. *AB* bisects *PQ* at point *N*. The statement that is true of *N* is:
   [A] *N* is the midpoint of AB.
   [B] *N* is the midpoint of *AB* and the midpoint of *PQ*.
   [C] *N* is the midpoint of *PQ*.
   [D] *N* divides *PQ* in the ratio 2:1.
2. Point *P* is the midpoint of *AB*. Complete the statement: $PB = 7$ cm, *AB* is equal to:
   [A] 7 cm   [B] 14 cm   [C] 3.5 cm   [D] none of the above
3. The diagram(s) in the figure below that demonstrate(s) the correct way of constructing an angle of 60° is:

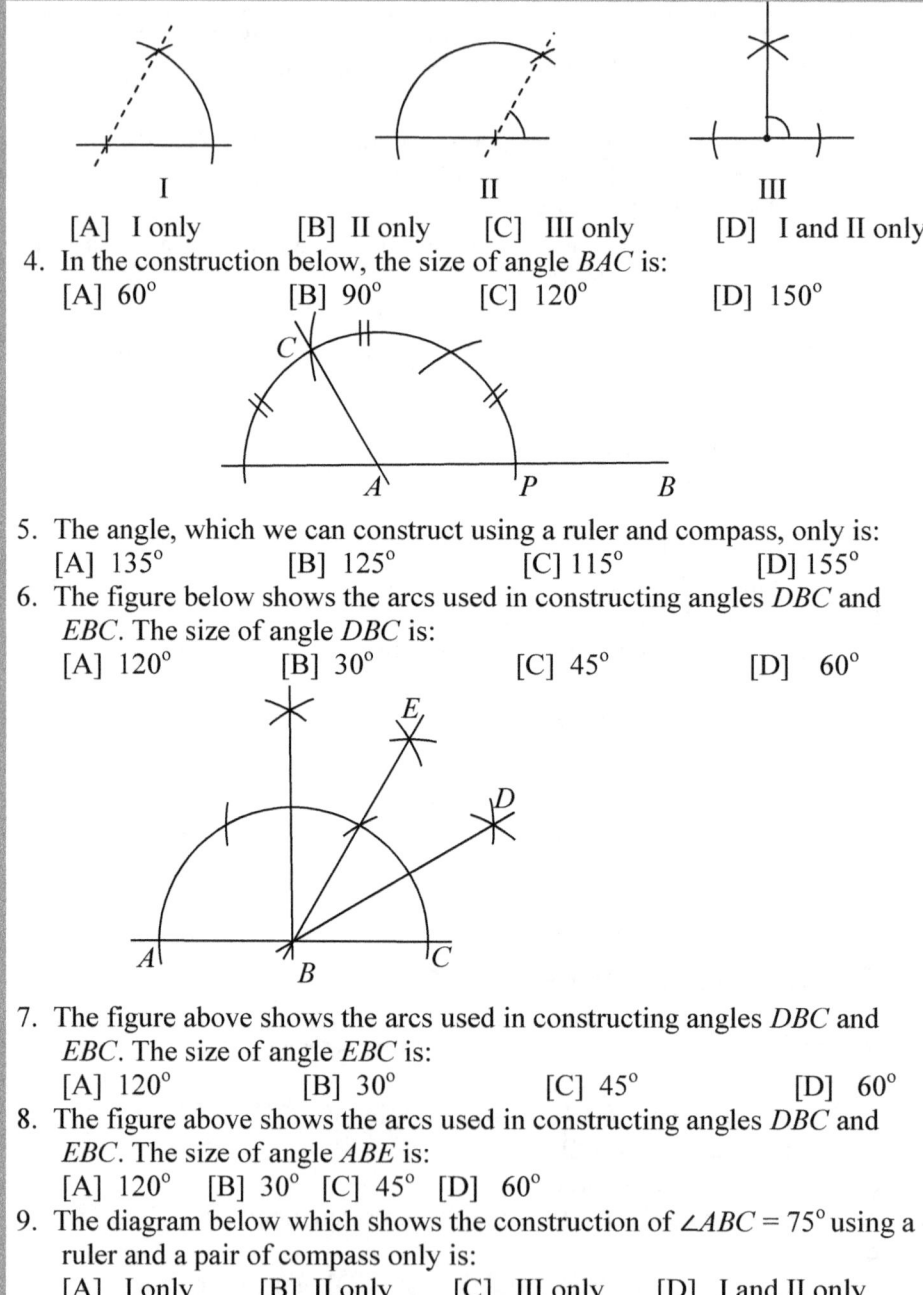

[A] I only [B] II only [C] III only [D] I and II only

4. In the construction below, the size of angle *BAC* is:
   [A] 60° [B] 90° [C] 120° [D] 150°

5. The angle, which we can construct using a ruler and compass, only is:
   [A] 135° [B] 125° [C] 115° [D] 155°

6. The figure below shows the arcs used in constructing angles *DBC* and *EBC*. The size of angle *DBC* is:
   [A] 120° [B] 30° [C] 45° [D] 60°

7. The figure above shows the arcs used in constructing angles *DBC* and *EBC*. The size of angle *EBC* is:
   [A] 120° [B] 30° [C] 45° [D] 60°

8. The figure above shows the arcs used in constructing angles *DBC* and *EBC*. The size of angle *ABE* is:
   [A] 120° [B] 30° [C] 45° [D] 60°

9. The diagram below which shows the construction of $\angle ABC = 75°$ using a ruler and a pair of compass only is:
   [A] I only [B] II only [C] III only [D] I and II only

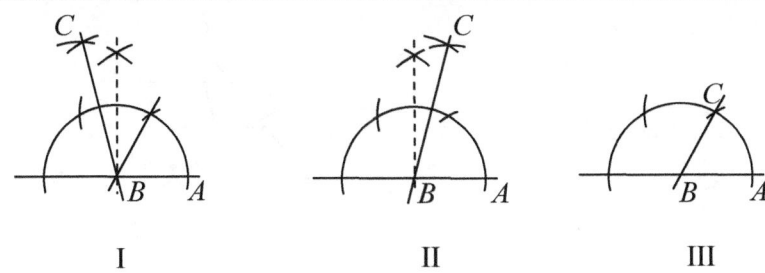

I II III

13. The locus of a point, which is moving such that its distance d from a fixed line l is always constant, is:
   [A] a mediator of *l* [B] a line parallel to *l* [C] a parabola [D] a circle
14. The locus of a point, which is moving such that its distance from a fixed point is always, constant (the same) is:
   [A] a mediator. [B] a line parallel to the point.
   [C] the mediator constant from the point. [D] a circle
15. The locus of a point, which is moving such that its distance from two fixed points is always equal, is:
   [A] a perpendicular bisector *l*. [B] a line parallel to *l*.
   [C] the angle bisector from the two points. [D] a circle.
16. The locus of a point, which is equidistant from two intersecting lines, is:
   [A] the perpendicular bisector of the intersecting lines.
   [B] a line parallel to the intersecting lines.
   [C] a circle.
   [D] the angle bisector between the intersecting lines
17. The diagram in the figure below that shows the construction of a perpendicular bisector is:

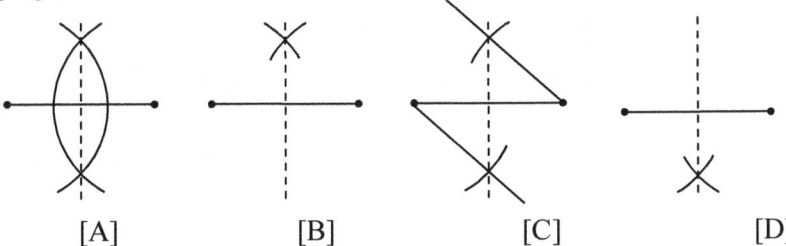

[A]   [B]   [C]   [D]

18. The figure below which shows the correct construction of the bisector of the angle *BAC* is:

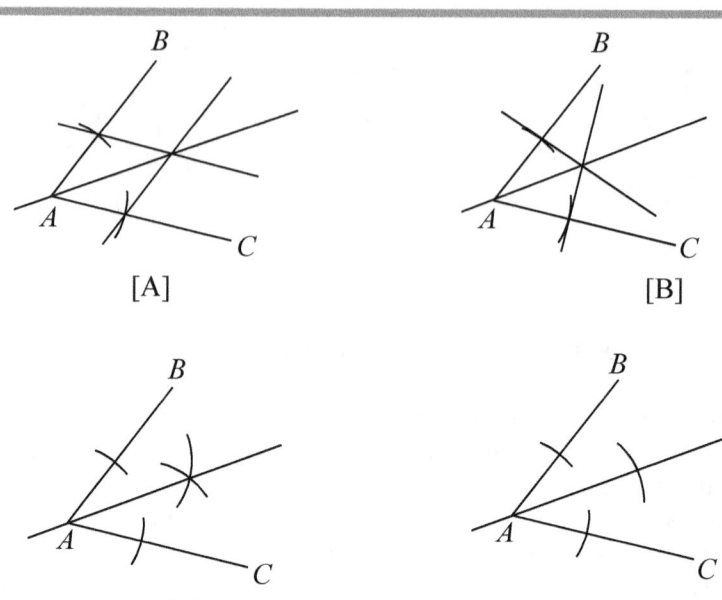

19. Another name for perpendicular bisector is:
    [A] Altitude   [B] Midpoint   [C] Angle bisector   [D] Mediator
20. In the figure below, *AM* is said to be the:
    [A] median   [B] mediator   [C] altitude   [D] angle bisector

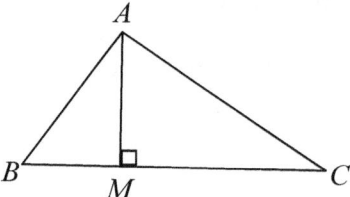

21. In the following figure, given that M is the midpoint of *BC*, *AM* is called the:
    [A] median   [B] mediator   [C] altitude   [D] angle bisector

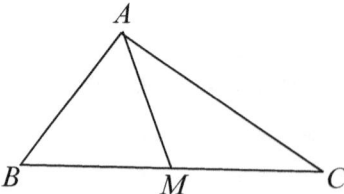

22. In the following figure, given that, *M* is the midpoint of *BC*, we call *NM*:
    [A] median   [B] mediator   [C] altitude   [D] angle bisector

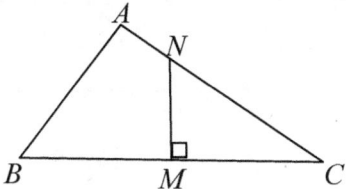

23. The figure below shows the construction of:
    [A] A congruent segment [B] A congruent angle
    [C] A perpendicular bisector [D] An angle bisector

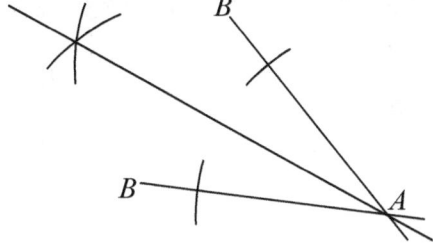

24. The following figure shows the construction of:
    [A] A mediator [B] A perpendicular to AB
    [C] An angle bisector [D] Intersecting lines

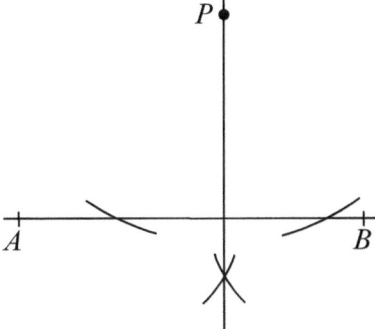

25. In the figure below, the correct construction of a perpendicular bisector, which is equal in length to the line segment [XY] is:

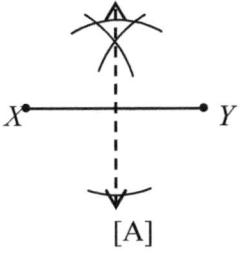 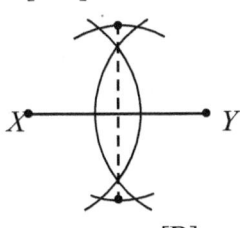

[A] [B]

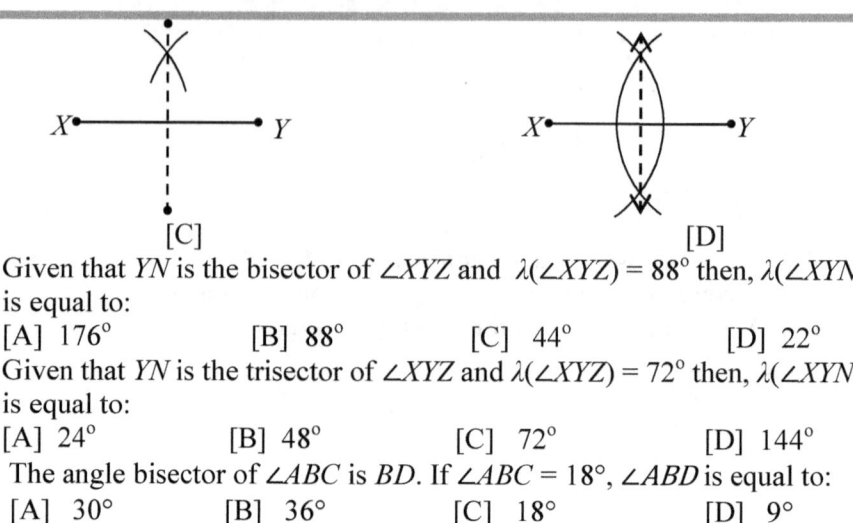

[C] [D]

26. Given that YN is the bisector of ∠XYZ and λ(∠XYZ) = 88° then, λ(∠XYN) is equal to:
    [A] 176°   [B] 88°   [C] 44°   [D] 22°

27. Given that YN is the trisector of ∠XYZ and λ(∠XYZ) = 72° then, λ(∠XYN) is equal to:
    [A] 24°   [B] 48°   [C] 72°   [D] 144°

28. The angle bisector of ∠ABC is BD. If ∠ABC = 18°, ∠ABD is equal to:
    [A] 30°   [B] 36°   [C] 18°   [D] 9°

Topic 7

TRIGONOMETRY

Objectives

At the end of this topic, the learner should be able to:

1. Obtain the sine and cosine of acute angles.
2. Define radian measure.
3. Give angles in radians.
4. Derive the trigonometric identity $\sin^2 x + \cos^2 x = 1$.
5. Draw trigonometric (unit) circle.
6. Develop some trigonometric identities.
7. Draw graphs of trigonometric functions.
8. Determine the trigonometric ratios of angles in the range $0° \leq \theta \leq 360°$.
9. Solve simple trigonometric equations of the first order within $0° \leq \theta \leq 90°$, algebraically and graphically.

7.1 Review and Revision

In module 11, topic 6 of book 3, we introduced trigonometry. The learner is advised to revise that topic and do the following review and revision exercise before continuing.

Review and Revision

1. In the triangle below, find AB.

   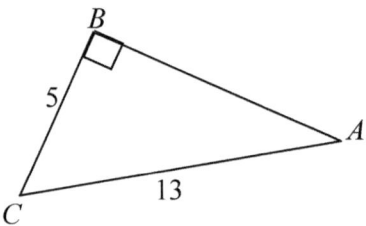

2. In the figure below, find
   (a) $\sin x$ (b) $\cos x$ (c) $\tan x$ (d) $\tan y$ (e) $\sin y$ (f) $\cos y$

   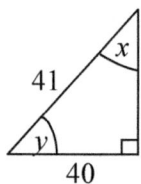

3. Use a calculator to find the following
   (a) $\sin 47°26'$ (b) $\cos 53°47'$ (c) $\tan 72°13'$
4. Use a calculator to find the following
   (a) $\sin^{-1} 0.1564$ (b) $\cos^{-1} 0.342$ (c) $\tan^{-1} 01.9251$
5. Use a tables to find the following
   (a) $\sin 72°20'$ (b) $\cos 63°43'$ (c) $\tan 82°16'$
6. Use a calculator to find the following
   (a) $\sin^{-1} 0.2564$ (b) $\cos^{-1} 0.542$ (c) $\tan^{-1} 01.2251$
7. If $\tan A = \dfrac{4}{3}$, find the values of $\sin A$ and $\cos A$ without using tables or calculators.
8. Find $\cos \emptyset$ from the following figure.

   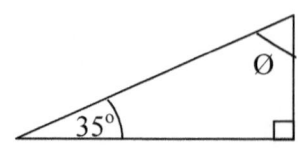

9. Find $\cos^2 60° + \tan^2 45°$ leaving your answer in surd form.
10. The figure below shows the cross-section of roof. How long must the carpenter cut the plank support *BN*? What is the length of the base *AC*?

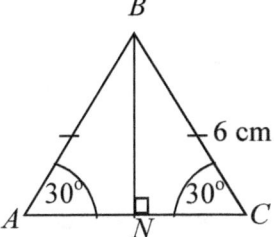

7.2 Reciprocal Trigonometric Ratios

Sometimes it is easier to solve problems using the reciprocals of the familiar trigonometric ratios sine, cosine and tangent studied in module 11, topic 6 of book 3. These reciprocals are as follows.

1. The cosecant (cosec) is the reciprocal of sine,

$$\operatorname{cosec}\theta = \frac{1}{\sin\theta} \Rightarrow \operatorname{cosec}\theta = \frac{\text{hyp}}{\text{opp}}$$

2. The secant (sec) is the reciprocal of cosine,

$$\sec\theta = \frac{1}{\cos\theta} \Rightarrow \sec\theta = \frac{\text{hyp}}{\text{adj}}$$

3. The cotangent (cot) is the reciprocal of tangent,

$$\cot\theta = \frac{1}{\tan\theta} \Rightarrow \cot\theta = \frac{\text{adj}}{\text{opp}}$$

 Exercise 7:1

1. In figure (i) and (ii) below, find
 (a) cot *x* (b) cosec *x* (c) sec *x* (d) cot *y* (e) cosec *y* (f) sec *y*

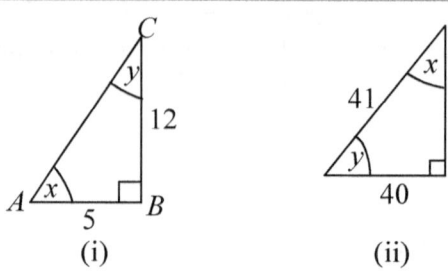

(i)   (ii)

2. In the figure below, find the value of
   (a) $\cot x$ (b) $\operatorname{cosec} x$ (c) $\sec x$ (d) $\cot y$ (e) $\operatorname{cosec} y$ (f) $\sec y$

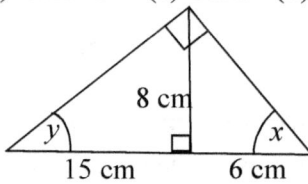

3. If α is acute and $\cot \alpha = \dfrac{24}{7}$, find the value of
   (a) $\operatorname{cosec} \alpha$ (b) $\sec \alpha$

4. Using figure (i) below, write down the values of
   (i) $\operatorname{cosec} \theta$ (ii) $\sec \theta$ (iii) $\cot \theta$

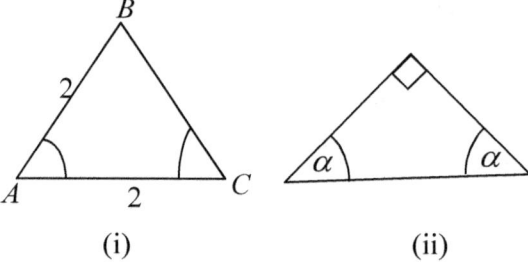

(i)   (ii)

5. Using figure (ii) above, write down the value of
   (i) $\operatorname{cosec} \alpha$ (ii) $\sec \alpha$ (iii) $\cot \alpha$

6. Given that x is an acute angle and that $\operatorname{cosec} x = \dfrac{m}{n}$, find $\sec x$ and $\cot x$ in terms of m and n.

7. If α is acute and $\sec A = \dfrac{4}{5}$, find the value of
   (a) $\operatorname{cosec} A$ (b) $\cot A$

8. In Figure (i) and (ii) below, find
   (a) $\sec w$ (b) $\operatorname{cosec} w$ (c) $\cot w$ (d) $\sec x$
   (e) $\operatorname{cosec} x$ (f) $\cot x$ (g) $\sec y$ (h) $\operatorname{cosec} y$

Module 16, Topic 7: Trigonometry

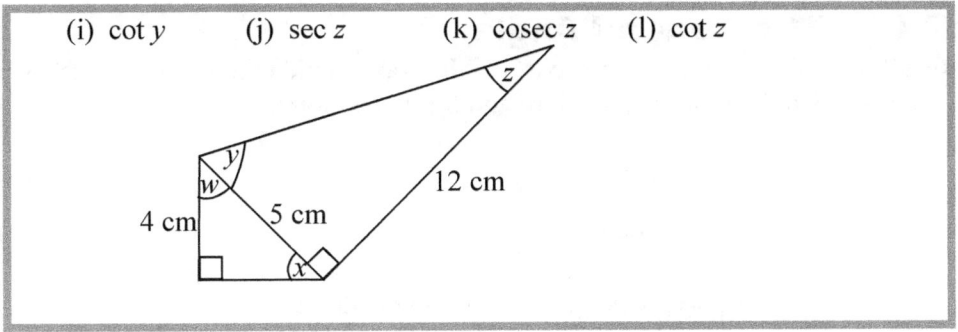

(i) cot y   (j) sec z   (k) cosec z   (l) cot z

7.3   Reciprocal Trig Ratios from Tables

It was earlier mentioned that when using tables to refer the trigonometric ratios of sine, cosine and tangent, the difference is added for both sine and tangent because these are increasing functions, while the difference is subtracted for cosine because it is a decreasing function.

Using tables to refer the trigonometric ratios cosecant, secant and cotangent is similar noting that, the cosecant and the cotangent are decreasing functions while the secant is an increasing function. Therefore, the difference is added for secant and subtracted for cosecant and cotangent.

Generalizing,

> To refer trigonometric ratios from tables, the difference is;
> 1. Subtracted for all trigonometric ratios starting with the prefix **co-**
> 2. Added for all other trigonometric ratios.

 Exercise 7:2

1. Use tables to find the following trigonometric ratios
   (a) cosec 34°   (b) cosec 72°18'   (c) cosec 81°
   (d) sec 52°   (e) sec 54°36'   (f) sec 63°43'
   (g) cot 48°   (h) cot 27°6'   (i) cot 82°16'
2. Use tables to find the angles whose cosecants are given below.
   (a) 1.6243   (b) 3.5843   (c) 1.0803   (d) 5.9554
   (e) 1.0105   (f) 2.6260   (g) 1.4746   (h) 4.1824
3. Use tables to find the angles whose secants are given below.
   (a) 1.2062   (b) 1.0589   (c) 4.3684   (d) 1.0033
   (e) 2.1116   (f) 1.1262   (g) 1.1753   (h) 11.1045
4. Use tables to find the angles whose cotangents are given below.
   (a) 1.3270   (b) 2.4506   (c) 0.5147   (d) 6.5606
   (e) 0.4935   (f) 2.2062   (g) 1.4994   (h) 0.1257

7.4 The General Angle

Recall from Topic 19 that, the *x*-axis and the *y*-axis divide the *x-y* plane into four sections called the first, second, third and fourth quadrants.

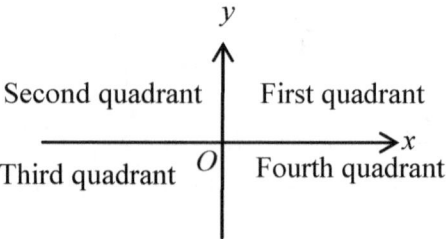

To find the trigonometric ratios of angles greater than 90°, consider a point *P* (figure below), which starts from the line *Ox*, where *O* is the origin and moves in the anticlockwise sense on a circle of radius 1 unit, called a unit circle. The coordinates of *P* when *P* is in each of the four quadrants will be either positive or negative though the radius *OP* remains positive as shown.

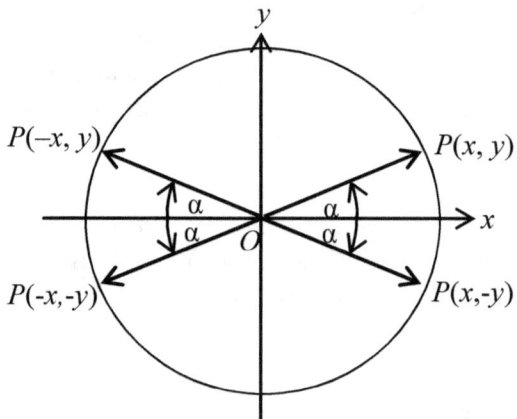

The angle α which *OP* makes with the *x*-axis remains acute (i.e. $0° \leq \alpha \leq 90°$), no matter the angle of rotation θ of *P* from *Ox*. We call α the **associated angle**. The values of α in the first, second, third and fourth quadrants are θ, $180° - \theta$, $\theta - 180°$ and $360° - \theta$ respectively. Mathematicians can prove (though beyond the scope of this book) that in each position, the trigonometric ratios of the angle θ are numerically equal in magnitude to the trigonometric ratios of the associated angle α.

In the first quadrant $0° \leq \theta \leq 90°$ and $\alpha = \theta$ (figure below).

Module 16, Topic 7: Trigonometry

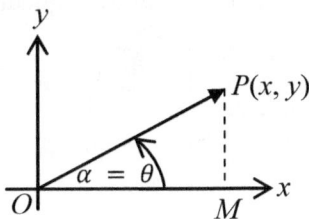

$$\sin\theta = \frac{y}{1} = y = \sin\alpha$$

$$\cos\theta = \frac{x}{1} = x = \cos\alpha$$

$$\tan\theta = \frac{y}{x} = \tan\alpha$$

In the second quadrant $90° \leq \theta \leq 180°$ and $\alpha = 180° - \theta$ (figure below).

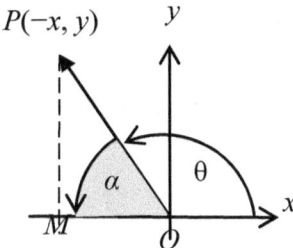

$$\sin\theta = \frac{y}{1} = y = \sin\alpha$$

$$\cos\theta = \frac{-x}{1} = -x = -\cos\alpha$$

$$\tan\theta = \frac{y}{-x} = -\tan\alpha$$

In the third quadrant $180° = \theta \leq 270°$ and $\alpha = \theta - 180°$ (figure below).

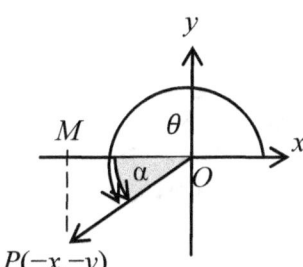

$$\sin\theta = \frac{-y}{1} = -y = -\sin\alpha$$

$$\cos\theta = \frac{-x}{1} = -x = -\cos\alpha$$

$$\tan\theta = \frac{-y}{-x} = \tan\alpha$$

In the fourth quadrant $270° = \theta \leq 360°$ and $\alpha = 360° - \theta$ (figure below).

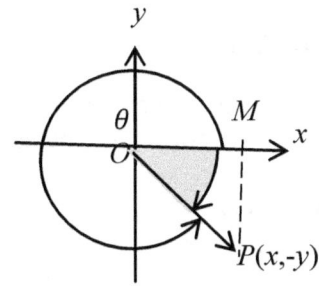

$$\sin\theta = \frac{-y}{1} = -y = -\sin\alpha$$

$$\cos\theta = \frac{x}{1} = x = \cos\alpha$$

$$\tan\theta = \frac{-y}{x} = -\tan\alpha$$

The figure below summarizes the positive trigonometric ratios, in each quadrant.

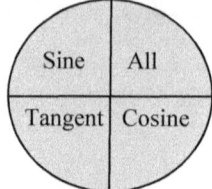 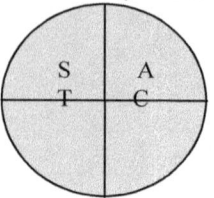

Thus,
In the first quadrant, ALL the three are positive.
In the second quadrant, only the SINE is positive.
In the third quadrant, only the TANGENT is positive.
In the fourth quadrant, only the COSINE is positive.

The mnemonic **ACTS** may help to recall these facts to memory. Note that the mnemonic is read from the fourth quadrant.

Exercise 7:3

1. Evaluate the following without using tables or calculators.
   (a) cos 300°   (b) tan 210°   (c) tan 120°
   (d) sin 270°   (e) sin 135°   (f) cos 150°
2. Evaluate the following using tables
   (a) cos 100°   (b) tan 235°   (c) tan 125°
   (d) sin 260°   (e) sin 160°   (f) cos 170°
3. Evaluate the following without using tables or calculators.
   (a) cosec 300° (b) cot 210°   (c) cot 120°
   (d) sec 270°   (e) sec 135°   (f) cosec 150°
4. Evaluate the following using tables
   (a) cosec 100° (b) cot 235°   (c) cot 125°
   (d) sec 260°   (e) sec 160°   (f) cosec 170°

7.5 The Meaning of Negative Angles

Conventionally angles measured in the anticlockwise sense are considered positive while angles measured in the clockwise sense are considered negative.

Consider again the point *P*, which is moving on the circle.

Module 16, Topic 7: Trigonometry

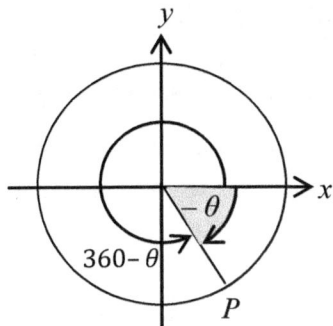

Suppose P is on any position, which is not along Ox. P could attain this position by either moving clockwise or anticlockwise. It follows that the angle $-\theta$ is equivalent to the angle $(360 - \theta)$ as shown in the figure above.
Hence the trigonometric ratios of $-\theta$ are equal to the trigonometric ratios of $(360 - \theta)$.

$$-\theta \equiv (360 - \theta)$$

From the above explanations,
$$-30° \equiv (360 - 30)° \equiv 330°$$
$$-120° \equiv (360 - 120)° \equiv 240°$$
$$-45° \equiv (360 - 45)° \equiv 315°$$
$$-270° \equiv (360 - 270)° \equiv 90°$$

Generalizing,
$$(-\theta) \equiv (360 - \theta)$$
$$\Leftrightarrow \sin(-\theta) \equiv \sin(360 - \theta) = -\sin\theta$$
$$\Leftrightarrow \cos(-\theta) \equiv \cos(360 - \theta) = \cos\theta$$
$$\Leftrightarrow \tan(-\theta) \equiv \tan(360 - \theta) = -\tan\theta$$

 Example

Evaluate the following
(a) $\sin 330°$ (b) $\tan 315°$ (c) $\cos 240°$

Solution

(a) $\sin 330° = \sin(360 - 330)° = -\sin 30° = -\dfrac{1}{2}$

(b) $\tan 315° = \tan(360 - 315)° = -\tan 45° = -1$

(c) $\cos 240° = \cos(360 - 240)° = \cos 120°$

Since $120°$ is in the second quadrant, the associated angle is $180° - 120° = 60°$ and the cosine is negative. $\Rightarrow \cos 240° = -\cos 60° = -\dfrac{1}{2}$

Exercise 7:4

1. Evaluate the following without using tables or calculators.
   (a) $\cos(-300)°$ (b) $\tan(-210)°$ (c) $\tan(-120)°$
   (d) $\sin(-270)°$ (e) $\sin(-135)°$ (f) $\cos(-150)°$
2. Evaluate the following using tables.
   (a) $\cos(-100)°$ (b) $\tan(-235)°$ (c) $\tan(-125)°$
   (d) $\sin(-260)°$ (e) $\sin(-160)°$ (f) $\cos(-170)°$
3. Evaluate the following using calculators.
   (a) $\cos(-100)°$ (b) $\tan(-235)°$ (c) $\tan(-125)°$
   (d) $\sin(-260)°$ (e) $\sin(-160)°$ (f) $\cos(-170)°$
4. Evaluate the following without using tables or calculators.
   (a) $\text{cosec}(-100)°$ (b) $\cot(-235)°$ (c) $\cot(-125)°$
   (d) $\sec(-260)°$ (e) $\sec(-160)°$ (f) $\text{cosec}(-170)°$
5. Evaluate the following using tables.
   (a) $\text{cosec}(-100)°$ (b) $\cot(-235)°$ (c) $\cot(-125)°$
   (d) $\sec(-260)°$ (e) $\sec(-160)°$ (f) $\text{cosec}(-170)°$
6. Evaluate the following using calculators.
   (a) $\text{cosec}(-100)°$ (b) $\cot(-235)°$ (c) $\cot(-125)°$
   (d) $\sec(-260)°$ (e) $\sec(-160)°$ (f) $\text{cosec}(-170)°$

7.6 Radian Measure

A radian is the angle subtended at the centre of a circle by an arc whose length is equal to the radius of the circle.

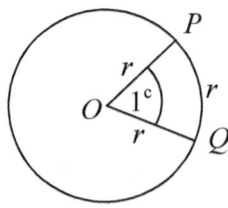

$$\Rightarrow 360° = 2\pi \text{ or } 180° = \pi$$

We can use the above relation to convert degrees to radians and vice versa.

Module 16, Topic 7: Trigonometry

 Example

Convert (a) 135° to radians   (b) $\frac{7\pi}{5}$ to degrees.

Solution

(a) $135° = \frac{135°}{360°} \times 2\pi = \frac{3\pi}{4}$ (b) $\frac{7\pi}{5} = \frac{\left(\frac{7\pi}{5}\right)}{2\pi} \times 360 = 252°$

 Exercise 7:5

1. Covert the following to radians.
    (a) 30° (b) 120° (c) 240° (d) 300° (e) 22.5°
2. Covert the following to dgrees.
    (a) $\frac{3\pi}{2}$ (b) $\frac{5\pi}{4}$ (c) $\frac{7\pi}{6}$ (d) $\frac{\pi}{12}$ (e) $\frac{5\pi}{3}$

7.7 Basic Trigonometric Identities

 Investigative Activity

In the following triangle, write down
(a) $\sin\theta$ (b) $\cos\theta$ (c) $\tan\theta$ (d) $\frac{\sin\theta}{\cos\theta}$ (e) $\sin^2\theta + \cos^2\theta$

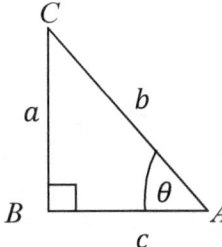

From the above investigation, we see that
$$\tan\theta = \frac{\sin\theta}{\cos\theta} \quad \ldots\ldots\ldots\ldots\ldots\ldots \text{①}$$
$$\sin^2\theta + \cos^2\theta = 1 \quad \ldots\ldots\ldots\ldots\ldots\ldots \text{②}$$

We also saw earlier under the section, reciprocal tri ratios that,

$$\csc\theta = \frac{1}{\sin\theta} \quad \ldots\ldots\ldots\ldots\ldots\ldots\ldots ③$$

$$\sec\theta = \frac{1}{\cos\theta} \quad \ldots\ldots\ldots\ldots\ldots\ldots\ldots ④$$

$$\cot\theta = \frac{1}{\tan\theta} \quad \ldots\ldots\ldots\ldots\ldots\ldots\ldots ⑤$$

The above ① and ⑤ are examples of trigonometric identities.

7.8 Graphs of Trigonometric Functions

To draw the graph of a trigonometric function $y = f(\theta)$, we make a table of values of y against θ. Next we plot the graph with θ on the horizontal axis and y on the vertical axis.

? Brainstorming Exercise

1. Use your calculator to complete the following table to 1 decimal place.

θ in °	0	30	60	90	120	150	180	210	240	270	300	330	360
$\sin\theta$													
$2\sin\theta$													

2. Use your table to sketch the graphs of $y = \sin\theta$ and $y = 2\sin\theta$ on the same Cartesian axes.
3. Using your graph, state the greatest and least values of
   (a) $\sin\theta$ (b) $2\sin\theta$
4. What remarks can you make concerning the two graphs?

The following shows the typical nature of the graphs of $y = \sin\theta$ and $y = 2\sin\theta$.

Take note of the following properties of the graph.
1. The maximum value of $\sin\theta$ is 1 and the minimum value is -1.
2. The maximum value of $2\sin\theta$ is 2 and the minimum value is -2.
3. The shape of both graphs from $\theta = 0$ to $\theta = 360°$, repeats indefinitely if the graph is extended. We say the period of the sine curve is 360°.

We can make a table of values and draw the graph of multiples of a trig ratio.

Module 16, Topic 7: Trigonometry

$y = \sin \theta$

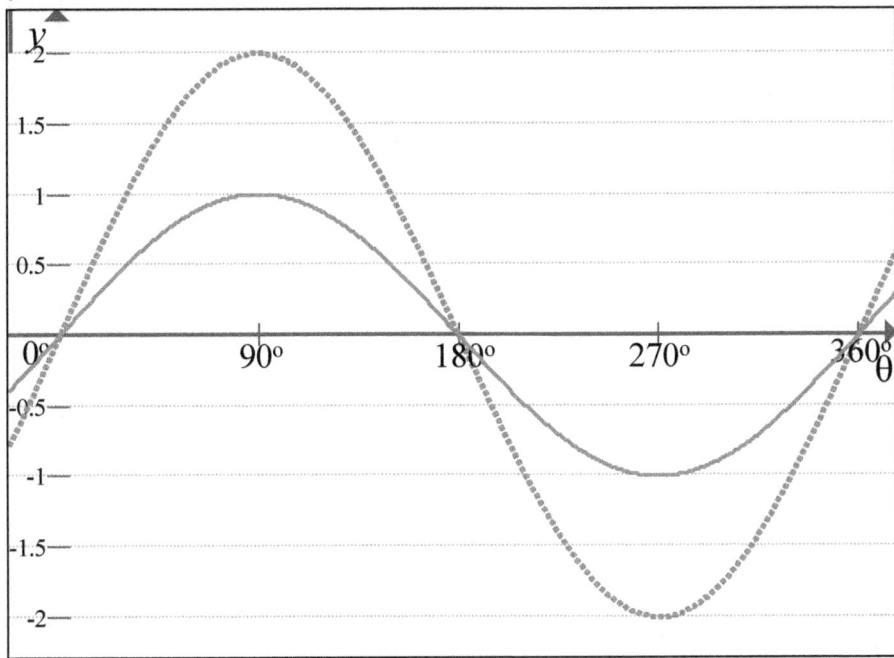

Example

Make a table of values to 1 decimal place of y against θ for θ between 0° and 360° for $y = \tan \theta$ and $y = 2 \tan \theta$.
  (a) Use your table to draw the graphs of $y = \tan \theta$ and $y = 2 \tan \theta$.
  (b) What conclusions do you draw concerning these graphs?

Solution

θ in °	0	30	60	90	120	150	180	210	240	270	300	330	360
$\tan \theta$	0	0.6	1.7	∞	−1.7	−0.6	0	0.6	1.7	∞	−1.7	−0.6	0
$2 \tan \theta$	0	1.2	3.5	∞	−3.5	−1.2	0	1.2	3.5	∞	−3.5	−1.2	0

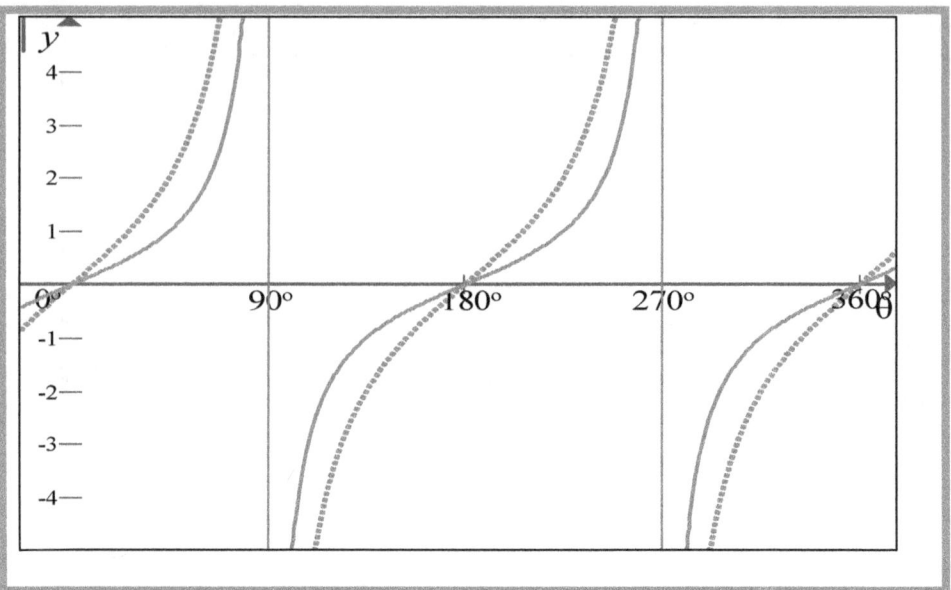

 Exercise 7:6

Make tables of values for each of the following and draw the graph of the trigonometric function.

(a) $y = \cos x$ (b) $y = 2\cos x$ (c) $y = -\cos x$
(d) $y = -2\cos x$ (e) $y = -2\sin x$ (e) $y = -2\tan x$.

7.9 Simple Trigonometric Equations

A trigonometric equation is an equation that involves trigonometric ratios. We can solve trigonometric equations algebraically or graphically.

 Example

1. Given that $0° \leq \theta \leq 90°$. Use an algebraic method to find the value of θ for which $\sin \theta = \cos 60°$.

   Solution
   $\sin \theta = \cos 60° \Rightarrow \sin \theta = \frac{1}{2}$ and $\theta = 30°$.

2. Solve the equation $\sin 2\theta° = \cos \theta°$ using a graphical method.

Solution

To solve this equation using a graphical method, we plot the graphs of $y = \sin 2\theta°$ and $y = \cos \theta°$. The horizontal coordinate of their point of intersection gives the solution.

θ	0°	9°	18°	27°	36°	45°	54°	63°	72°	81°	90°
$\sin 2\theta$	0	0.31	0.59	0.81	0.95	1	0.95	0.81	0.59	0.31	0
$\cos \theta$	1	0.99	0.95	0.89	0.81	0.71	0.59	0.45	0.31	0.16	0

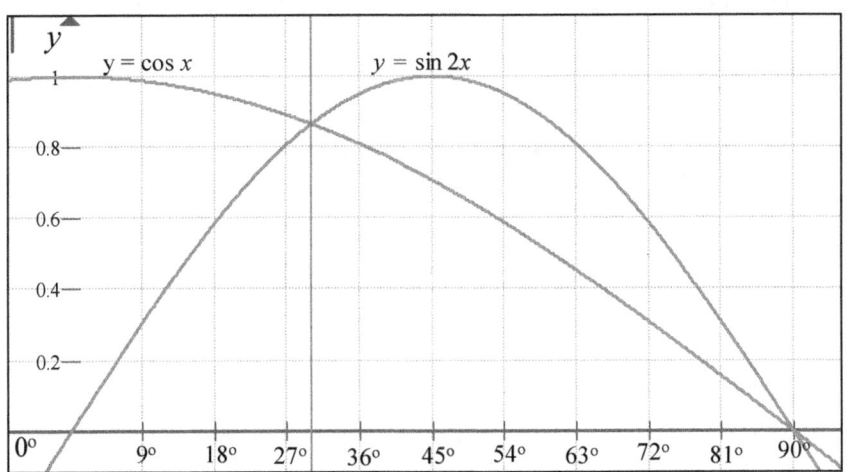

From the graph, $\theta = 30°$ or $\theta = 90°$.

 Exercise 7:7

Solve the following trigonometric equation in the range $0 \leq x \leq 90°$.
1. $\sin 2x° = 0$
2. $\sin x° = 2 \cos x° \sin x°$
3. $3 \tan^2 x° - \sec^2 x° = 1$
4. $\cos x° - \sin x° = 1$
5. $\tan x° = \sec x°$

 Multiple Choice Exercise 7

1. The pair of trigonometric ratios such that one is the inverse of the other is:
   [A] $\sin \theta$ and $\csc \theta$
   [B] $\cos \theta$ and $\cot \theta$
   [C] $\sin \theta$ and $\sec \theta$
   [D] $\cos \theta$ and $\csc \theta$

2. Given that $\tan x = \dfrac{5}{12}$. The value of $\sin x + \cos x$ is:

   [A] $\dfrac{5}{13}$ [B] $\dfrac{7}{13}$ [C] $\dfrac{17}{13}$ [D] $\dfrac{5}{12}$

3. Given that $\sin P = \dfrac{5}{13}$ where P is acute. The value of $\cot P - \tan P$ is:

   [A] $\dfrac{79}{156}$ [B] $\dfrac{95}{156}$ [C] $\dfrac{5}{13}$ [D] $\dfrac{13}{12}$

4. If $\sin(x+30)° = \cos 40°$, the value of $x°$ is:
   [A] 15° [B] 20° [C] 60° [D] 90°
5. If $10 \tan 60° = 20 \tan x$. Correct to the nearest degree x is equal to:
   [A] 30° [B] 40° [C] 41° [D] 60°
6. If $\sin x = \cos 70°$, x equals:
   [A] 110° [B] 70° [C] 30° [D] 20°
7. If $\sin x = \dfrac{12}{13}$, where $0° < x < 90°$. The value of $1 - \cos^2 x$ is:

   [A] $\dfrac{25}{169}$ [B] $\dfrac{64}{169}$ [C] $\dfrac{105}{169}$ [D] $\dfrac{144}{169}$

8. Using Mathematical tables $\cos 40° - \sin 30°$ equals:
   [A] -0.2660 [B] 0.2660 [C] -0.0266 [D] 0.0266
9. If $\cos 60° = \dfrac{1}{2}$, The angle whose cosine is equal to $\dfrac{1}{2}$ is:
   [A] 120° [B] 150° [C] 210° [D] 300°
10. If $\cos x$ is negative and $\sin x$ is negative, it is true that:
    [A] $0° < x < 90°$ [B] $90° < x < 180°$ [C] $180° < x < 270°$ [D] $270° < x < 360°$
11. If $\sin \theta = \dfrac{\sqrt{3}}{2}$ and $\cos \theta = -\dfrac{1}{2}$. The value of θ is:

    [A] 30° [B] 60° [C] 90° [D] 120°
12. The value of $\tan 315°$ is:

    [A] -1 [B] $-\dfrac{\sqrt{2}}{2}$ [C] 0 [D] 1

13. The value of $\sin 210°$ is:

    [A] $\dfrac{1}{2}$ [B] $-\dfrac{\sqrt{3}}{2}$ [C] $-\dfrac{1}{2}$ [D] $\dfrac{\sqrt{3}}{2}$

14. $\cos 75°$ has the same value as:
    [A] $\cos 115°$ [B] $\cos 255°$ [C] $\cos 285°$ [D] $-\cos 255°$
15. If $\sin\theta = \cos\theta$ for $0° \leq \theta \leq 360°$. The possible values of θ are:
    [A] 45°, 225° [B] 135°, 315° [C] 45°, 315° [D] 135°, 225°
16. $\cos 57°$ has the same value as:
    [A] $\sin 213°$ [B] $\cos 303°$ [C] $\cos 127°$ [D] $\cos 137°$
17. If $\sin\theta = -\dfrac{1}{2}$. The values of θ between 0° and 360° are:

    [A] 120°, 240° [B] 120°, 180° [C] 210°, 330° [D] 210°, 300°
18. $\cos 65°$ has the same value as:

[A] sin 65°   [B] cos 25°   [C] cos 205°   [D] cos 295°

19. In $\triangle PQR$, $\angle PQR$ is a right angle $|QR| = 2$ cm and $\angle PRQ = 60°$. $|PR|$ is equal to:

   [A] $4\sqrt{3}$ cm   [B] 4 cm   [C] $2\sqrt{3}$ cm   [D] 1 cm

20. If $\cos x = \dfrac{5}{8}$ for $0° \leq x \leq 180°$, the value of x is:

   [A] 141.3°   [B] 128.7°   [C] 51.3°   [D] 48.7°

21. In surd form $\sin 45° \cos 30° + \cos 45° \sin 30°$ is equal to:

   [A] $\dfrac{\sqrt{2}}{2}$ [B] $\dfrac{\sqrt{3}}{2}$ [C] $\sqrt{2}$ [D] $\dfrac{\sqrt{6}+\sqrt{2}}{4}$

Topic 8

APPLICATIONS OF TRIGONOMETRY

Objectives
At the end of this topic, the learner should be able to:

1. Apply trigonometry to solve problems involving right-angled triangles.
2. Distinguish between angle of elevation and angle of depression and solve problems involving angle of elevation and angle of depression.
3. Understand the convention for denoting bearings and solve problems on bearings in two dimensions.
4. State and use the Sine and Cosine Formulae to find the unknown sides or angles of any triangle.

Module 16, Topic 8: Applications of Trigonometry

8.1 Solutions to Triangles

The trigonometric ratios studied in Topic 7 are used to solve trigonometric problems involving right-angled triangles. There are three cases to consider.

Given One Side and One Acute Angle

Example

1. In the figure below, find to 2 decimal points (a) AB (b) AC

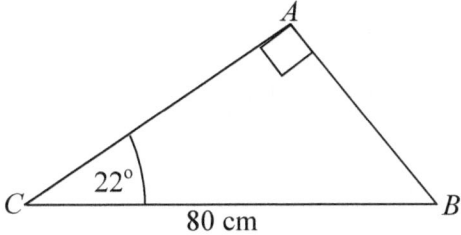

Solution
(a) $AB = 80\sin 22 = 80(0.3746) = 29.97$ cm
(b) $AC = 80\cos 22 = 80(0.9272) = 74.18$ cm

2. In the triangle ABC, angle BAC is equal to $23°35'$. Given that $BC = 60$ cm, find
 (a) AB (b) AC.

Solution

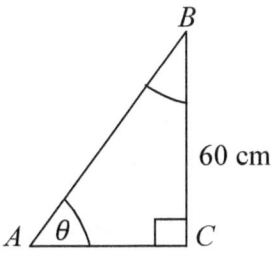

(a) $AB = \dfrac{BC}{\sin\theta}$, where $\theta = 23°35'$

$\Rightarrow AB = \dfrac{60}{\sin 23°35'} = \dfrac{60}{0.4}$ cm $= 150$ cm

(b) $AC = \dfrac{BC}{\tan \theta}$, where $\theta = 23°35'$

$\Rightarrow AC = \dfrac{60}{\tan 23°35'} = \dfrac{60}{0.4369}$ cm $= 136$ cm

In solving problems involving triangles, there are often a number of alternative methods as demonstrated below.
$AB = BC \operatorname{cosec} 23°35' = 60(2.5) = 150$ cm
$AC = BC \cot 23°35' = 60(2.2937) = 137.3$ cm.

We can find AB and AC using $\angle B$ as follows.
$\angle B = 90° - \theta = 90° - 23°35' = 66°25'$.
$AB = BC \sec 66°25' = 60(2.5) = 150$ cm
$AC = BC \tan 66°25' = 60(2.29) = 137.4$ cm

Exercise 8:1

1. Find the length of the two unknown sides of the triangles in Figure (a) and (b) below to two decimal places.

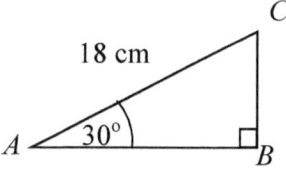

(a)

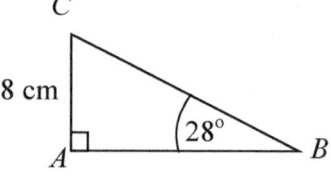

(b)

2. Find to the nearest degree each acute angle of a triangle whose sides are in the ratio;
   (a) 5:12:13 (b) 8:15:17 (c) 7:24:25 (d) 11:60:61
3. A ladder leans against a vertical wall and makes an angle of 70° with the horizontal ground. The foot of the ladder is 3 m from the wall. How far up the wall to the nearest m does the ladder reach?
4. The figure below shows the cross-section of roof. How long must the carpenter cut the plank support BN? What is the length of the base AC?

Module 16, Topic 8: Applications of Trigonometry

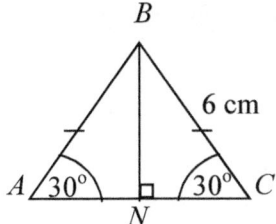

5. A road rises 105 m for every x m of the horizontal. If the angle between the slope and the horizontal is $6°$, find the value of x.
6. The roof of a bamboo hurt is a triangular prism. If the altitude of the roof from the ceiling is, 3 m and the equal slanting sides make each an angle of $30°$ with the horizontal. Calculate:
   (a) The length of the slanting sides
   (b) The angle made by the slanting sides at the upper edge.
   (c) The length of the base between the two slanting sides.
7. A plane takes off at A and ascends at a fixed angle of $22°$ with the level ground. After it flies 3000 m, find to the nearest metre;
   (a) The altitude of the plane   (b) The distance covered horizontally.

Given Two Sides of a Right-Angled Triangle

Given two sides of a right-angled triangle, we can find the third side using the Pythagoras theorem and can find the angles using the trig ratios of the sides.

 Example

1. Given the triangle below, find
   (i) a, leaving your answer in surd form.   (ii) Angle A (iii) Angle C

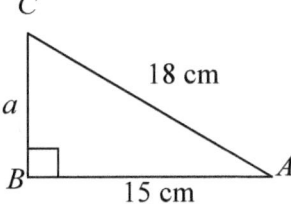

Solution
(i) Using the Pythagoras theorem,
   $a^2 = 18^2 - 15^2 \Rightarrow a = 3\sqrt{11}$ cm

(ii) $\cos A = \dfrac{15}{18} \Rightarrow A = \cos^{-1} 0.55278 = 33.6°$

(iii) Angle $C = 90 - 33.6 = 56.4°$

2. Given that $\hat{B} = 90°, a = 8.1$ cm and $c = 7.5$ cm.
   Find (a) angle A (b) angle C (c) AC

   Solution

   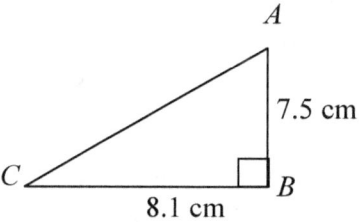

   (a) $\tan A = \dfrac{8.1}{7.5} \Rightarrow \angle A = \tan^{-1} 1.08 = 47.2°$

   (b) $\angle C = 90 - 47.2 = 42.8°$

   (c) By the Pythagoras theorem, $AC = \sqrt{8.1^2 + 7.5^2} = 11.04$ cm

 Exercise 8:2

1. In figure (a) to (f) below find,
   (i) The unknown side leaving your answer in surd form.
   (ii) The angles X and Y to 1 decimal place.

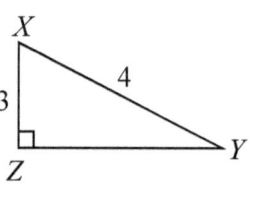
(a)

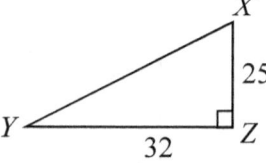
(b)

Module 16, Topic 8: Applications of Trigonometry

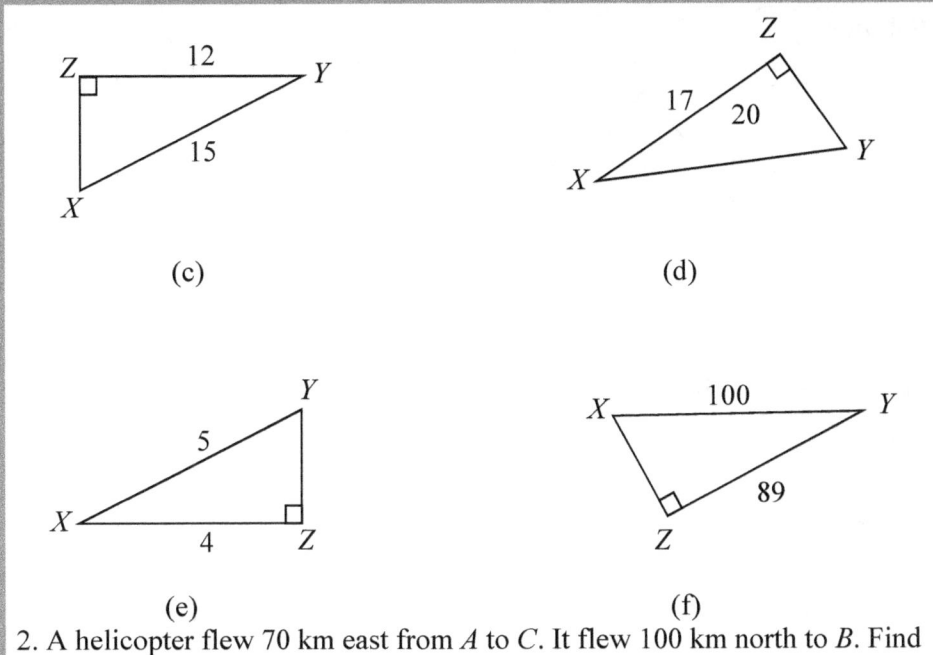

(c) (d)

(e) (f)

2. A helicopter flew 70 km east from A to C. It flew 100 km north to B. Find the angle of turn, to the nearest degree that it must make at B to return to A.

3. In the figure below, find the side marked y and determine the values of $\angle X$ and $\angle Z$

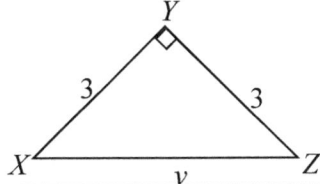

Given an Isosceles Triangle with Given Sides

To solve an isosceles triangle with given sides, draw a perpendicular from the vertex between the equal sides to the opposite side.

 Example

Find the angles of the triangle ABC with $AB = 9.6$ cm and $AC = BC = 8.2$ cm.

141

Solution

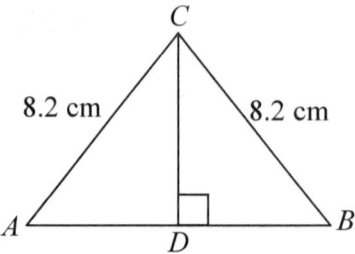

Since CD bisects AB,

$AD = \dfrac{9.6}{2} = 4.8$ cm

From $\triangle ADC$ $\cos A = \dfrac{4.8}{8.2}$

$\Rightarrow \angle A = \cos^{-1}\dfrac{4.8}{8.2} = 54°10'$, $\angle B = 54°10'$ and
$\angle C = 180° - 2(54°10') = 71°40'$

 Exercise 8:3

1. In the figure below, find AC to 3 significant figures.

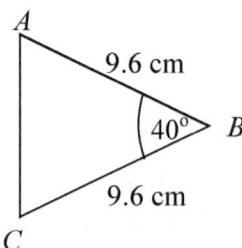

2. An isosceles triangle ABC has AB = BC = 6 cm and a base 4 cm long. Find the altitude of the triangle, leaving your answer in surd form. Also find the angles of the triangle to 3 significant figures.

3. Given that all the triangles in figure (a) to (c) are isosceles, find the perpendicular distance from B to AC, leaving your answer in surd form. Also find the angles A, B and C of the triangles.

Module 16, Topic 8: Applications of Trigonometry

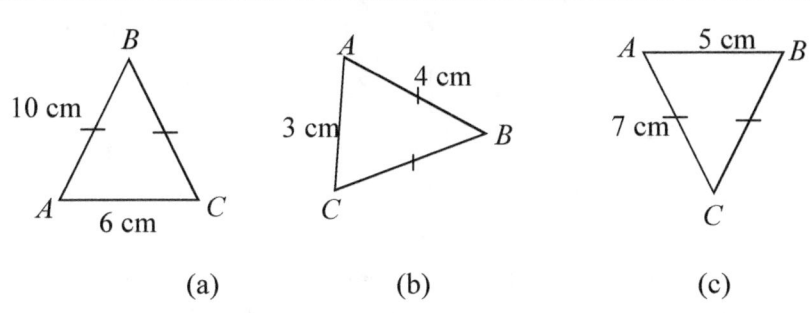

    (a)   (b)   (c)

4. The base angle of an isosceles triangle is 28° and each arm is 45 cm. Find to the nearest cm
   (a) The altitude of the line drawn to the base   (b) The length of the base
5. An isosceles triangle ABC has $AB = AC = 8$ cm, and $BC = 5$ cm. Calculate the size of its angles.

8.2 Angles Of Elevation And Depression

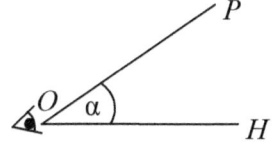

α is the angle of elevation.

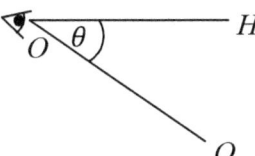

θ is the angle of depression.

The angles of elevation and depression give the direction of one point with respect to another in a vertical plane. Thus α is the angle of elevation of P with respect to the horizontal OH, while θ is the angle of depression of Q with respect to the horizontal OH.

 Example

1. The angle of elevation of the top of a house is 30° to a man lying 11 m away from the base of the house. Calculate the height of the house.

   Solution
   Let OT be the tree of height h and OM the distance of the man from the tree. Then from the figure below, $h = 11\tan 30° = \dfrac{11\sqrt{3}}{3}$ m

143

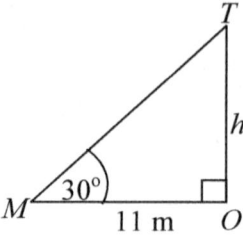

2. A road surveyor measures the difference in altitude of two points to be 2 m. If the distance between these two points is 12 m, calculate the angle of elevation of the higher point from the lower one. Hence, find the angle of depression of the lower point from the higher one.

Solution
Let the lower point be P, the higher point be Q and the difference in altitude between the points be OQ as shown in the figure below. Then the angle θ of elevation of Q from P is equal to the angle α of depression of P from Q [alternate angles between parallel lines].

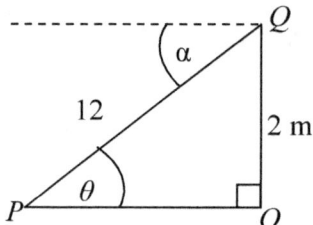

$\sin\theta = \dfrac{2}{12} = \dfrac{1}{6} \Rightarrow \theta = \sin^{-1}\left(\dfrac{1}{6}\right) \approx 9.6°$. Hence $\alpha = 9.6°$

3. Two points A and B 15 m apart are in line and on the same level with a tree. If the angles of elevation of the top of the tree from the points A and B are 16° and 25° respectively, calculate the height of the tree and the distance of each of the points from the tree.

Solution
Let the height OT of the tree be h.
The point B with the greater angle of elevation is closer to the tree than the point A.

Module 16, Topic 8: Applications of Trigonometry

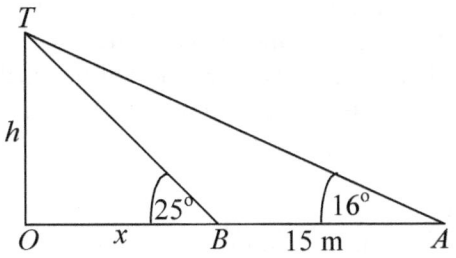

$h = x\tan 25°$ .... ①
$h = (x+15)\tan 16°$ .... ②
② − ①: $(x+15)\tan 16° - x\tan 25° = 0$
$$x = \frac{15\tan 16°}{\tan 25° - \tan 16°} = \frac{15(0.29)}{0.47 - 0.29} = 23.95$$
Therefore, $h = 23.95 \tan 25° = 11.17$ m.
Hence, distance of B from the tree = 23.95 m and that of A = 23.95 + 15 = 38.95 m.

 Exercise 8:4

1. The angle of elevation of the top of a pole from a point 50 m (away from the top) is 21°. Find to the nearest metre.
   (a) The height of the pole
   (b) The distance between the pole and the point
2. An electrician climbed on a straight pole 13 m tall and sighted his bag at an angle of depression of 30°. Calculate,
   (a) The distance of the bag from the bottom of the pole.
   (b) The distance of the bag from the face of the electrician.
3. From the top of a cliff 18 m high, the angle of depression of a fish in the sea is 24°. Find to the nearest metre the distance of the fish from the bottom of the cliff.
4. An observer at the top of a hill 75 m above the level of a lake sighted two fishermen directly in line. Find to the nearest metre the distance between the fishermen, if the angles of depressions noted by the observer were:
   (a) 20° and 15°   (b) 35° and 24°   (c) 9° and 6°
5. The angle of elevation of the sun at a certain time is 42°. Calculate to the nearest metre;
   (a) The height h of a tree whose shadow is 25 m long.
   (b) The shadow of the tree along the level ground if its height is 35 m.
6. Standing on a platform 200 m away from a building, Ndoh found that the angle of elevation of the top of a building is 21°. Given that, the heights of the platform and that of Ndoh are 3.3 m and 1.7 m respectively. Find the

height of the building to the nearest metre.
7. Sitting at the top of a tree 20 m tall, a boy sighted a bird and a goat directly beneath the bird. Given that, the angle of elevation of the bird is 25° and the angle of depression of the goat is 32°. Find to the nearest metre;
   (a) The distance of the goat from the tree
   (b) The height of the bird above the ground
8. A man standing at the top of a telephone transmitter pole 50 m tall, sights two cars on a horizontal road. Find the distance between the two cars if the angles of depression noted by the man were 21° and 36°.

8.3 Bearings In Two Dimensions

Bearings deal with the angular direction of one point from another. There are main two ways of expressing the bearing of one point from another. These are the cardinal point or compass bearing and the three digit bearings.

Cardinal Points or Compass Bearing

This method uses the four cardinal points North, South, West and East. In this method bearings are measured from North or South to East or West. Notice that the angle from north to east or north to west is only 90°. Similarly from south to east or south to west is only 90°. Thus N 35° E means an angle of 35° from the North towards the East, S 42° W means an angle of 42° from the South towards the West.

Three Digit Bearings

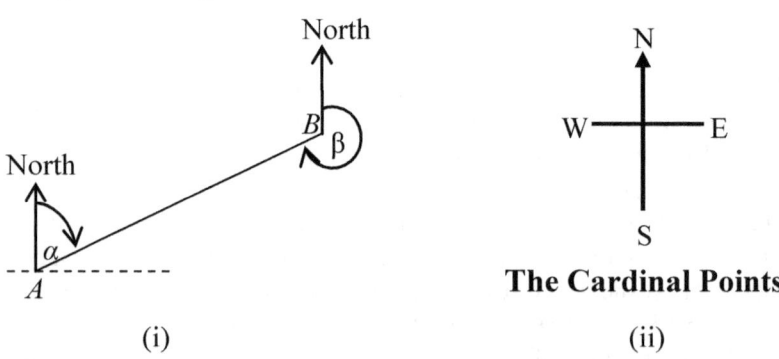

(i)   **The Cardinal Points**
                                   (ii)

Conventionally and for analytical purposes bearings are measured in the clockwise direction from the north which is considered 0°. In this method N 35° E will be written as 035°, S 30° W will be written as 210°. By this convention, all bearings are given to three digits. This minimizes errors. Thus, east will be represented by 090°, South West by 225°, North West by 315°, west by 270° etc. We can convert three digit bearings to compass bearings and vice versa.

Module 16, Topic 8: Applications of Trigonometry

 Example

1. Convert the following 3 digit bearings to compass bearings.
   (a) 075°   (b) 324°   (c) 138°   (d) 249°

   Solution

   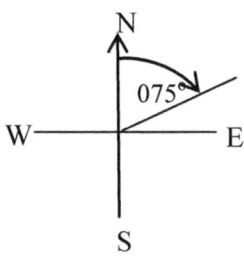

   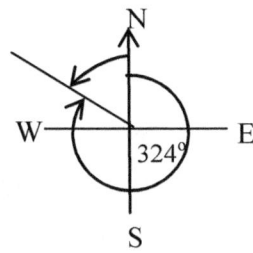

   (a)   075° = N 75° E   (b)   324° = 360° − 324° = N 36° W

   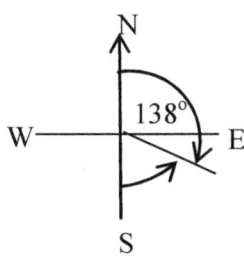

   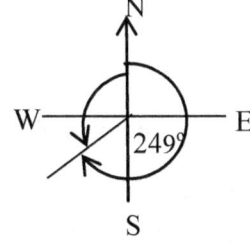

   (c)   138° = 180° − 138° = S 21° W   (d)   249° = 270° − 249° = S 42° E

2. Convert the following compass bearings 3 digit bearings to
   (a) N 85° E   (b) S 28° W

   Solution

   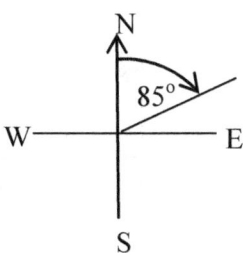

   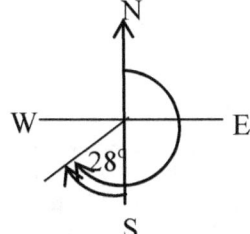

   (a)   N 85° E = 085°   (b)   S 28° W = 180° + 28° = 208°

   In the examples above, α is the bearing of the point B from the point A, and β is the bearing of the point A from the point B. Thus if the bearing of a point B from another point A is θ, then the bearing of the point A from B will be $(180° + \theta)$.

3. A boy starts from a point A and moves on a bearing of $020°$ to a point B, which is 5 km away from A. He then changes his course to a bearing of $110°$ and moves to a point C, which is 12 km from B. Find the distance and bearing of the point C from the point A, giving your answer to one decimal place.

Solution

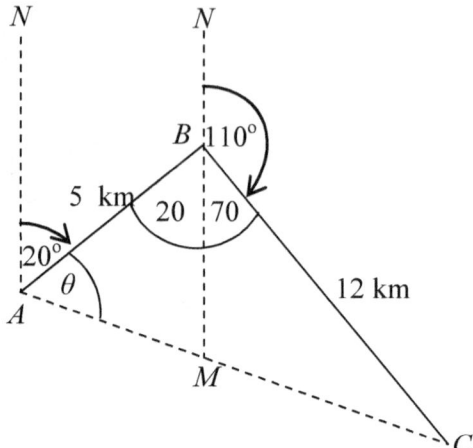

From the figure above, $\angle ABM = 20°$ (alternate angles)
$\angle CBM = 180° - 110°$ (angles on straight line) $= 70°$
$\angle ABC = 70° + 20° = 90°$
Using the Pythagoras theorem,
$AC = \sqrt{AB^2 + BC^2} = \sqrt{5^2 + 12^2} = \sqrt{169} = 13$ m
$\sin\theta = \dfrac{12}{13} \Rightarrow \theta = \sin^{-1}\left(\dfrac{12}{13}\right) \approx 67.38°$
Bearing of C from A $= 20° + 67.38° \approx 87.4°$, and the distance of C from A is 13m.

 Exercise 8:5

1. Anye walks 3 km from a point P on a bearing of $023°$. He then walks 4 km on a bearing of $113°$ to a point Q. Calculate
   (a) The distance of Q from P
   (b) The bearing of Q from P
2. From a point A, a boy walks 2 km on a bearing of $017°$. He then walks 3 km on a bearing of $107°$ to C. What is the distance and bearing of C from A?
3. A mail van travels from its head office in Yaounde to a town B 210 km

away on a bearing of 055° to deliver a mail. It then changes course and moves to its branch office in Yaounde on a bearing of 220°. If the branch office is directly east of the head office, calculate correct to 3 significant figures
(i) The distance between the branch office and the head office;
(ii) How far is town B from the branch office?

4. A King sent two messengers to deliver a message to two of his notables Shey and Nformi. Shey lives 20 km away from the palace on a bearing of 205° while Nformi lives 15 km from the palace on a bearing of 060°. Calculate to the nearest whole number
   (a) The distance apart of Nformi's home from Shey's home.
   (b) The bearing of Nformi's home from Shey's home.

5. In Figure 29:20, |XY| = 8 cm, |YZ| = 13 cm, the distance of Y from X is 050° and the bearing of Z from Y is 130°.
   (a) Calculate correct to 3 significant figures,
       (i) |XZ| (ii) the bearing of Z from X.
   (b) (i) Calculate the shortest distance between point Y and XZ. Hence, find the area of triangle XYZ.

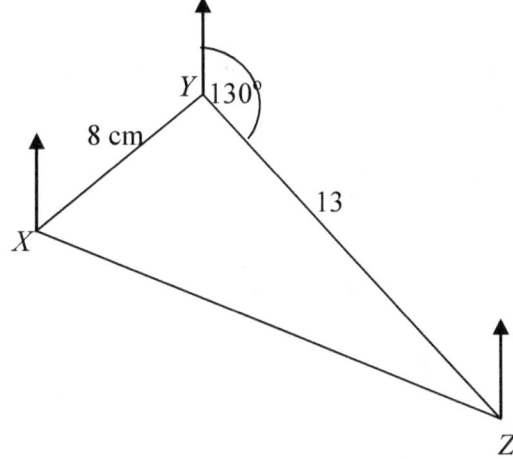

6. Three hills P, Q and R are such that Q is 24 km away from P on bearing of 150° and R is 32 km away from P, and P is on a bearing of 015° from Q. Calculate to one decimal place,
   (i) The distance between the hills Q and R.
   (ii) The bearing of hill R from hill Q.

8.4 Sine And Cosine Formulae

Sometimes problems in trigonometry involve triangles that are not right-angled. In such a case, the sine or the cosine formulae (sometimes referred to as the sine and cosine rules) are used.
Consider the triangles in the figure below.

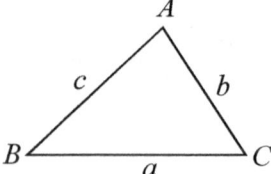 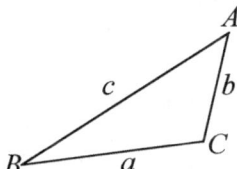

Each side of a triangle is proportional to the opposite side. This fact is expressed by the **sine formula**, which states that in any triangle

$$\frac{a}{\sin A} = \frac{b}{\sin B} = \frac{c}{\sin C}$$

This form is used when a side is required. When an angle is required, the following equivalent form is used

$$\frac{\sin A}{a} = \frac{\sin B}{b} = \frac{\sin C}{c}$$

The **cosine formula** states that in any triangle

$$a^2 = b^2 + c^2 - 2bc \cos A$$
$$b^2 = a^2 + c^2 - 2ac \cos B$$
$$c^2 = a^2 + b^2 - 2ab \cos C$$

The sine rule is used when:
(i) Two angles and one side opposite one of the given angles are given.
(ii) Two sides and an angle opposite one of the given sides are given.

The cosine rule is used when:
(i) Two sides and the included angle are given.
(ii) All the three sides are given.

Module 16, Topic 8: Applications of Trigonometry

 Example

1. In the figure below, find AC and AB.

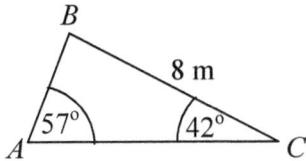

Solution
$$\angle B = 180 - (57 + 42) = 81°$$
Using the sine rule,
$$\frac{a}{\sin A} = \frac{b}{\sin B} = \frac{c}{\sin C}$$
$$\Rightarrow \frac{c}{\sin 42} = \frac{b}{\sin 81} = \frac{8}{\sin 57}$$
$$\Rightarrow b = \frac{8 \sin 81}{\sin 57} = 9.4 \text{ m and } c = \frac{8 \sin 42}{\sin 57} = 6.4 \text{ m}$$

2. In the figure below, find BC.

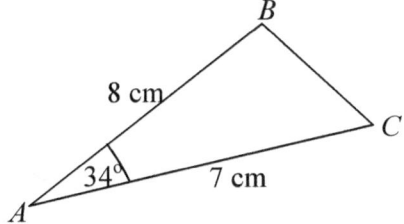

Solution
$$a^2 = b^2 + c^2 - 2bc \cos A$$
$$a^2 = 7^2 + 8^2 - 2(7)(8) \cos 34°$$
$$\Rightarrow a = \sqrt{20.15} = 4.5 \text{ cm}$$

Sometimes both the sine and the cosine rules are required to solve some problems.

3. In the figure below, calculate
   (a) AC (b) $\angle BAC$ (c) $\angle ACB$

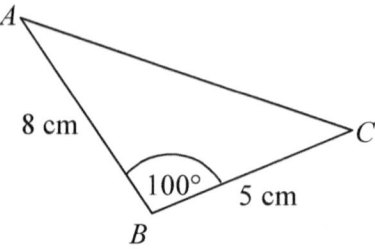

Solution
(a) By the cosine rule,
$$b^2 = a^2 + c^2 - 2ac\cos B$$
$$b^2 = 5^2 + 8^2 - 2(5)(8)\cos 100°$$
$$\Rightarrow b = \sqrt{102.89} = 10.14 \text{ cm}$$

(b) By the sine rule,
$$\frac{\sin A}{5} = \frac{\sin 100°}{10.14}$$
$$\Rightarrow \sin A = \frac{5\sin 100°}{10.14} = 0.4856$$
$$\Rightarrow \angle BAC = \sin^{-1} 0.4856 = 29.1°$$

(c) $\angle ACB = 180 - (100 + 29.1°) = 50.9°$

 Exercise 8:6

1. In the following figure, find to 1 decimal place BC and AC.

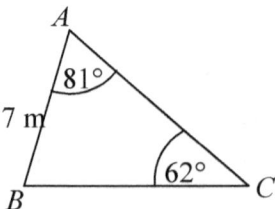

2. A ship steams 3 km from a port P on a bearing of $080°$ and then 4 km on a bearing of $047°$. Find to 1 decimal place its distance and bearing from P.
3. A ship sails 2 km due north and then 3 km on a bearing of $060°$. Find its distance and bearing from the original position.
4. A ship sails 110 km from a port X on a bearing of $035°$ and then 116 km

on a bearing of 105° to another point Y. Calculate to 1 decimal place:
(a) the distance XY   (b) the bearing of Y from X.

5. The figure below shows the position of two schools G and H the inspectorate I. G is 8 km on a bearing of 290° from I and G is from H 10 km on a bearing of 050°.

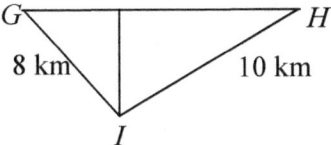

(a) Determine ∠GIH
(b) Calculate correct to 1 decimal place,
(i) The distance of GH.   (ii) The bearing of H from G.

Multiple Choice Exercise 8

1. In the triangle shown below the length of the side marked x is:
   [A] 7sin 56°   [B] 7tan 56°   [C] 7tan 34°   [D] 7cos 34°

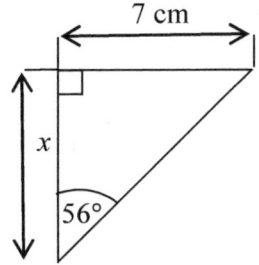

2. In the following figure, $AC = 42$ cm, $AD = 12$ cm and $BD = 16$ cm. As a vulgar fraction tan A is:

   [A] $\dfrac{4}{5}$ [B] $\dfrac{3}{5}$ [C] $\dfrac{5}{6}$ [D] $\dfrac{4}{3}$

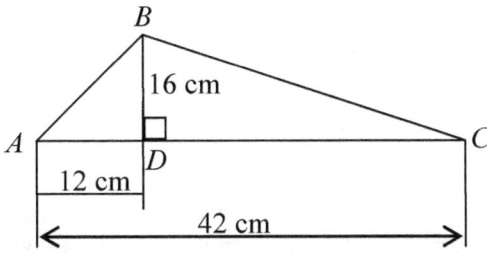

3. In the figure above, $AC = 42$ cm, $AD = 12$ cm and $BD = 16$ cm. Sin A as a decimal fraction is:

[A] 0.8   [B] 1.3   [C] 1.6   [D] 0.6

4. In the figure above, $AC = 42$ cm, $AD = 12$ cm and $BD = 16$ cm. The size of C to the nearest degree is:
   [A] 27°   [B] 32°   [C] 28°   [D] 58°

5. In the figure above, $AC = 42$ cm, $AD = 12$ cm and $BD = 16$ cm. The perimeter of triangle BDC in cm is:
   [A] 70   [B] 80   [C] 90   [D] 92

6. The figure below is a trapezium $ABCD$.

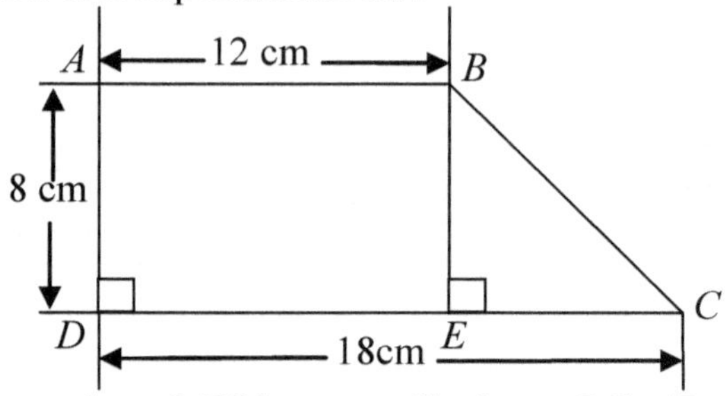

   Angle D is a right angle and BE is perpendicular to DC. $AB = 12$ cm and $AD = 8$ cm. The value of $DC = 18$ cm. The value of $\tan C$ is:
   [A] $\dfrac{4}{9}$ [B] $\dfrac{2}{3}$ [C] $\dfrac{4}{3}$ [D] $\dfrac{4}{5}$

7. A ladder 9 m long leans against a vertical wall, making an angle of 64° with the horizontal ground. To one decimal place, the distance of the foot of the ladder from the wall is:
   [A] 3.9 m   [B] 5.8 m   [C] 7.9 m   [D] 8.1 m

8. When helicopter is 800 m above the ground, its angle of elevation from a point P on the ground is 30°. The distance of the helicopter from P by the line of sight is:
   [A] 400 m   [B] 800 m   [C] 1600 m   [D] 1700 m

9. The angle of elevation of X from Y is 30°. If $XY = 40$ m the height of X above the level of Y is:
   [A] 10 m   [B] 20 m   [C] 40 m   [D] 50 m

10. If the shadow of a pole 7 m high is equal to half its length, the angle of elevation of the sun; correct to the nearest degree is:
    [A] 63°   [B] 0°   [C] 60°   [D] 26°

11. The angle of elevation of the top of a tree 39 m away from the point on the ground is 30°. The height of the tree is:
    [A] $39\sqrt{3}$ m   [B] $13\sqrt{3}$ m   [C] $\dfrac{13}{\sqrt{3}}$ m   [D] $\dfrac{\sqrt{3}}{13}$ m

12. A ladder leans against the wall at an angle 60° to the wall. If the foot of the ladder is 5 m away from the wall, the length of the ladder is:
    [A] $\dfrac{5\sqrt{3}}{3}$ m   [B] 5 m   [C] $5\sqrt{3}$ m   [D] $\dfrac{10\sqrt{3}}{3}$ m

Module 16, Topic 8: Applications of Trigonometry

13. The angle of elevation of the top of a tower from a point on the horizontal ground, 40 m away from the foot of the tower is 30°. The height of the tower is:

    [A] 20 m   [B] $\dfrac{40\sqrt{3}}{3}$ m   [C] $20\sqrt{3}$ m   [D] $40\sqrt{3}$ m

14. At a point 500 m from the base of a water-tank, the angle of elevation of the top of the tank is 45°. The height of the tank is:
    [A] 500 m   [B] 353 m   [C] 354 m   [D] 250 m

15. A ladder 6 m long leans against a vertical wall, so that it makes an angle of 60° with the wall. The distance of the foot of the ladder from the wall is:

    [A] 3 m   [B] 6 m   [C] $2\sqrt{3}$ m   [D] $3\sqrt{3}$ m

16. The angle of elevation of the top X of a vertical pole from a point P on a level ground is 60°. The distance from P to the foot of the pole is 55 m. Without using tables, the height of the pole is:

    [A] $\dfrac{50}{3}$ m   [B] 50 m   [C] $55\sqrt{3}$ m   [D] 60 m

17. The angle of elevation of a point T on a tower from a point U on the horizontal ground is 30°. If $TU = 54$ m, the height of T above the ground is:
    [A] 108 m   [B] 72 m   [C] 31.2 m   [D] 27 m

18. In Figure 29:29, $\angle PQR$ is a right angle $|QR| = 2$ cm and $\angle PRQ = 60°$. $|PR|$ is equal to:

    [A] 1 m   [B] 4 m   [C] $2\sqrt{3}$ m   [D] $\dfrac{4}{3}$ m

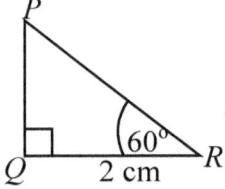

19. A ladder 5 m long rest against a wall so that it foot makes an angle of 30° with the horizontal. The distance of the foot of the ladder from the wall is:

    [A] $\dfrac{5\sqrt{3}}{3}$ m   [B] $2\dfrac{1}{2}$ m   [C] $\dfrac{5\sqrt{3}}{2}$ m   [D] $\dfrac{10\sqrt{3}}{3}$ m

20. A pole of length l leans against a vertical wall so that it makes an angle of 60° with the horizontal ground. If the top of the pole is 8 m above the ground, l must be:

[A] $16\sqrt{3}$ m   [B] $\dfrac{\sqrt{3}}{16}$ m   [C] 16 m   [D] $\dfrac{16\sqrt{3}}{3}$ m

21. The angle of depression of a point on the ground from the top of a building is 20.3°. If the distance of the point from the foot of the building is 40 m, the height of the building, correct to one decimal place is:
     [A] 13.9 m   [B] 28.1 m   [C] 27.8 m   [D] 14.8 m

22. From the top of a cliff 20 m high, surveyor sights a boat at sea 75 m from the foot of the cliff. The angle of depression of the boat from the top of the cliff is:
     [A] 14.9°   [B] 15.5°   [C] 74.5°   [D] 75.1°

23. From the top of a building 10 m high, the angle of depression of a stone lying on the ground is 69°. Correct to one decimal place, the distance of the stone from the foot of the building is:
     [A] 3.6 m   [B] 3.8 m   [C] 6.0 m   [D] 9.3 m

24. A cliff on the bank of a river is 300 metres high. If the angle of depression of a point on the opposite side of the river is 60°, the width of the river is:
     [A] 100 m   [B] $75\sqrt{3}$ m   [C] $100\sqrt{3}$ m   [D] $200\sqrt{3}$ m

25. From the top of a cliff, the angle of depression of a boat on the sea is 60°; if the top of the cliff is 25 m above the sea level, the horizontal distance from the bottom of the cliff to the boat is:
     [A] $\dfrac{\sqrt{3}}{25}$ m   [B] $25\sqrt{3}$ m   [C] $\dfrac{25\sqrt{3}}{3}$ m   [D] $\dfrac{25}{3}$ m

26. The angle of depression of a point Q from a vertical tower PR, 30 m high, is 40°. If the foot P of the tower is on the same level ground as Q, correct to two decimal places, $|PQ|$ should be:
     [A] 35.75 m   [B] 25.00 m   [C] 22.98 m   [D] 19228 m

27. In the figure below, AB is a vertical pole and BC is horizontal. If $|AC| = 10$ m and $|BC| = 5$ m. The angle of depression of C from A is:

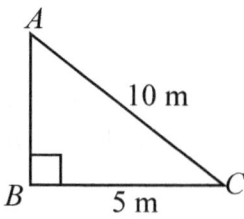

     [A] 63°   [B] 60°   [C] 45°   [D] 30°

28. In the figure below, QRT is a straight line. If $\angle PTR = 90°$, $\angle PRT = 60°$, $\angle PQR = 30°$ and $|PQ| = 6\sqrt{3}$ m. $|RT|$ is equal to:

[A] 0.3 m   [B] $\frac{\sqrt{3}}{2}$ m   [C] 3 m   [D] $3\sqrt{3}$ m

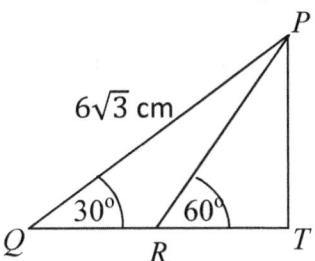

29. The true bearing of 250° as a compass bearing is:
    [A] S 70° E   [B] N 70° E   [C] S 70° W   [D] N 70° W
30. The bearing, which is equivalent to S 50o, W is:
    [A] 230°   [B] 220°   [C] 130°   [D] 040°
31. The bearing S 40° E is the same as:
    [A] 040°   [B] 050°   [C] 130°   [D] 140°
32. The bearing S 50° W is the same as:
    [A] 050°   [B] 130°   [C] 140°   [D] 230°
33. Town P is 150 km from a town Q in the direction 050°. The bearing of Q from P is:
    [A] 050°   [B] 150°   [C] 230°   [D] 310°
34. The bearing of P from Q is x, where $270° < x < 360°$. The bearing of Q from P is:
    [A] $(x-90)°$ [B] $(x+180)°$ [C] $(x-135)°$ [D] $(x-180)°$
35. The figure below shows the position of three ships A, B and C at sea. B is due north of C such that $|AB|=|BC|$ and the bearing of B from A is 040°. The bearing of A from C is:
    [A] 040°   [B] 070°   [C] 110°   [D] 290°

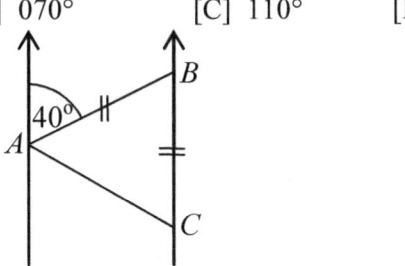

36. A tree is 8 km due south of a building. Ambe is standing 8 km west of the tree. The distance of Ambe from the building is:
    [A] $4\sqrt{2}$ km   [B] 8 km   [C] $8\sqrt{2}$ km   [D] 16 km
37. A tree is 8 km due south of a building. Ambe is standing 8 km west of the tree. The bearing of Ambe from the building is:

[A] 315°   [B] 270°   [C] 225°   [D] 135°

38. Three observation posts P, Q and R are such that Q is due east of P and R is due north of Q, if $|PQ| = 5$ km and $|PR| = 10$ km, $|QR|$ equals:
    [A] 50 km   [B] 9.5 km   [C] 7.6 km   [D] 8.7 km

39. From the figure below, the bearing of Q from P is:
    [A] 236°   [B] 214°   [C] 146°   [D] 124°

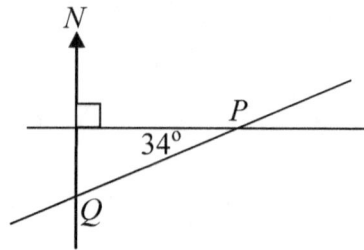

40. The bearing of a town Q from a town P is 215°. P is 80 km north of R while R is due east of Q. The distance between Q from R correct to the nearest km is:
    [A] 46 km   [B] 56 km   [C] 38 km   [D] 98 km

41. From the diagram below, the bearing of C from B is:
    [A] 060°   [B] 090°   [C] 120°   [D] 240°

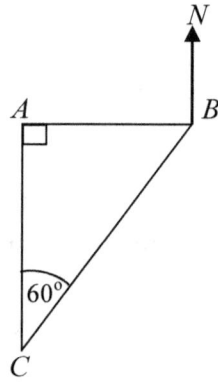

Topic 9

CIRCLES

Objectives

At the end of this topic, the learner should be able to:

1. Define each of the vocabularies associated with a circle.
2. Identify an arc, a chord, a tangent.
3. Establish the different properties of a circle.
4. Calculate the area of a sector and that of a segment.
5. Establish and apply tangent properties of a circle.
6. Find arc length, length of a tangent to a circle and the length of intersecting chords.
7. Establish, be able to state and apply the different circle theorems so as to determine other measures.
8. State the properties of a cyclic quadrilateral.

9.1 Vocabularies Associated with Circles

Much work was done on circles in modules 2, topic 11 of book 1. In this topic, we shall review and revise this material before continuing with new material.

Revision Exercise

In the following figure,
(i) Name the points O.
(ii) Name the set of points on the line:
 (a) AC (b) OC (c) FD (d) ABC
 (e) ACX (f) EY (g) AEC (h) AX
(iii) Name the set of points in the region.
 (a) DEFD (b) CODC
 (c) ABCA (d) AECA
(iv) State the name of the figure formed by connecting the points ADEF.
(v) State the name of the distance ABCDEF round the figure.

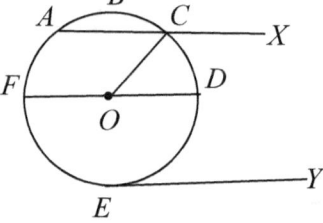

9.2 Mensuration of the Circle

In modules 2, topic 11 of book 1, we saw that:
1. The circumference C of a circle is given by $C = \pi d$.
2. The area A of a circle is given by $A = \pi r^2$.

Examples

1. Calculate the circumference of a circle whose radius is 21 cm, taking $\pi = \frac{22}{7}$.

   Solution
   $$C = 2\pi r = 2\left(\frac{22}{7}\right)(21) = 132 \text{ cm}$$

2. Find the radius of a circle whose circumference is 11 cm, taking π as 3.142.

   Solution
   $$r = \frac{C}{2\pi} = \frac{11}{2(3.142)} = 1.75 \text{ cm}$$

3. Calculate the area of a circle whose radius is 14 cm. Take $\pi = \frac{22}{7}$.

Solution

$$A = \pi r^2 = \frac{22}{7}(14)^2 = 616 \text{ cm}^2$$

4. Calculate to the nearest centimeter the radius of a circle whose area is 2464 cm². (Take =3.142)

   ### Solution

   $$r = \sqrt{\frac{A}{\pi}} = \sqrt{\frac{2464}{3.142}} = 28 \text{ cm}$$

5. Find the area of a circle with circumference 44 cm. Take $\pi = \frac{22}{7}$.

   ### Solution

   $$A = \pi r^2 \text{ and } C = 2\pi r \Rightarrow r = \frac{C}{2\pi}$$

   $$\Rightarrow A = \frac{C^2}{4\pi} = \frac{(44)^2}{4\left(\frac{22}{7}\right)} = \frac{44^2(7)}{4(22)} = 154 \text{ cm}^2$$

Exercise 9:2

Where necessary in this exercise take $\pi = \frac{22}{7}$.

1. Find the area of a circle whose radius is 3.5 cm.
2. Calculate the maximum area a goat will graze if it is tied to a pole by a rope 7 m from its neck.
3. The area of a circle is 38.5 cm². What is its radius?
4. Find the diameter of a circle whose area is 616 m².
5. Calculate the circumference of a circle whose radius is $3\frac{1}{2}$ cm.
6. Find the circumference of a circle whose radius is 14 cm.
7. Calculate the radius of a circle whose circumference is 44 cm.
8. Calculate the area of a circle whose circumference is 154 cm.
9. Find the distance covered by a bike, which runs once round a circular track of radius 21 m.
10. Find the distance between two concentric circles with areas 804 cm² and 1661 cm².
11. Calculate to the nearest whole number the area of the shaded portion in the figure below.

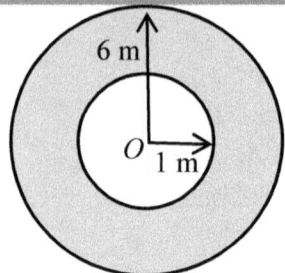

12. Two circles have radii 3.5 cm and 7 cm respectively. Find the area between them given that they have the same centre.

9.3 Arc Length

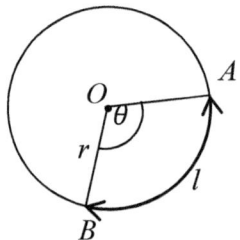

Since a revolution is 360°, the length l of an arc AB of a circle subtending an angle θ at the centre of the circle is the fraction $\dfrac{\theta}{360}$ of the circumference of the circle.

Hence, arc length l is given by $l = \dfrac{\theta}{360} \times C$, where C is the circumference of the circle. Since $C = 2\pi r$, the length l of arc AB is

$$l = \dfrac{\theta}{360} \times 2\pi r \implies r = \dfrac{360l}{2\pi\theta} \text{ and } \theta = \dfrac{360l}{2\pi r}.$$

 Example

In these examples, take $\pi = \dfrac{22}{7}$.
1. An arc of a circle of radius 21 cm subtends an angle of 120° at the centre of a circle. Calculate the length of the arc.

Module 16, Topic 9: Circles

Solution

$$l = \frac{\theta}{360} \times 2\pi r = \frac{120}{360} \times 2(\tfrac{22}{7})(21) = 44 \text{ cm}$$

2. Given that the length of the major arc of a circle of radius 14 cm is 70 cm, calculate the angle subtended at the centre by the arc.

   Solution

   $$\Rightarrow \theta = \frac{360 l}{2\pi r} = \frac{360(70)(7)}{2(22)(14)} = 286.4°$$

3. An arc of length 48 cm subtends an angle of 55° at the centre of a circle. Find the radius of the circle.

   Solution

   $$\Rightarrow r = \frac{360 l}{2\pi \theta} = \frac{360(48)(7)}{2(22)(55)} = 50 \text{ cm}$$

9.4 Area of a Sector

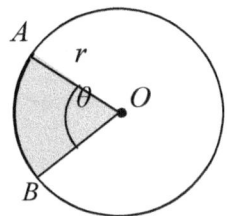

In the same way as the length of an arc of a circle, the area of a sector of a circle subtending an angle θ at the centre of the circle is the fraction $\frac{\theta}{360}$ of the area of the circle.

Therefore the area S of sector OAB will be

$$S = \frac{\theta}{360} \times \text{area of the circle}$$

$$\Rightarrow S = \frac{\theta}{360} \times \pi r^2 \quad \Rightarrow \quad r = \sqrt{\frac{360 S}{\pi \theta}} \text{ and } \theta = \frac{360 S}{\pi r^2}.$$

Example

In these examples, take $\pi = \frac{22}{7}$.

1. A sector of a circle of radius 21 cm subtends an angle of 120° at the centre of the circle. Calculate the area of the sector.

   Solution
   $$S = \frac{\theta}{360} \times \pi r^2$$
   $$\Rightarrow S = \frac{120°}{360} \times \tfrac{22}{7} \times 21^2 = 462 \text{ cm}^2$$

2. The length of an arc of a circle of radius 14 cm is 21 cm. Calculate the area of the sector of the circle.

   Solution
   $$S = \frac{\theta}{360} \times \pi r^2 \quad \text{...........①}$$
   $$l = \frac{\theta}{360} \times 2\pi r \quad \text{...........②}$$
   ① ÷ ② : $\frac{S}{l} = \frac{r}{2}$
   $$\Rightarrow S = \frac{rl}{2} = \frac{14(21)}{2} = 147 \text{ cm}^2$$

3. The radius of a sector of area 176 cm² is 14 cm. Calculate the angle subtended by the sector at the centre of the circle $\left(\text{Take } \pi = \frac{22}{7}\right)$.

   Solution
   $$S = \frac{\theta}{360} \times \pi r^2$$
   $$\Rightarrow \theta = \frac{360S}{\pi r^2} = \frac{360(176)(7)}{22(14)^2} = 102.9°$$

4. The area of a sector, which subtends an angle of 108° at the centre, is 198 cm². Find the radius of the sector. Take $\pi = \frac{22}{7}$.

   Solution
   $$S = \frac{\theta}{360} \times \pi r^2$$
   $$\Rightarrow r = \sqrt{\frac{360S}{\pi \theta}} = \sqrt{\frac{360(198)(7)}{22(108)}} = 14.5 \text{ cm}$$

9.5 Area of a Segment

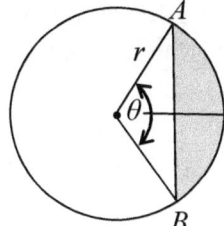

Area of Segment = Area of sector − Area of triangle

$$A_s = \frac{\theta}{360} \times \pi r^2 - \frac{1}{2} r^2 \sin \theta$$

 Example

A chord subtends an angle of 135° at the centre of a circle of radius 21 cm. Calculate the area of the minor segment of the circle to the nearest square centimetre. $\left(\text{Take } \pi = \frac{22}{7}\right)$.

Solution

$$A_s = \frac{\theta}{360} \times \pi r^2 - \frac{1}{2} r^2 \sin \theta$$
$$= \frac{135}{360} \times \frac{22}{7} (21)^2 - \frac{1}{2} (21)^2 \sin 135$$
$$\Rightarrow A_s = 519.75 - 155.92 = 364 \text{ cm}^2$$

 Exercise 9:3

Where necessary in this exercise take $\pi = \frac{22}{7}$.

1. The arc of a circle subtends an angle of 57° at the centre of the circle. Find the length of the arc if the diameter of the circle is 7 cm.
2. Calculate the angle subtended by an arc of length 21 cm at the centre of a circle of radius 14 cm.
3. Find the perimeter of a sector, which subtends an angle of 108° at the centre of a circle of radius 7 cm.

4. Calculate the perimeter of a sector of a circle of radius 14 cm, which subtends an angle of 140° at the centre.
5. The arc length of a sector, which subtends an angle of 210° at the centre, is 88 cm. Find the radius of the sector.
6. In the figure below, AB is a chord of a circle centre O. Given that $|AB| = 24.2$ cm and that the perimeter of $\triangle AOB$ is 52.2 cm. Calculate angle AOB correct to the nearest degree.

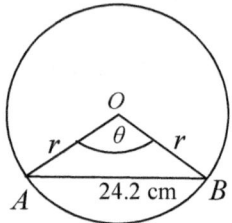

7. Calculate the length of a chord of a circle of radius 10 cm, given that the chord is 6 cm from the centre of the circle.
8. Calculate the area of the shaded segment of the sector shown below.

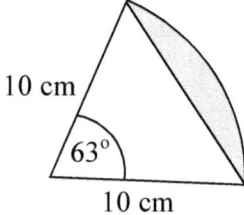

9. The figure below represents a miniature fan, which is in the shape of a third of a circle of radius 14 cm. The shaded part is painted and the remainder is a third of a circle of radius 10.5 cm. Calculate the area of the shaded part

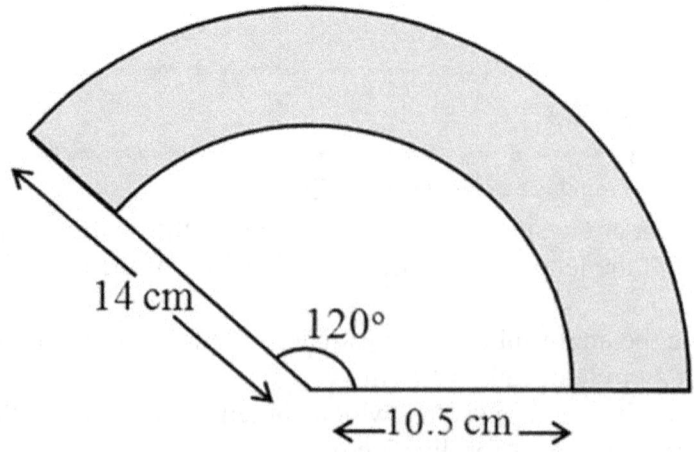

Module 16, Topic 9: Circles

10. The area of a semi circle is 616 cm². Find to the nearest whole number the radius and the perimeter of the semi circle.
11. The circumference of a circle is 52.8 cm. Find to the nearest cm², the area of a semi circular arc of the circle.
12. The following figure shows the actual area of a sports stadium. Given that the ends are semi-circular, find the total area in m² to 3 significant figures of the stadium.

13. Figure (a) below is made up of a rectangle $ABCD$ of sides $2x$ metres by y metres and a semi-circle on AB as diameter. Write down, in terms of x, y and π an expression for the perimeter p of the figure. Hence, find x in terms of p, y and π.

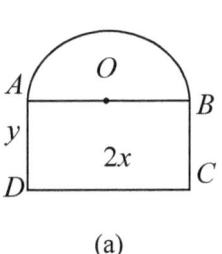 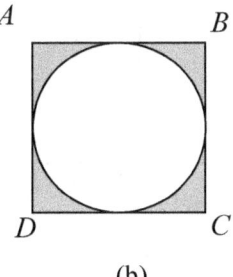

(a)   (b)

14. Figure (b) above shows a circle of radius $3\frac{1}{2}$ cm inscribed in a square $ABCD$.
    (i) Find the length of a side of the square.
    (ii) Find the area of the shaded region.
15. Calculate the area of the shaded portion in the figure (a) below.

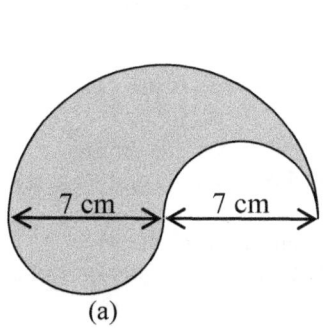 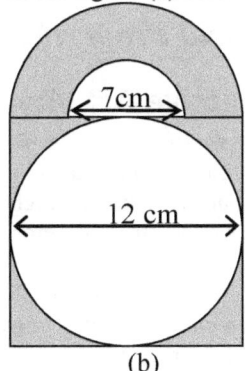

(a)   (b)

16. Calculate the area of the shaded portion in figure (b) above.

17. Figure (a) below shows a circle of radius 6 cm inscribed in a square, also inscribed in a larger circle. Calculate the area of the un-shaded portion.

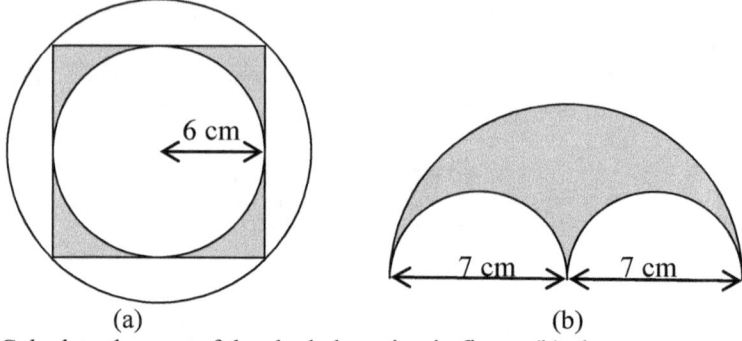

(a) (b)

18. Calculate the area of the shaded portion in figure (b) above.

19. The following figure shows, sectors, which subtend angles of, 90° pinned at the vertices *A, B, C* and *D* of a square of length 2 cm.

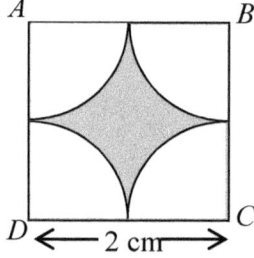

(a) Find the perimeter of the shaded portion.
(b) Find the area of the shaded portion.

9.6   Circle Theorems

Symmetrical Properties of a Circle

 Investigative Activity

1. Draw a large circle and mark the centre *O*.
2. Draw two non-parallel chords which are not diameters of the circle.
3. Construct the perpendicular bisectors of these chords and extend them until they meet.
4. Where is the point of intersection of these perpendicular bisectors?
5. What conclusion can you draw concerning the perpendicular bisector of a chord and the centre of the circle?
6. Draw another large circle and mark the centre *O*.

> 7. Draw two equal non-parallel chords which are not diameters of the circle.
> 8. Construct the perpendicular bisectors of these chords.
> 9. Measure the distance of each chord from the centre of the circle and record your result.
> 10. What conclusion do you draw concerning the distance of equal chords from the centre of the circle?

From the above investigation, we can see that:
(1) The perpendicular bisector of a chord passes through the centre (Figure (a)).

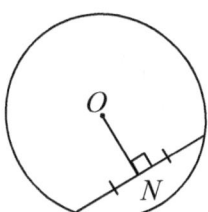 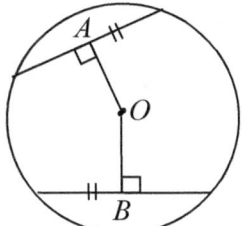

(2) Equal chords of a circle are equidistant from the centre (Figure (b)).

Angles in a Semi-circle

Investigative Activity

> 1. Draw a large circle and mark the centre O.
> 2. Draw a diameter AD of the circle.
> 3. Mark a point B on the circle and draw chords AB and DB to meet at B.
> 4. Measure the angle ABD.
> 5. Mark another point C on a different point on the circle and draw the chords AC and DC to meet at C.
> 6. Mark another point C on a different point on the circle and draw the chords AC and DC to meet at C.
> 7. Measure the angle ACD.
> 8. What conclusion do you draw concerning the angle subtended at the circumference by a diameter of a circle?

From the above investigation, we can see that:
(3) The angle in a semi-circle is 90° (as shown below)

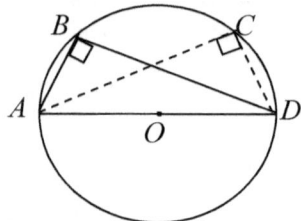

Example

In the figure below, AOD is a diameter of a circle centre O and radius 10 cm; P is a variable point on the circumference of the circle. Find the length of AP when $\triangle APD$ is isosceles.

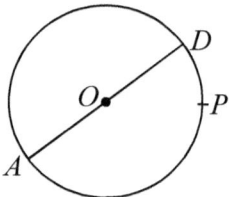

Solution

$\triangle APD$ is isosceles $\Rightarrow AP = PD$
$\angle APD = 90°$ [Angle in a semi-circle]
$\angle DAP = ADP = 45°$ [APD is an isosceles $\triangle$]

$$AP = AD\sin 45° = 20\left(\frac{\sqrt{2}}{2}\right) = 10\sqrt{2} \text{ cm}$$

Alternatively,
Let the other two sides of the isosceles triangle be x cm.
Using the Pythagoras theorem,
$$x^2 + x^2 = 20^2 \Leftrightarrow x = 10\sqrt{2}$$

Angles in the Same Segment and at the Centre

 Investigative Activity

1. Draw a large circle and mark the centre O.
2. Mark three points P, Q and R on the circle.
3. Draw the chords PR and QR and measure the angle PRQ.
4. Mark another point S on a different point in the same segment and draw the chords PS and QS.
5. Measure the angle PSQ.
6. What conclusion do you draw concerning the angle subtended by the same chord on the same segment of a circle?
7. Draw the radii OP and OQ.
8. Measure the angle POQ and record your result.
9. Compare your result with angles PRQ and PSQ.
10. What conclusion do you draw about the relationship between the angle subtended by an arc at the centre and on the circle?

From the above investigation, we can see that:

(4) The angles subtended in the same segment by the same arc are equal.
   In the figure below the ∠ PSQ and ∠ PRQ, are both angles subtended by the arc PQ, in the same segment. Hence, they are equal.

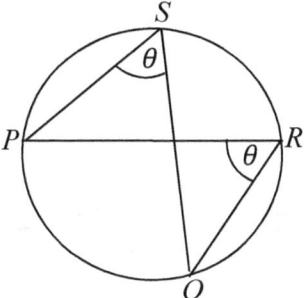

(5) The angle subtended by an arc PQ at the centre of a circle is twice the angle subtended at the circumference by the same arc.

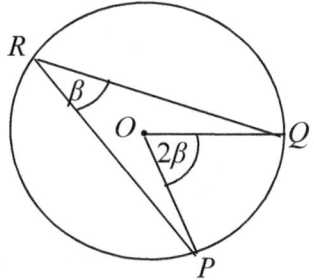

Example

In the figure below, A, B and C are points on the circle whose centre is O and $O\hat{C}A = 25°$. Calculate (i) Angle AOC (ii) Angle CBA.

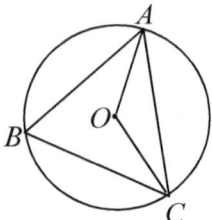

Solution
(i) $OA = OC$ [radii of same circle]

$\Rightarrow \triangle AOC$ is isosceles

Hence, $\angle OAC = \angle OCA = 25°$ [base $\angle$s of isosceles $\triangle$]
$\angle OAC + \angle OCA + \angle AOC = 180°$ [angles in a $\triangle$]
$\angle AOC = 180° - 2(25°) = 130°$

(ii) $\angle CBA = \dfrac{1}{2} \angle AOC$ [$\angle$at centre is twice $\angle$ at circumference]

$\Rightarrow \angle CBA = \dfrac{1}{2}(130°) = 65°$

Angles in Opposite Segments and in a Cyclic Quadrilateral

 Investigative Activity

1. Draw a large circle and mark the centre O.
2. Draw any quadrilateral whose vertices $ABCD$ lie on the circle.
3. Measure the angles of the quadrilateral and record them.
4. Sum the opposite angles of the quadrilateral.
5. What conclusion do you draw concerning the opposite angles of a cyclic quadrilateral?
6. Draw a chord AC to join A and C on your circle.
7. Are B and D on the same segment or are they in opposite segments?
8. What conclusion can you draw about opposite angles of a cyclic quadrilateral?
9. Produce DC to F. What can you say about the angles BCD and BCF?
10. Compare the size of the angles BCF and BAD.
11. What conclusion can you draw about the exterior angle of a cyclic quadrilateral and the interior opposite angle of the quadrilateral?

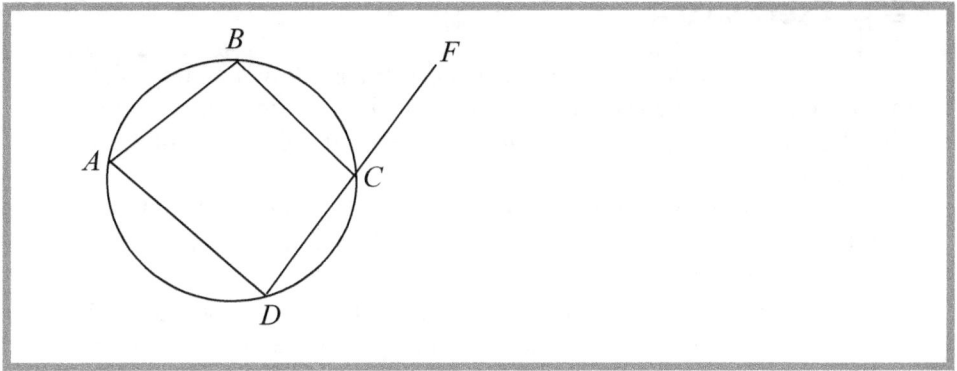

From the above investigations, we can see that:
(6) Opposite angles of a cyclic quadrilateral are supplementary i.e. their sum is 180°.
(7) The angles subtended by a chord in opposite segments are supplementary (Figure (a) below).

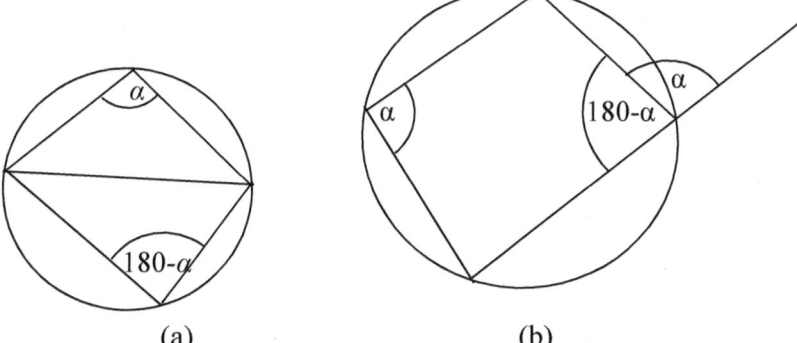

        (a)   (b)

(8) The exterior angle of a cyclic quadrilateral is equal to the opposite interior angle.

The Cyclic Quadrilateral Test

To test whether or not a quadrilateral is a cyclic quadrilateral, it suffices to show that its opposite angles are supplementary as in theorem (6).

Exercise 9:4

1. Two parallel chords of length 6 cm and 8 cm are drawn on opposite sides of the centre O of a circle of radius 5 cm. Calculate the distance between the chords. What would the distance have been if the two parallel chords had been on the same side of the centre?
2. P is the point of contact of a tangent TP to a circle, centre R and radius 8 cm. If $TR = 17$ cm, calculate the area of $\triangle TPR$ and the length of the perpendicular from P to TR.
3. AB is a chord of a circle and C is the midpoint of the minor arc AB. If $AC = 5$ cm and $AB = 8$ cm, calculate the length of the diameter of the circle.
4. In figure (a) below, the chord PQ and RS intersect at X inside the circle, centre T. If $\angle PTR = 64°$ and the minor arc PR is twice the minor arc QS, calculate $\angle QXS$.

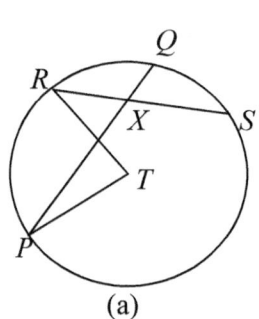
(a)

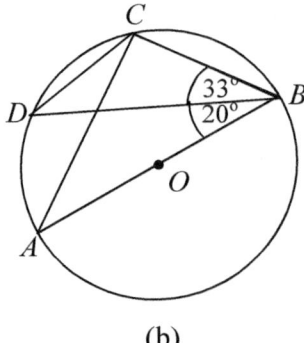
(b)

5. Figure (b) above shows a circle with centre O angle $DBC = 33°$ and $\angle DBA = 20°$. Calculate angle CDB.
6. The figure below shows a cyclic quadrilateral $ABCD$ with $AD = AB$, $BC = DC$ and angle $BCE = 126°$. Determine, in degrees, the values of
   (a) angle ACD
   (b) angle BAD

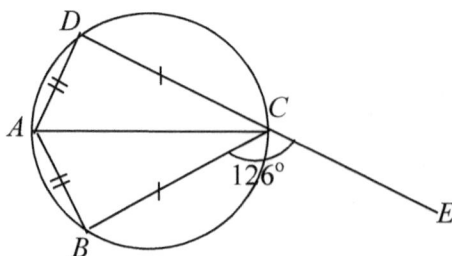

7. Find the values of x and y in figure (a) below. Show your, working clearly.

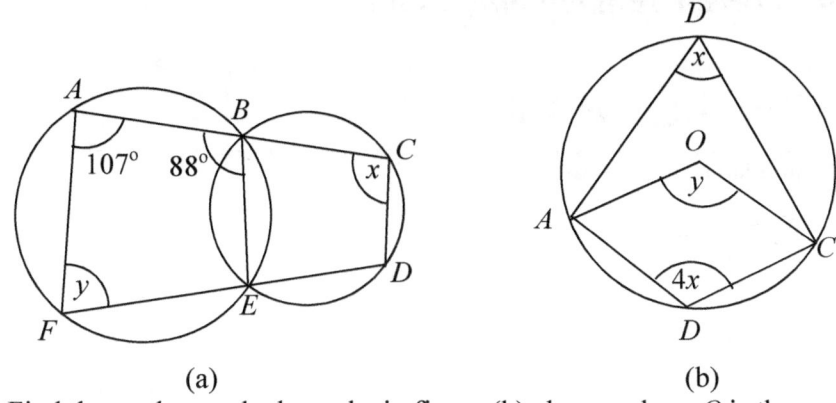

(a)   (b)

8. Find the angles marked *x* and *y* in figure (b) above, where O is the centre of the circle.
9. In the figure below, $ABCD$ is a cyclic quadrilateral in which AB is a diameter and DC is parallel to AB. The side CD is produced to E and $\angle BDE = 63°$. Calculate angle DBC.

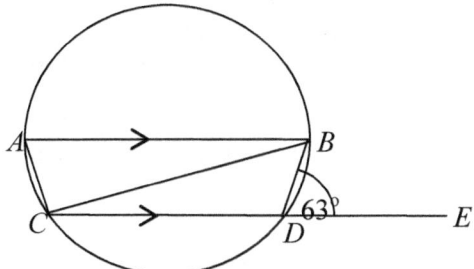

10. The figure below, (not drawn to scale) shows a circle with lines PS and QT passing through the centre O. Given that $\angle RQT = 30°$ and $\angle RPS = 15°$, find giving reasons for your arguments, the values of
    (a) $\angle PRS$ (b) $\angle RSP$ (c) $\angle RST$ (d) $\angle TOR$

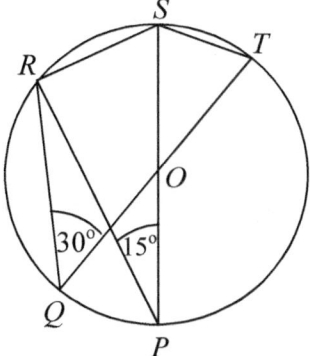

Angle between Tangent and Radius

Investigative Activity

1. Draw a large circle centre O.

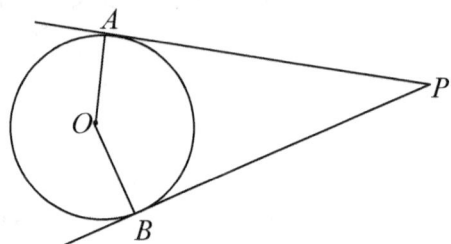

2. Draw a straight line PA to touch the circle at one point A.
3. Draw a radius to meet the tangent PA at A the point of contact.
4. Measure the angle OAP and record the value.
5. Draw another straight line PB to touch the circle at another point B.
6. Draw a radius to meet the tangent PB at B the point of contact.
7. Measure the angle OBP and record the value.
8. What conclusion do you draw concerning the angle between a tangent and a radius of a circle?
9. Measure the lengths AP and BP and record your result.
10. What conclusion do you draw concerning the length of a tangent to a circle from the same external point?

From the above investigation, we see that:

(9) A tangent drawn to a circle is perpendicular to the radius of the circle at the point of contact (as shown below).

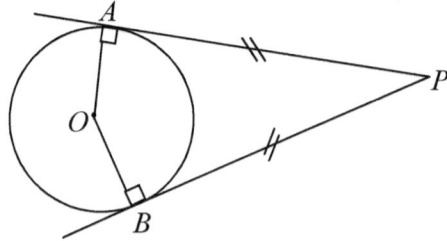

(10) Tangents to a circle from the same external point are equal in length (as shown above).

Module 16, Topic 9: Circles

Alternate Segment Theorem

Opposite (alternate) segment to angle *x*

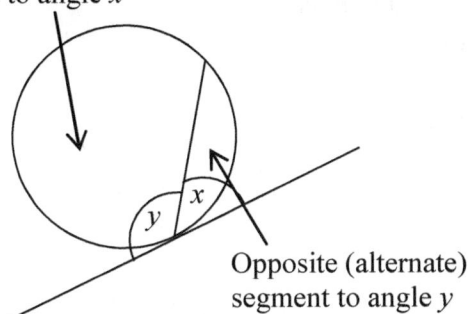

Opposite (alternate) segment to angle *y*

(11) The **alternate segment theorem** states that, the angle between a tangent and a chord at the point of contact is equal to the angle in the alternate (opposite) segment (as shown below).

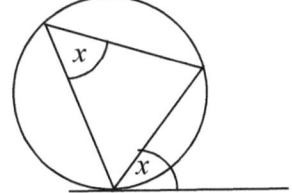

 Example

TA and *TB* are tangents drawn to a circle from a point *T*. *X* is a point on the major arc of the circle. If ∠*AXB* = 63° calculate ∠*ATB*.

Solution

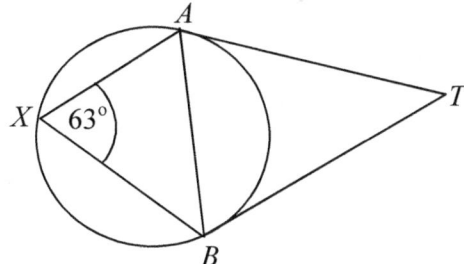

$AT = TB$ (tangents from same external point)

$\angle TAB = \angle TBA$ ($\triangle ABT$ is isosceles)

Also $\angle TAB = \angle AXB = 63°$ ($\angle$ in alt segment)

$\angle TAB + \angle TBA + \angle ATB = 180°$ ($\angle$s in a $\triangle$)

$\Rightarrow 63° + 63° + \angle ATB = 180°$

$\Rightarrow \angle ATB = 180° - 126° = 54°$

Exercise 9:5

1. In the figure below, ATM is a tangent to the circle at T. The secant ABC cuts the circle at B and C. The diameter through B cuts the circle again at D. If $\angle DBC = \angle ATB = 40°$, calculate the measures of angles BAT, CBT, and DTM and hence say why:
   (i) The line AC is parallel to TD.   (ii) The line CT is a diameter.

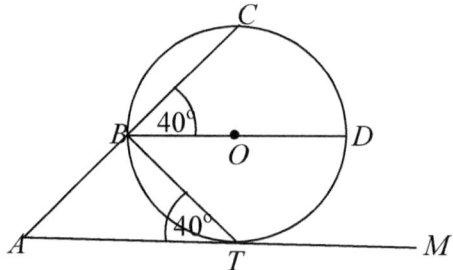

2. In the figure below, TA and TC are tangents to a circle whose centre is O. A, B and C are points on the circle and angle $OCA = 25°$. Given that the radius of the circle is 7 cm, calculate
   (i) $\angle OAC$ (ii) $\angle CBA$ (iii) $\angle ACT$ (iv) the length of AC
   (v) the length of OT (vi) the area of $OATC$.

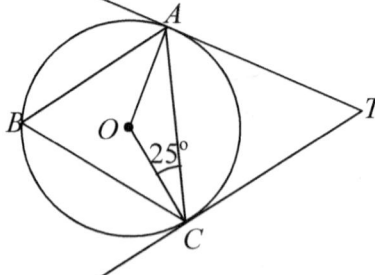

3. Find giving reasons, the values of the angles marked x, y and z in the figure below.

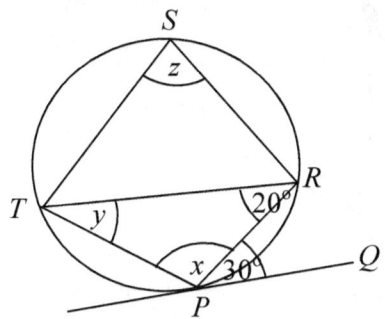

4. The following shows triangle ABC with ∠CBA = 50° and ∠BCA = 70°; inscribed in a circle centre O of radius 5 cm. DT is a tangent. Find
   (a) ∠CAB   (b) Find ∠CEA
   (c) Show that ∠CAE = 30°, giving reasons.
   (d) Calculate to one decimal place, the lengths of DE and DB.

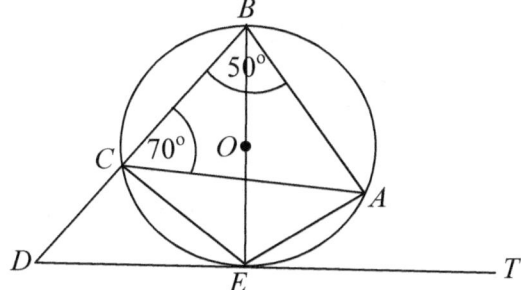

5. In the figure below, ABC is a tangent to the circle. Given that ∠DBA = 50°, ∠FBC = 64° and ED = EF. Find the value of ∠EDF.

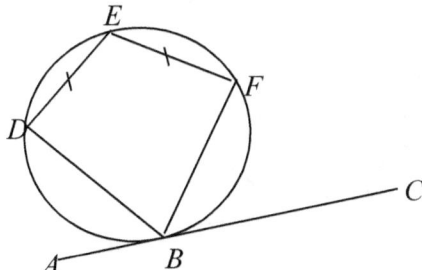

6. In the figure below, the lines TA and TB are tangents to the circle with centre O. Given that angle ATB = 30° find the angles denoted by the letters a, b and c.

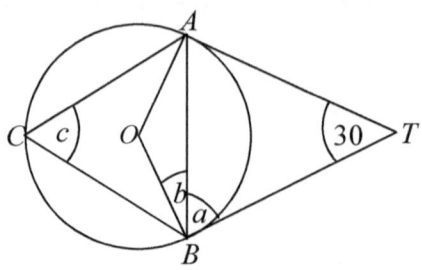

7. The figure below, is a circle with centre O, tangent PTQ, and angle below ATQ = 58°. Calculate the values of x and y where x = ∠TBP and y = ∠TPB.

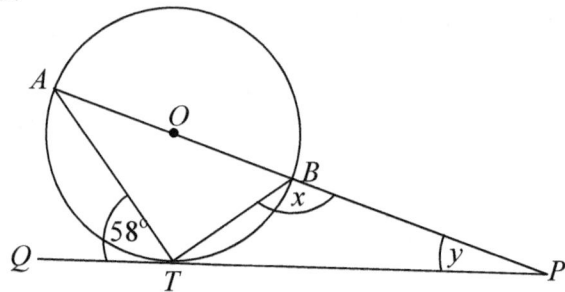

8. Given that PQ is parallel to BA, find the values of the angles x, y and z in the figure below.

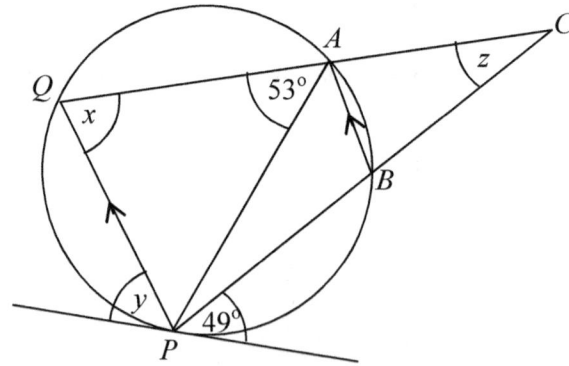

Intersecting chord theorem

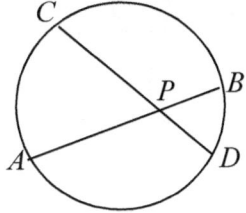
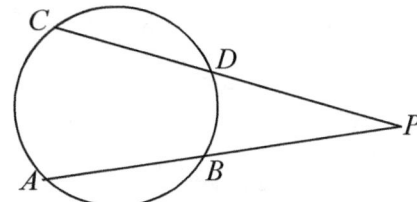

(i): **Internal intersection**   (ii): **External intersection**

In both cases, in the figures above, $PA \cdot PB = PC \cdot PD$

If in (ii) PC is a tangent then, $PA \cdot PB = PC^2$

Note that in both cases we reckon all distances from P.

 Example

1. Find the value of x in the figure below.

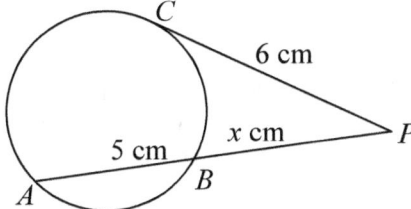

Solution

Using the intersecting chord theorem
$$x(x+5) = 6^2$$
$$\Leftrightarrow x^2 + 5x - 36 = 0$$
$$(x+9)(x-4) = 0$$
$$x = -9 \text{ or } x = 4$$

Since $x \geq 0$, $x = 4$ cm.

2. In the figure below, ABC is a tangent to the circle. Given that, EB is perpendicular to AC, $DN = 7.2$ cm, $NF = 6$ cm, $BN = 9$ cm and $NE = x$ cm. Find x.

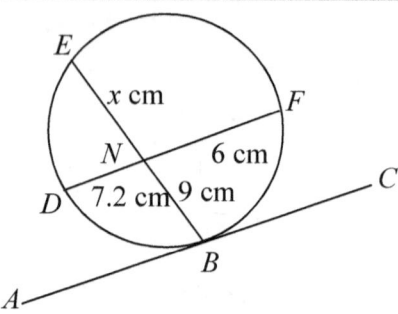

Solution
Using the intersecting chord theorem
$$9x = 6(7.2)$$
$$x = \frac{6(7.2)}{9}$$
$$\Rightarrow x = 4.8 \text{ cm}$$

 Exercise 9:6

1. Two chords BA and DC, of a circle when produced meet at O. If $AB = 5$ cm, $OA = 4$ cm and $OC = 3$ cm, calculate the length of the chord CD.
2. In figure (a) below, find (i) The length of NC, (ii) The angle ABC

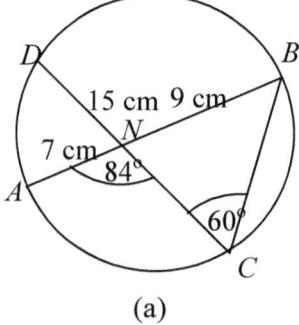   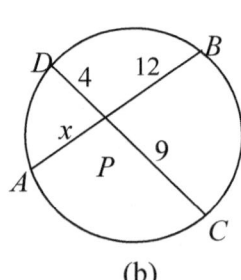

(a)   (b)

3. Find the value of x in figure (b).
4. Given that O is the center of the circle, find the value of y in the figure below.

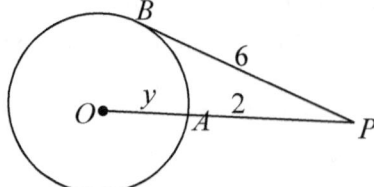

Module 16, Topic 9: Circles

5. The figure below shows the chords *DC* and *FE* produced to meet at *A*, and *AB* drawn to touch the circle at *B*. If *AC* = 7 cm, *CD* = 21 cm and *AE* = 8 cm, calculate the lengths of *AB* and *EF*.

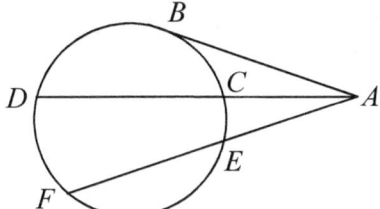

6. A chord *AB* of a circle is of length 48 cm, and 7 cm from the centre. *AB* is produced to *P* so that *BP* is of length 16 cm. Calculate the radius of the circle and the length of the tangent from P to the circle.
7. In the figure below, calculate the value of *x*.

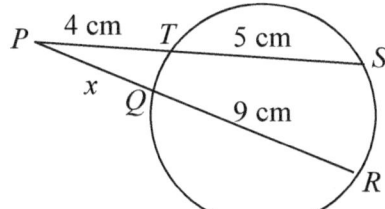

Multiple Choice Exercise 9

In this exercise where necessary, take $\pi = \dfrac{22}{7}$.

1. If *O* is the centre of the circle in the figure below, the area of the shaded part is:

   [A] $\dfrac{3\pi r^2}{8}$ [B] $\dfrac{\pi r^2}{2}$ [C] $\dfrac{\pi r^2}{4}$ [D] $\dfrac{\pi r}{2}$

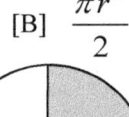

2. The area of a circular field is 154 m². The perimeter of the field is:
   [A] 44 m   [B] 49 m   [C] 88 m   [D] 176 m
3. The area of a circle is 38.5 cm², its diameter is?
   [A] 22 m   [B] 14 m   [C] 7 m   [D] 6 m
4. The diameter of a circular field whose area is 616 cm² is:
5. [A] 98.00 m   [B] 28.00 m   [C] 49.00 m   [D] 24.82 m
6. 29. In the figure below, *PXR* and *PYO* are two semi-circles with diameters 14

cm and 7 cm respectively. The area of the enclosed region *PXROY* correct to the nearest whole number is:
7. [A] 96 cm² [B] 116 cm² [C] 154 cm² [D] 192 cm²
8. 30. In the figure below, *PXR* and *PYO* are two semi-circles with diameters 14 cm and 7 cm respectively. The perimeter of the region is:
[A] 2 cm [B] 33 cm [C] 40 cm [D] 66 cm

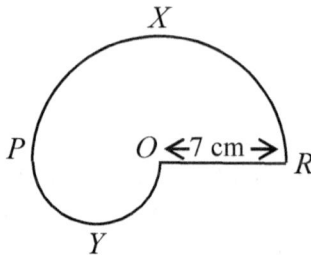

9. An arc of a circle of radius 7 cm is 14 cm long. The angle the arc subtends at the centre of the circle is:
[A] 44° [B] 51.43° [C] 98° [D] 114.55°
10. Correct to three significant figures the length of an arc, which subtends an angle of 70° at the centre of a circle of radius 4 cm is:
[A] 2.44 cm [B] 4.89 cm [C] 9.78 cm [D] 25.1 cm
11. An arc of length 22 cm subtends an angle θ at the centre of a circle. If the radius of the circle is 15 cm, the value of θ will be:
[A] 84° [B] 70° [C] 96° [D] 156°
12. In the figure below, *O* is the centre of the circle with radius 10 cm and $ABC = 30°$. The length of the arc *AC*, correct to one decimal place is:
[A] 5.2 cm [B] 13.2 cm [C] 10.5 cm [D] 20.6 cm

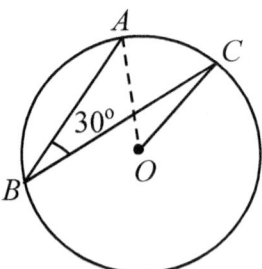

13. To 2 significant figures, the length of the arc of a circle of radius 3.5 cm that subtends an angle of 75° at the centre of the circle is:
[A] 2.3 cm [B] 4.6 cm [C] 8 cm [D] 16 cm
14. The angle of a sector of a circle of diameter 8 cm is 135°. The area of the sector is:
[A] $9\frac{3}{7}$ cm² [B] $12\frac{4}{7}$ cm² [C] $18\frac{6}{7}$ cm² [D] $25\frac{1}{7}$ cm²
15. A sector of a circle of radius 7 cm has an area of 44 cm². The angle of the sector correct to the nearest degree is:
[A] 103° [B] 26° [C] 6° [D] 206°

Module 16, Topic 9: Circles

16. A sector of a circle of radius 9 cm subtends an angle of 120° at the centre of the circle. The area of the sector to the nearest cm² is:
    [A] 75 cm² [B] 84 cm² [C] 85 cm² [D] 86 cm²

17. A circle has radius x cm. The area of a sector of the circle with angle 135°, in terms of x and π is:
    [A] $\dfrac{\pi x^2}{8}$ [B] $\dfrac{3\pi x^2}{8}$ [C] $\dfrac{5\pi x^2}{8}$ [D] $\dfrac{\pi x}{8}$

18. The area of the minor sector POQ in the figure below is:
    [A] $148\dfrac{1}{2}$ cm² [B] $32\dfrac{1}{12}$ cm² [C] $6\dfrac{5}{12}$ cm² [D] $1\dfrac{5}{6}$ cm²

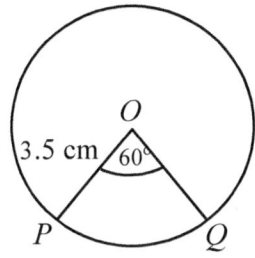

19. The length of an arc of a circle of radius 5 cm is 4 cm. The area of the sector is:
    [A] 2 cm² [B] 8 cm² [C] 10 cm² [D] 20 cm²

20. Correct to three significant figures, the area of the minor sector, OPQ in the figure (a) below is approximately equal to:
    [A] 3.41 cm² [B] 157 cm² [C] 10.9 cm² [D] 5.35 cm²

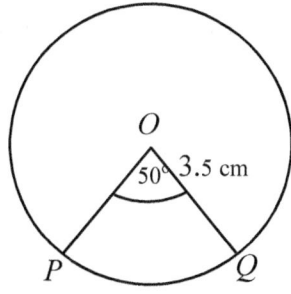 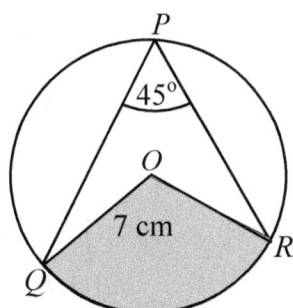

21. In the figure (b) above, PQR is a circle with centre O. $|OQ| = |OR| = 7$ cm. The area of the shaded portion is:
    [A] 11 cm² [B] 22 cm² [C] 38.5 cm² [D] 77 cm²

22. In the figure below, O is the centre of the circle. Given that $OA = 5$ cm, $OD = 3$ cm and $\angle AOD = \angle BOD$, the length of the chord AB is:
    [A] 8 cm [B] 5 cm [C] 3 cm [D] 15 cm

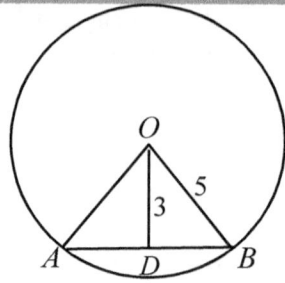

23. The figure below, shows the shaded segment of a circle of radius 7 cm. The area of the triangle OXY is cm². The area of the segment is:

[A] $\frac{5}{12}$ cm [B] $\frac{7}{12}$ cm [C] $1\frac{1}{6}$ cm [D] $2\frac{1}{3}$ cm

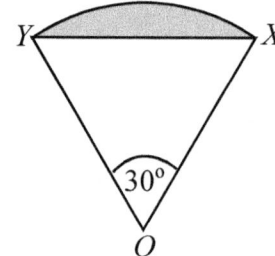

24. In the figure below, $PQRO$ is one quarter of a circle with centre O. $|RQ| = |PR| = 7$ cm. Correct to two decimal places, the area of the shaded portion is:
 [A] 57.70 cm²   [B] 38.50 cm²   [C] 27.00 cm²   [D] 19.25 cm²

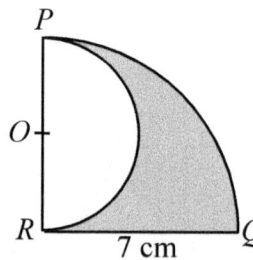

25. The area of a square is equal to that of a triangle of base 9 cm and altitude 32 cm. The length of the side of the square must be:
 [A] 6 cm   [B] 6.2 cm   [C] 12 cm   [D] 2.2 cm

26. The length of the side of a square, which is equal in area to a rectangle measuring 45 cm by 5 cm, is:
 [A] 25 cm   [B] 23 cm   [C] 16 cm   [D] 15 cm

27. In the figure (a) below, O is the centre of the circle through points L, M, and N. If $\angle MLN = 74°$ and $\angle MNL = 39°$. The value of $\angle LON$ is:
 [A] 100°   [B] 113°   [C] 126°   [D] 134°

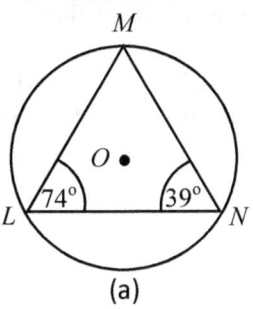
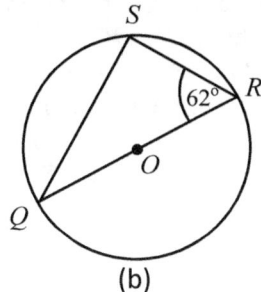

(a)   (b)

28. In the figure (b) above, O is the centre of the circle. If $\angle QRS = 62°$, the value of $\angle SQR$ is:
    [A] 14°   [B] 28°   [C] 31°   [D] 90°

29. In the figure (a) below, $PQRS$ is a circle. $\angle SPR = p°$ and $\angle SQR = 2x°$. The value of x in terms of p is:

    [A] $x = 2p$ [B] $x = p - 2$ [C] $x = p^2$ [D] $x = \dfrac{p}{2}$

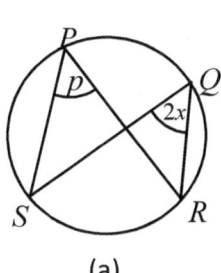

(a)   (b)

30. In figure (b) above, O is the centre of the circle. If $\angle PAQ = 75°$, the value of $\angle PBQ$ is:
    [A] 51°   [B] 75°   [C] 105°   [D] 150°

31. In figure (a) below, O is the centre of the circle QRT and PT is the tangent to the circle at T. The angle x is:
    [A] 40°   [B] 35°   [C] 25°   [D] 20°

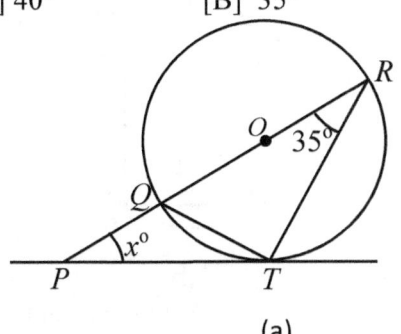
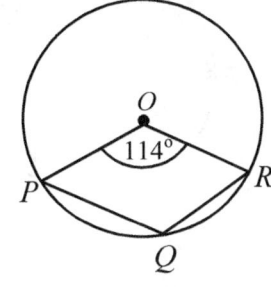

(a)   (b)

32. In figure (b) above, O is the centre of the circle. Given that ∠POR = 114° the value of ∠PQR is:
   [A] 123°   [B] 118.5°   [C] 117°   [D] 114°

33. In the figure below, O is the centre of the circle. If ∠POQ = 39° and ∠PRQ = 5x° the value of x is:
   [A] 4   [B] 8   [C] 16   [D] 20

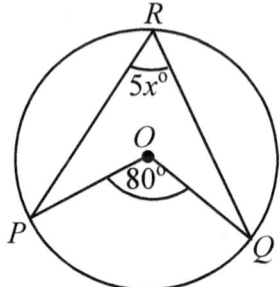

34. In the figure below, SQ is the tangent to the circle at P. XP∥YQ, ∠XPY = 56° and ∠PXY = 80°. The value of angle PQY is:
   [A] 34°   [B] 36°   [C] 44°   [D] 46°

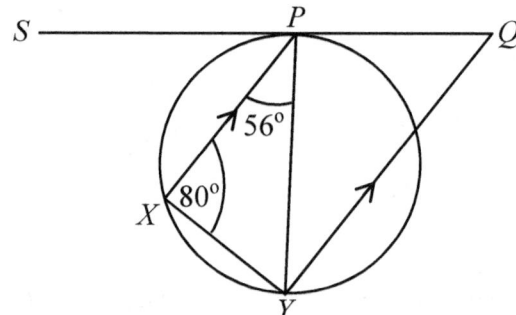

35. In figure (a) below, O is the centre of the circle. It is true to say that:
   [A] a = b   [B] b + c = 100   [C] a + b = c   [D] a = b and b + c = 100

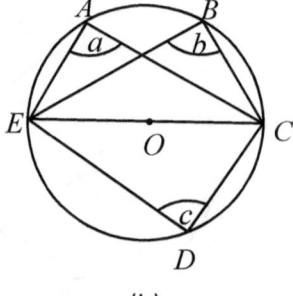

(b)   (b)

36. In figure (b) above, O is the centre of the circle, ∠SOR = 64° and ∠PSO = 36°. The value of ∠PQR is:

[A] 100°   [B] 96°   [C] 94°   [D] 86°

37. In figure (a) below, $ABCD$ is a circle. The value of x is:

   [A] $\dfrac{20}{9}$ [B] $\dfrac{36}{5}$ [C] 3   [D] $\dfrac{45}{4}$

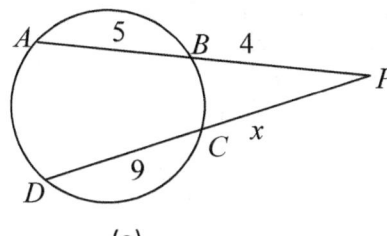
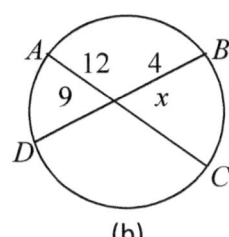

(a)   (b)

38. In figure (b) above, $ABCD$ is a circle. The value of x is:

   [A] $\dfrac{48}{9}$ [B] 3   [C] 36   [D] $\dfrac{16}{3}$

39. Given that in figure (a) below, $PQ = 6$ cm, $TR = 5$ cm and $RQ = 7$ cm. The radius of the circle with centre O is:
   [A] 8 cm   [B] 6 cm   [C] 2 cm   [D] 4 cm

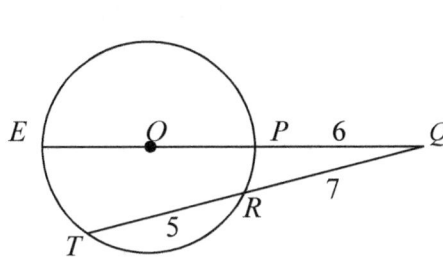
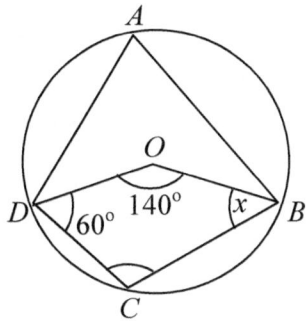

(a)   (b)

40. O is the centre of the circle in figure (b) above. The size of the angle marked x is:
   [A] 120°   [B] 40°   [C] 50°   [D] 70°

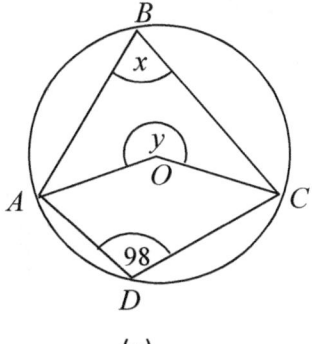
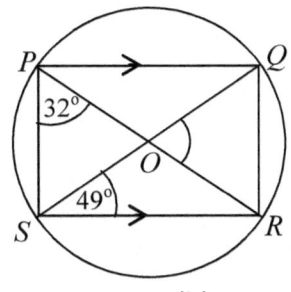

(a)   (b)

189

41. In figure (a) above, O is the centre of the circle. The value of $y - x$ is:
   [A] 164° [B] 114° [C] 66° [D] 16°
42. In figure (b) above, ∠RPS = 32° and ∠QSR = 49°. The value of ∠QOR is:
   [A] 64° [B] 82° [C] 98° [D] 116°
43. In figure (a) below, O is the centre of the circle. ∠BAO = 30° and ∠BCO = 20°. The value of reflex angle AOC is:
   [A] 330° [B] 300° [C] 270° [D] 260°

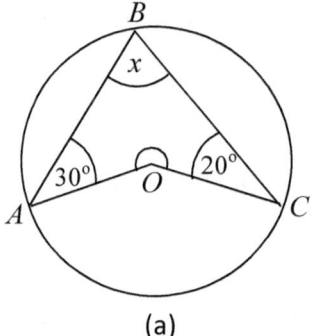

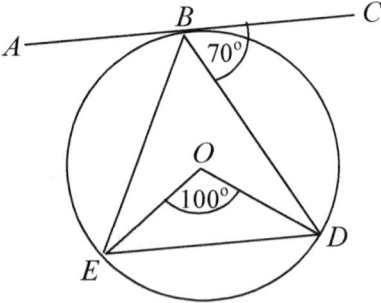

(a)   (b)

44. In figure (b) above, X, Y and Z are points on a circle centre O. WZ is a tangent to the circle at the point Z and ∠XYZ = 22°. The value of θ is:
   [A] 112° [B] 68° [C] 46° [D] 22°
45. In the figure below, O is the centre, ∠DOE = 100° and ∠CBD = 70°. The value of ∠BEO is:
   [A] 20° [B] 30° [C] 40° [D] 60°

Topic 10

THE EARTH AS A SPHERE

Objectives

At the end of this topic, the learner should be able to:

1. Draw a sphere to represent the earth and indicate the great circles, small circles, the equator and the meridian.
2. Locate a place on the surface of the earth in terms of latitudes and longitudes.
3. Tell time using GMT as reference point.
4. Find the distance between two points on the earth's surface.
5. Find the shortest distance between two points on the earth's surface.
6. Calculate time in relation to longitude.

10.1 Surface Area and Volume of a Sphere

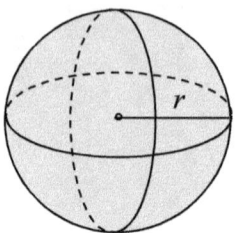

The surface area S of a sphere with radius r is four times the area of a circle with radius r.

Area of sphere, $S = 4 \times$ Area of circle $= 4\pi r^2$

The volume V of the sphere is given by, $V = \dfrac{4}{3}\pi r^3$

 Example

Calculate the surface area of and the volume of a sphere whose radius is 14 cm.

Solution

$S = 4\pi r^2 = 4\left(\dfrac{27}{7}\right)(10.5)^2 = 1386$ cm^2

$V = \dfrac{4}{3}\pi r^3 = \dfrac{4}{3}\left(\dfrac{22}{7}\right)(10.5)^3 = 4851$ cm^3

 Exercise 10:1

1. Find the surface area and volume of the spheres with the following radius.
   (a) 21 cm (b) 10.5 cm
2. The surface area of a sphere is 616 cm^2, what is the volume of the sphere correct to two significant figures?
3. Find the ratio of volumes of two spheres whose radii are in the ratio 1:7.
4. Calculate to the nearest whole number the volume of air required to fill completely a ball which when fully expanded has a diameter of 15 cm.
5. A blacksmith melts a lead sphere of diameter 100 mm and uses it to make spherical balls of diameter 1 mm. If no lead is lost in the process, calculate the number of small spheres he makes.
6. A hemisphere has a diameter of 6 cm. Calculate
   (a) Its Volume (b) The area of its curved surface. (c) Its total surface area.

10.2 Volume and Surface Area of a Hemisphere

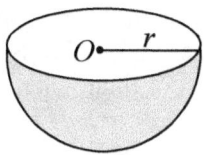

Since a hemisphere is half of a sphere, its volume V and curved surface area S_0 are half that of the sphere. Hence,

$$V = \frac{2}{3}\pi r^3 \quad \text{and} \quad S_0 = 2\pi r^2$$

In addition, a hemisphere has a circular part so the total surface area of a solid hemisphere is

$$S = \pi r^2 + 2\pi r^2 = 3\pi r^2$$

 Example

A hemispherical bowl with a flat lid has a radius of 7 cm. Calculate the
(a) Volume of the bowl
(b) The surface area of the bowl when (i) Open.   (ii) Closed.

Solution

Volume of bowl $V = \dfrac{2}{3}\pi r^3$

$$\Rightarrow V = \frac{2}{3}\left(\frac{22}{7}\right)(7)^3 = 718.7 \text{ cm}^3$$

(b) (i) $S_0 = 2\pi r^2 = 2\left(\dfrac{22}{7}\right)(7)^2 = 308 \text{ cm}^2$

(ii) $S = 3\pi r^2 = 3\left(\dfrac{22}{7}\right)(7)^2 = 462 \text{ cm}^2$

10.3 Position of a Place on the Surface of the earth

Geographers and scientists locate places on the earth surface using longitudes and latitudes as coordinates. Longitudes are imaginary lines on the earth surface which run from north to south, east or west of the Greenwich meridian. Latitudes on the other hand, are imaginary lines on the earth surface which run from east to west, north or south of the equator.

To measure points along the latitudes begin E or W of the Greenwich meridian.
To measure points along the longitudes begin N or S of the equator.
Longitudes and latitudes act like the coordinate plane.
Longitudes are measured from 0° to 90° N or S of the equator.
Latitudes are measured from 0° to 180° E or W of the Greenwich meridian.

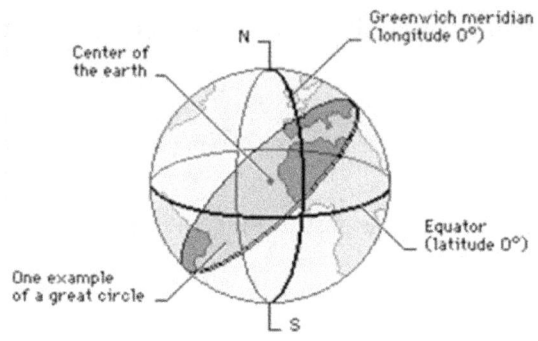

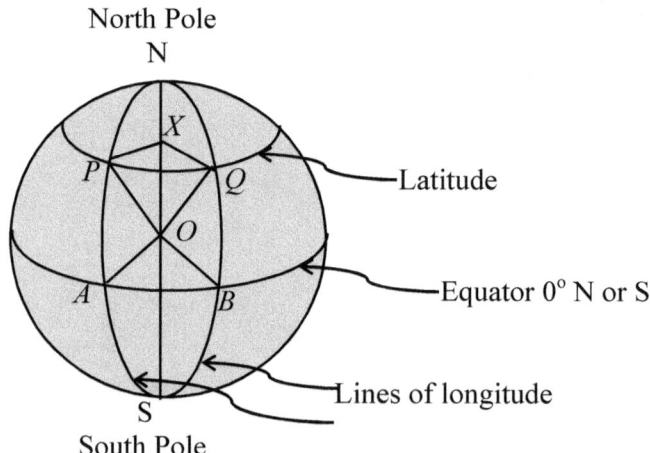

10.4 Great Circles

A great circle is a circular section of a sphere whose radius is equal to that of the sphere. Great circles of the earth include the equator and all the longitudes.
Using the formula for calculating arc length, treated in Topic 9, the distance D along a great circle is,

$$D = \frac{\theta}{360} \times 2\pi R$$

Module 16, Topic 10: The Earth as a Sphere

Where $R \approx 6370$ km, is the radius of the earth, and θ is the angular distance, which the two points subtend at the centre of the earth.

10.5 Small Circles

A small circle is any circle other than the equator and the longitudes drawn round the earth's surface.

Example

1. Calculate in km the shortest distance, measured over the earth's surface between the points $P(40°N, 20°W)$ and $Q(20°S, 20°W)$.

   Solution

   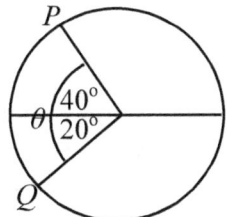

   $$\text{Arc length } PQ = \frac{\theta}{360} \times 2\pi R = \frac{60}{360} \times 2\left(\frac{22}{7}\right)(6370) = 6673.3 \text{ km}$$

2. Find in km the shortest distance, between the points $A(70° N, 50° W)$ and $B(20° N, 50° W)$.

   Solution

   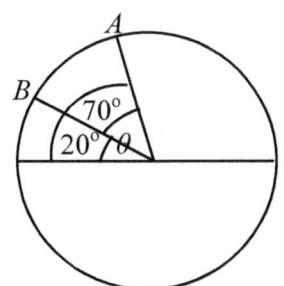

   $\theta = 70° - 20° = 50°$

   $$\text{Arc length } AB = \frac{\theta}{360} \times 2\pi R$$

   $$= \frac{50}{360} \times 2\left(\frac{22}{7}\right)(6370)$$

   $$= 5561.1 \text{ km}$$

Exercise 10:2

In this exercise take $\pi = \dfrac{22}{7}$ and R, the radius of the earth to be 6400 km.

1. Calculate in km the shortest distance, between the pair of points,
   (a) (50° N, 10° W) and (66° S, 10° W)   (b) (10° N, 97° E) and (65° N, 97° E)
   (c) (32° S, 30° W) and (84° S, 30° W)
2. Find in km the shortest distance, between the following pair of points, which are along the equator.
   (a) 150° W and 15° E   (b) 130° E and 92° E   (c) 40° W and 130° W
3. Calculate the angular distance between two points on the same line of longitude if their distance apart is   (a) 2816 km   (b) 10057 km

10.6 Time Zones in the World

Due to the rotation of the earth, when it is day in some places, it is night in others. Therefore, for legal, commercial, and social reasons, the earth is divided by longitudes into 24 geographic areas called time zones. Clocks within a given time zone are set to the same time. Each time zone is defined by its distance east or west of Greenwich, England. Time in each of the 12 zones east of Greenwich increases one hour for each zone. Time in each of the 12 zones to the west of Greenwich decreases one hour for each zone. The International Date Line divides the eastern and western time zones. The time difference between each side of the International Date Line is 24 hours. Thus, a traveler heading west across the date line loses one day while a traveler heading east gains a day.

 Example

A plane left London at 1:20 p.m. on Tuesday (London time) and arrived in Australia at 5:40 p.m. Wednesday (Australian time). How long did the plane take if Sydney time is 11 hours ahead of London time?

Solution

In terms of London time, time of arrival = Arrival time − 11 hours

$$= 5:40 - 11 \text{ hours} = 6:40 \text{ a.m.}$$

Therefore time taken = 6:40 a.m. − 1:20 p.m. = 17 hours 20 minutes.

Module 16, Topic 10: The Earth as a Sphere

 Exercise 10:3

1. State the following as 12 hour system time.
   (a) 17:25   (b) 23:36   (c) 14:20   (d) 11:10
2. State the following as 24 hour system time.
   (a) 8:34 p.m.   (b) 5:56 a.m.   (c) 12:45 p.m.   (d) 12:45 a.m.
3. Calculate the time between
   (a) 10:42 a.m. and 8:12 p.m.   (b) 10:42 p.m. and 8:12 a.m.
   (c) 8:16 a.m. and 7:20 p.m.   (d) 22:33 and 9:15
4. What is the meaning of (a) a.m.?   (b) p.m.?
5. A football match started at 4 p.m. Spanish local time. A city's time is 13 hours behind Spanish time. At what time did the match start in the local time of the city?
6. A plane left airport A at 8:30 p.m. on Tuesday and landed at airport B at 9:00 p.m. (local time) on the same day. If the time at airport B is 19 hours behind that at airport A. How long did the trip take?
7. When the time is 6:46 p.m. in Dakar, it is 10:46 p.m. in Nairobi. A plane takes 14 hours to travel from Dakar to Nairobi. If the plane leaves Dakar at 2:40 p.m. on Tuesday:
   (a) At what time (local time) will it arrive in Nairobi?
   (b) On what day will it arrive in Nairobi?

 Multiple Choice Exercise 10

In this exercises, where necessary, take $\pi = \frac{22}{7}$.

1. The surface area of a sphere of radius 7 cm is:
   [A] 86 cm^2 [B] 154 cm^2 [C] 616 cm^2 [D] 143 cm^2
2. Two solid spheres have volumes 250 cm^3 and 128 cm^3 respectively. The ratio of their radii is certainly:
   [A] 5:4   [B] 25:16   [C] 2:1   [D] 4:3
3. A hollow sphere has a volume of k cm^3 and a surface area of k cm^2. The diameter of the sphere is:
   [A] 3 cm   [B] 12 cm   [C] 9 cm   [D] 6 cm
4. A sphere has a surface area of 4312 cm^2. The radius of the sphere in cm correct to one decimal place is:
   [A] 18.0   [B] 18.5   [C] 19.0   [D] 19.5

5. The position of two countries P and Q are 15° N, 12° E and 65° N, 12° E respectively. Their difference in latitude is:
   [A] 100° [B] 80° [C] 50° [D] 24°
6. Cotonou and Niamey are on the same line of longitude and Niamey is 7° north of Cotonou. If the radius of the earth is 6400 km, the distance of Niamey north of Cotonou along the line of longitude correct to the nearest kilometre is:
   [A] 391 km [B] 503 km [C] 782 km [D] 1006 km
7. P and Q are two places on the same circle of latitude 79° S. P is on longitude 68° E, while Q is on longitude 22° W. The angular distance between P and Q is:
   [A] 12° [B] 45° [C] 48° [D] 90°
8. Two ships on the equator are on longitude 45° W and 45° E respectively. Their distance apart along the equator, correct to 2 significant figures is:
   [A] 3,200 km [B] 10,000 km [C] 6,400 km [D] 5,000 km
9. Two points P and Q are on longitude 67° W. Their latitudes differ by 90°. Taking the radius of the earth as 6400 km, their distance apart in terms of π is:
   [A] 6400π km [B] $\frac{6400}{\pi}$ km [C] 3200π km [D] $\frac{3200}{\pi}$ km
10. Two places are 2816 km apart on the same line of longitude. Taking $R = 6,400$ km, the angular difference between their latitudes is:
    [A] 25.2° [B] 26.1° [C] 51.3° [D] 63.9°
11. Abijan is 4° west of Accra and on the same circle of latitude. If the radius of this circle of latitude is 6370 km, the distance of Abijan west of Accra, correct to the nearest km is:
    [A] 222 [B] 445 [C] 890 [D] 5005

Topic 11

NETWORKS

Objectives

At the end of this topic, the learner should be able to:

1. Define the terms network, vertex or node, edge, arc or link and region.
2. Identify and count the number of regions, vertices and edges in a given network.
3. Use the relation $R + V - E = 2$ to find an unknown parameter, given 2 parameters.
4. List the set of vertices V and edges E given a diagrammatic network graph.
5. Distinguish between an ordered list and an unordered list.
6. Name and distinguish between the various types of network graphs.
7. Determine whether or not a given network is traversable.
8. Interpret simple networks in real life situations.

11.1 Network Terminology

Consider the diagram below. This diagram shows a set of objects or points 1, 2, 3, 4, 5, and 6, connected by lines. We call such a diagram a network. Thus, a **network** is a collection of points, called **vertices** or **nodes** and lines, called **edges** or **arcs** or **links**, connecting these points. The area bounded by the vertices and edges is called a **region**.

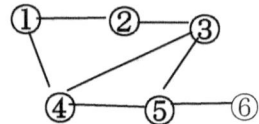

The network above has 6 vertices, 7 edges (arcs) and 3 regions. Notice that we count the area outside the network diagram as one region.

 Example

Find the number of vertices, regions, and edges in the network below.

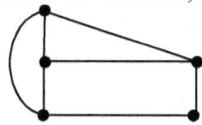

Solution
The network has 5 vertices, 4 regions, and 7 edges.

11.2 Euler's Formula

 Integration Activity

Given that R = number of regions,
V = number of vertices and
E = number of edges.
Use the networks in figures (a) to (f) to complete the table below. (a) has already been filled for network (a). What conclusion do you draw?

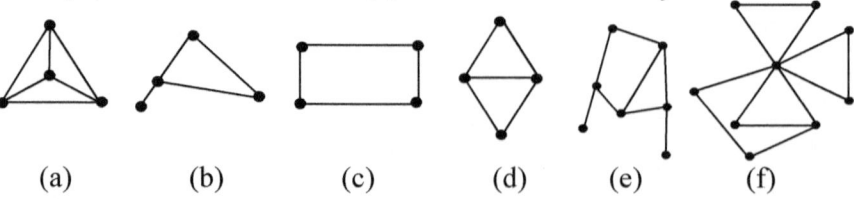

(a) (b) (c) (d) (e) (f)

Module 16, Topic 11: Networks

Network	R	V	E	R + V − E
(a)	4	4	6	2
(b)				
(c)				
(d)				
(e)				
(f)				

In any network, the relation $R + V - E = 2$ is always true.

We call this relation **Euler's formula** named after the Swiss mathematician Leonhard Euler who was the spark to network theory.

11.3 Ordered Lists and Unordered Lists

Consider a set X containing two elements a and b. We write this set using curly braces. Thus, $X = \{a, b\}$. The order in which we write the elements in any set is immaterial. Therefore, $X = \{a, b\} = \{b, a\}$. Since we can write the two elements in a set containing two elements in any order, a set of two elements is therefore an **unordered pair**.

It follows that, two sets $\{a, b\}$ and $\{c, d\}$ are equal if and only if either $a = c$ and $b = d$ or $a = d$ and $b = c$.

On the other hand, an **ordered pair** is a collection of two objects such that we can distinguish one as the *first element* and the other as the *second element*. We write ordered pairs using parentheses instead of curly braces, which are used for sets. Thus, the ordered pair (a, b) consists of the first element a and the second element b.

The fundamental property of ordered pairs is that two ordered pairs are equal if and only if their *first elements* are equal and their *second elements* are equal. This means that, two ordered pairs (a, b) and (c, d) are equal if and only if $a = c$ and $b = d$.

With an intuitive extension of the above notation, we can also write ordered lists for any finite number of elements. Thus,
- (1, 2, 3) is an **ordered list** and is not the same as (3, 2, 1) whereas the set {1, 2, 3} is an **unordered list** and is the same as {3, 2, 1}.

- {(1, 2), (3, 4)} is a set of two elements, each of which is an ordered pair. One of the elements of the set is (1, 2) and the other is (3, 4).
- ({1, 2}, {3}, {2, 4}) is an ordered list of three sets. The first element of the list is the set {1, 2}; the second element is the set {3}; the third element is the set {2, 4}.

11.4 Odd and Even Vertices

If the number of arcs meeting at a vertex is *even*, we call the vertex an **even vertex** otherwise it is an **odd vertex**.

 Example

In the network below, list the set:
(a) E of even vertices.   (b) O of odd vertices.

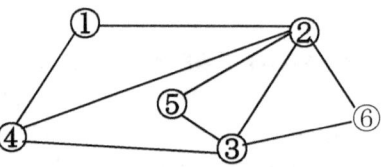

Solution
(a) $E = \{1, 3, 5, 6\}$ (b) $O = \{2, 4\}$.

 Exercise 11:1

In figure (a) to (g) list the set: (a) E of even vertices. (b) O of odd vertices.

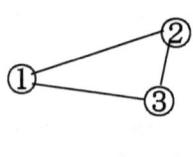   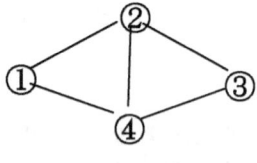   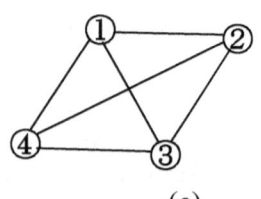

(a)   (b)   (c)

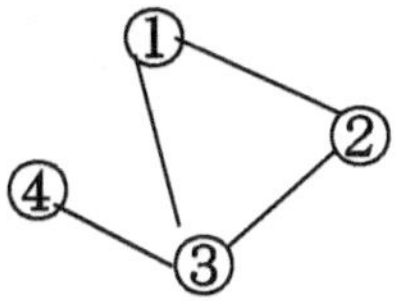

Module 16, Topic 11: Networks

(d) 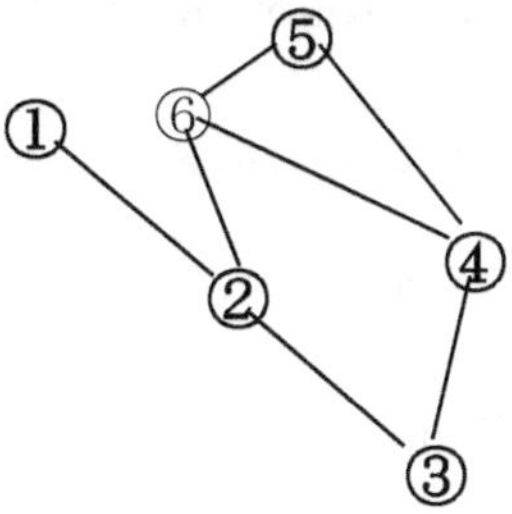   (e) 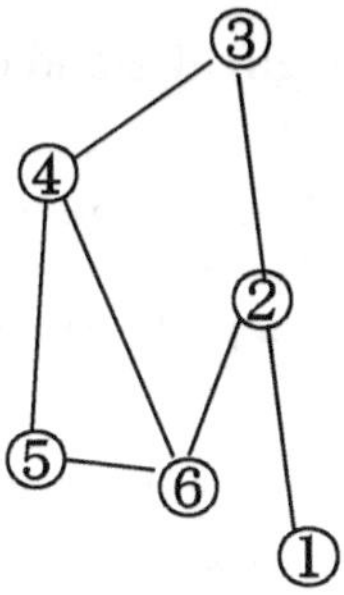   (f)

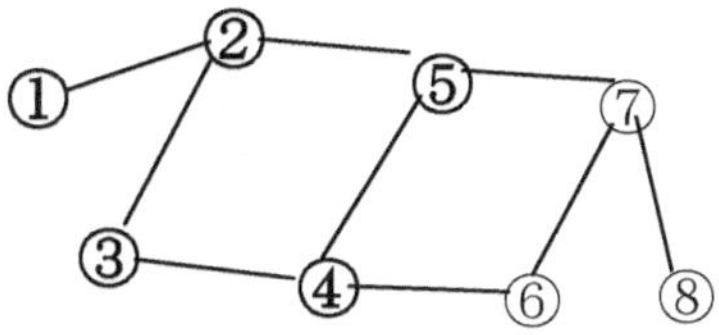

(g) 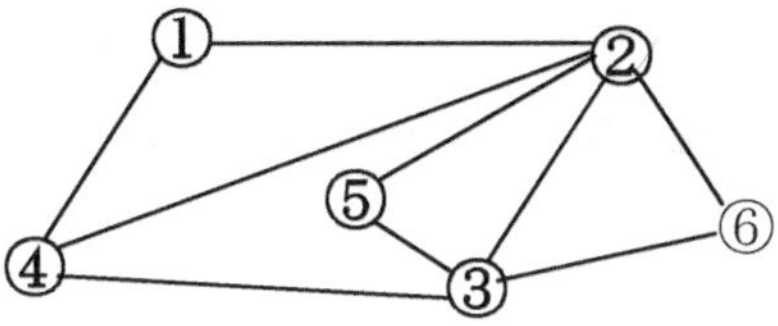   (h)

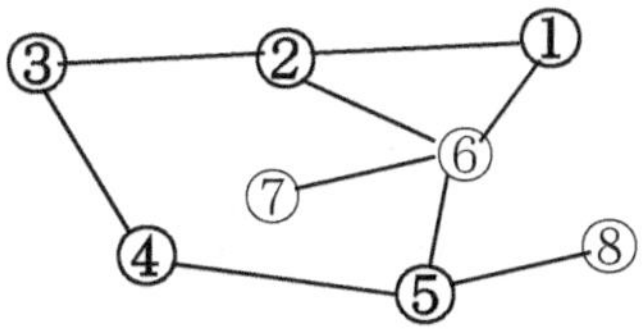

(i)

11.5 Traversable Networks

The figure (a) below shows the path which participants in a certain race must follow. The rules of the race are that each participant has to run through all the paths shown without passing more than once through any path. What are the possibilities of doing this? Figures (a) and (c) show the possibilities.

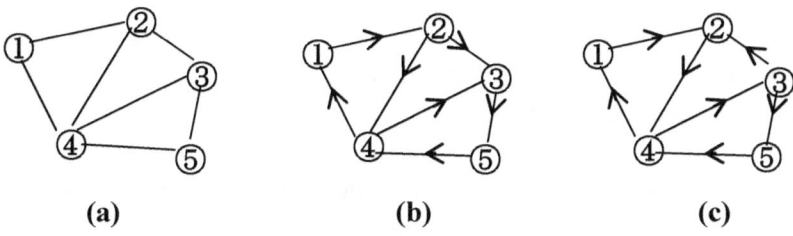

(a)   (b)   (c)

One can begin at 2 and end at 3. The following are two of the possible paths in this case.

2→4→1→2→3→5→4→3 or (2, 4, 1, 2, 3, 5, 4, 3)

2→3→5→4→1→2→4→3 or (2, 3, 5, 4, 1, 2, 4, 3)

Alternatively, one can begin at 3 and end at 2. Two of the possible paths in this case are shown below.

3→2→4→3→5→4→1→2 or (3, 2, 4, 3, 5, 4, 1, 2)

3→5→4→3→2→4→1→2 or (3, 5, 4, 3, 2, 4, 1, 2)

A **traversable** is a network, which we can trace exactly once beginning at some point without retracing any arc.

 Integration Activity

1. Examine each of the following networks.

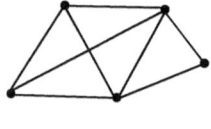   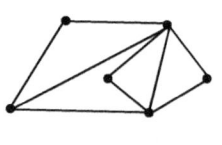   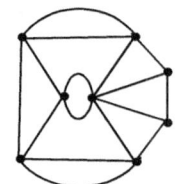

(a)   (b)   (c)

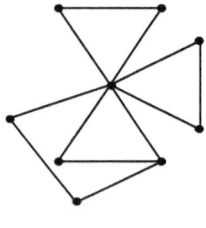   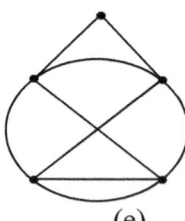   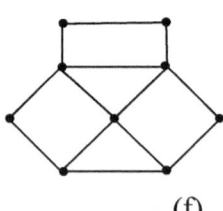

(d)   (e)   (f)

   (i) How many odd vertices does each network have?
   (ii) Is each network traversable?
   (iii) By tracing each of these networks on a separate paper, deduce a conjecture about the beginning and ending points of traversable networks with exactly the same number of odd vertices.
2. (i) Determine which of the networks in Figure 42:5 are traversable?
   (ii) For each traversable network, write down two different sequences of

nodes from the starting vertex to the ending vertex to show how we can traverse the network.
3. State the number of odd vertices in each of the traversable networks.
4. What is the relation between a traversable network and the number of odd vertices in the network?
5. Use the networks in Figure 42:5 to verify your conclusion.

11.6 Conditions for a Network to be Traversable

Whether a network is traversable depends on the number of odd vertices as follows.
1. A network, which has no odd vertices, is traversable. This means that a network whose vertices are all even is traversable. In this case, any vertex may be the beginning point, and the same vertex will be the ending point.
2. A network, which has exactly two odd vertices, is traversable. If one odd vertex is the beginning point, the other odd vertex is the ending point.
3. A network with more than two odd vertices is not traversable.

  Example

In figure (a) to (g), determine giving reasons for your answer which of the networks are traversable. Draw each of the traversable networks on a separate paper and mark the beginning point, the end-point and the direction using arrows in which we can traverse the network.

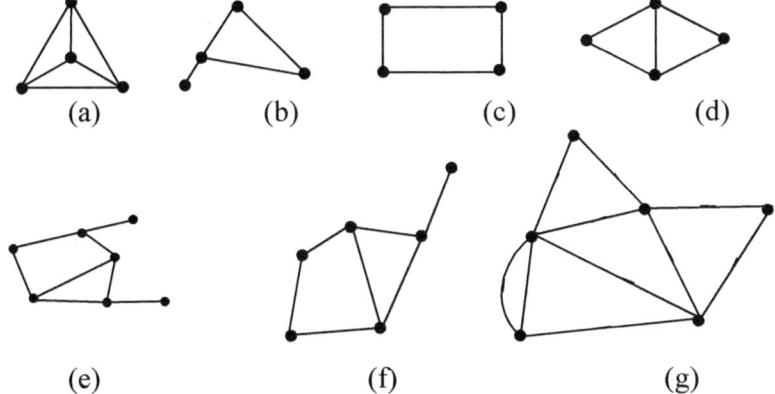

Solution
(a), (e) and (f) are not traversable because they each have more than 2 odd vertices.

(c) Traversable because all the vertices are even. In other words, it has no odd vertices.

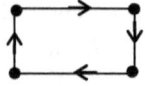

Any of the vertices may be the beginning or endpoint.

(b), (d) and (g) are traversable because each has exactly 2 odd vertices.

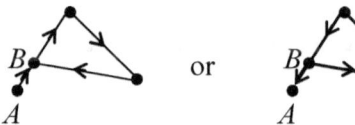

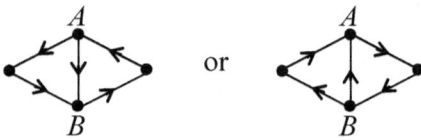

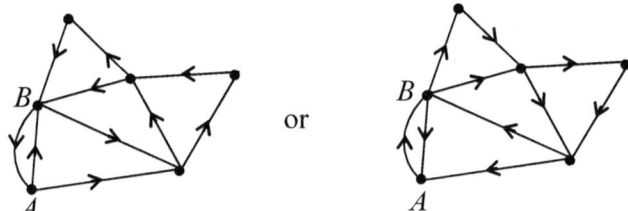

In all three cases, either of the odd vertices A or B can be the beginning point, and the other will be the endpoint.

 Exercise 11:2

Determine which of the networks in figures (a) to (c) is/are traversable.

(a)

(b)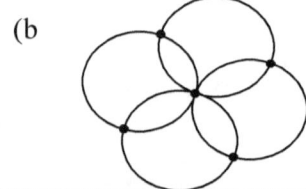

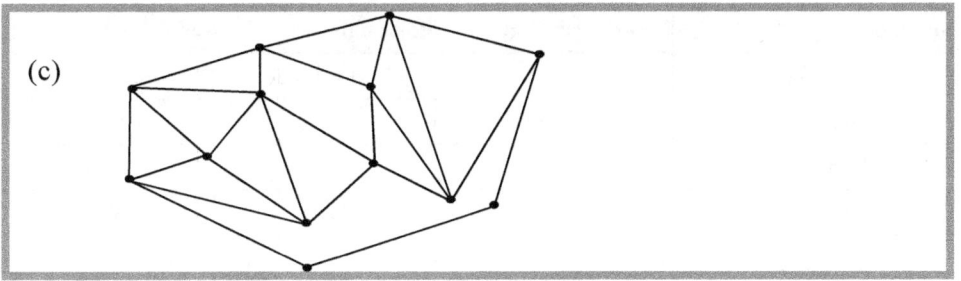

(c)

11.7 Networks in Real Life

We can use the concept of networks to describe many different systems (computer networks, technical, physical, biological, sociological etc, etc.).

Computer Networks are connections between groups of computers and associated devices that allow users to transfer information electronically. The local area network shown above is representative of the setup used in many offices and companies. Individual computers called workstations (W.S.), communicate to each other via cable or telephone lines linking to servers.

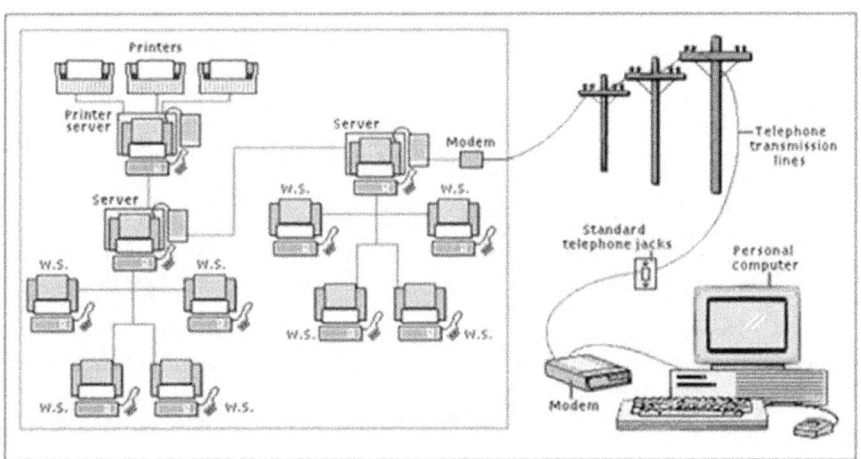

Computer Networking

The table below shows the existence of networks in different disciplines and the terminology used to refer to them, their points and their connection lines.

Discipline	Name used	Point name	Line name
Mathematics	Graph	Vertex (node)	Edge or arc
Computer science	Network	Node	Link
Physics	system	Site	Bond
Chemistry	Bonding	Atom	Bond
Sociology	Social network	Actor (individual)	Tie (friendship)
Communication	World wide web	Webpage Website	Link (d)
Communication	Internet	Network	Connection Bridge
Communication	Road system (road network)	Crossing (junction)	Road

The above table is not exhaustive. It illustrates the importance of networks and their applications in real life. Some applications of networks are in planning roads and museums. Museums planners often do it in such a way as to minimize congestion at doorways when so many visitors visit the museum. To realize this, visitors have to to move through each door only once. This implies that such a museum must be traversable.

Example

Figures (i) and (ii) below show the floor plans of two museums.
1. Draw network diagrams to represent each plan.
2. Use your network diagrams to determine giving reasons for your answer whether it is possible to visit all the rooms passing through each door exactly once.
3. If it is possible, confirm your results by sketching the plan and showing the route using arrows how to do this. Your diagram should indicate the starting and the finishing points.

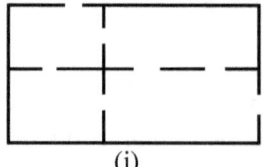
(i)

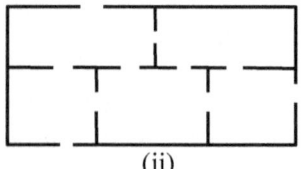
(ii)

Solution
1. Taking each room and the outside region to be nodes and the doors to be edges, we can draw the network diagrams as follows.

Module 16, Topic 11: Networks

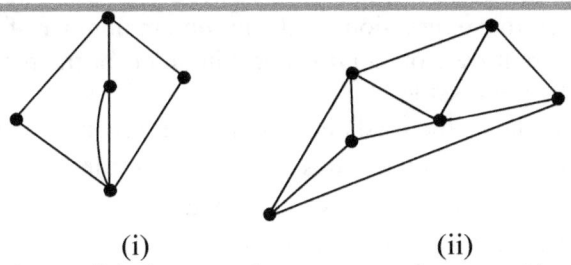

(i) (ii)

2. In (i) it is possible because there are exactly two odd nodes, meaning that the network is traversable. In (ii) it is impossible because there are four odd nodes, implying that the network is not traversable.

3.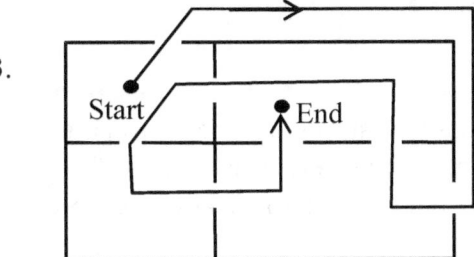

Exercise 11:3

1. (a) Mark the beginning point and the end point for traversing the following network.

   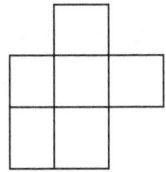

   (b) Draw shapes showing the minimum number of squares that we must remove from the grids in the figure below, in order for the remaining network to be traversable.

   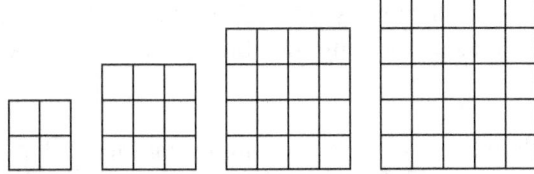

   (c) Using your answer in (b), find a pattern and predict the minimum number of squares that we must remove from a 12 × 12 grid of squares in order for the remaining network to be traversable.

(d) Write an algebraic expression for the minimum number of squares that we must remove from an $n \times n$ grid in order for the remaining network to be traversable.
2. The figure below shows the road network in a certain city. In order to reduce traffic, the city council decides to turn all roads into one way by building one more road linking two of the junctions in the city. Draw five different road network diagrams to show how the city council can do this.

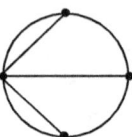

3. To organize a trade fair, a council builds a fence with gates as shown in the diagram below. In order to visit all the stands, a visitor must pass through all the gates. However, to reduce congestion at the gates the council passes an ultimatum that no one passes through any gate more than once.
   (a) Given that a visitor can begin either from inside or outside the fence, is it possible for a visitor to visit all the stands?
   (b) If so, draw a diagram showing the starting and finishing points. Use arrows to show the direction of movement.

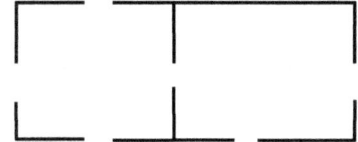

   (c) Using the apartments of the fence and the region outside the fence as nodes and the gates as arcs, draw a simple network diagram to represent the plan.

11.8 Network Graphs

A network graph is a set of objects connected by lines. We usually call objects in a graph *nodes* or *vertices* and the lines connecting the objects are called *links* or *edges* or *arcs*.

More formally, a graph is an ordered pair $G = (V, E)$ comprising the set V of vertices (or nodes) together with the set E of edges (or lines) such that each element of E is a doubleton (or paired) subset of V.

Module 16, Topic 11: Networks

 Example

1. Given the diagram in Figure 42:21 and the graph $G = (V, E)$. List the elements of V and E.

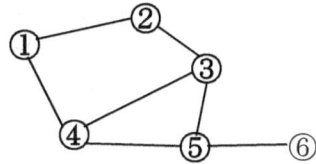

Solution
$V = \{1, 2, 3, 4, 5, 6\}$, $E = \{\{1,2\},\{1,4\},\{2,3\},\{3,4\},\},\{3,5\},\{4,5\},\{5,6\}\}$

2. A network is defined by $G = (V, E)$ where $V = \{a, b, c, d, e\}$ and $E = \{\{a, b\}, \{a, c\}, \{b, c\}, \{b, d\}, \{c, d\}, \{a, e\}\}$.
   (a) Sketch the network diagrammatically.
   (b) State the number of regions, vertices and edges of the network.

Solution
(a)

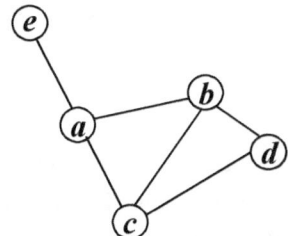

(b) Number of regions = 3, Number of vertices = 5, Number of edges = 6

11.1 Directed and Undirected Edges

An arc $a = (x, y)$ is conventionally considered to be directed from x to y and in its diagrammatic representation, the arrow points towards y.

In this case, we call x is the **tail** and y the **head**.
Since a path leads from x to y, y is said to be the successor of x and reachable from x and x is said to be the predecessor of y. To emphasize, y is the **direct**

successor of x and x is the **direct predecessor** of y. The arc (y, x) is called arc (x, y) **inverted**.

An arc $a = \{x, y\} = \{y, x\}$ is conventionally considered to be directed in either ways. Both x and y are heads and tails. Diagrammatically, we can represent this as

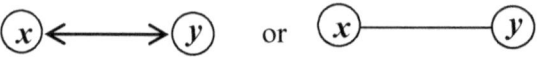

11.2 Properties of Network Graphs

(i) **Adjacent or Coincident Edges:** If two edges of a graph share a common vertex, we call them **adjacent** or **coincident edges**. In figure (a), the edges AX and BX are adjacent or coincident because they share the same vertex X.

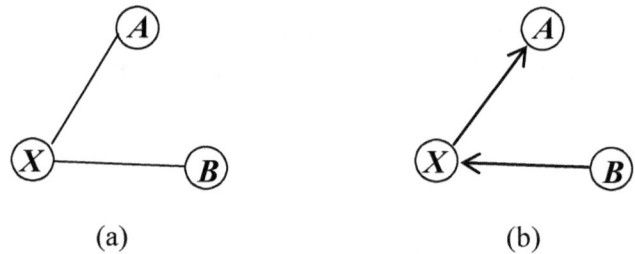

(a)   (b)

(ii) **Consecutive Arrows**

If the head of one arrow of a graph is at the nock (notch end) of another arrow of the graph, we call the two arrows **consecutive arrows**. In figure (b), the arrows BX and XA are consecutive the head of BX is at the tail of XA.

(iii) **Adjacent Vertices:** If two vertices of a graph share a common edge, we call them **adjacent vertices** and we say the common edge is joint the two vertices.

(iv) **Consecutive Vertices:** If two adjacent vertices are such that the head of the edge is on one vertex and the tail of the edge is on the other, we call the vertices **consecutive vertices**.

Module 16, Topic 11: Networks

(v) **Incident edge and Vertex:** We call an edge and a vertex on that edge **incident** edge and vertex.

Exercise 11:4

1. In the following figure, list
   (a) 8 pairs of coincident edges.
   (b) 3 pairs of consecutive arrows.
   (c) 7 pairs of adjacent vertices.
   (d) 5 pairs of consecutive vertices.
   (e) 8 pairs of incident edges and vertices.

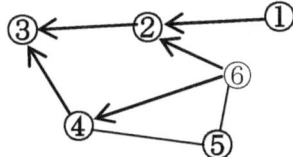

2. Write down the graph of the network in the figure above symbolically.
3. Draw the network structure for the graph $D = (V, A)$ where
   $V = \{1, 2, 3, 4, 5, 6\}$ and
   $A = \{(1, 2), (2,1), (3,1), (3, 2), (3,4), (3, 5), (4, 1), (4,5), (5,6)\}$.

11.9 Types of Network Graphs

1. *Directed Graphs*

A **directed graph** sometimes referred to simply as a **digraph** is an ordered pair $D = (V, A)$ such that V is the set of vertices and A is the set of ordered edges. In other words, A is a set of ordered pairs (a, b).

Example

1. The graph of a network structure is $D = (V, A)$, where $V = \{1, 2, 3\}$ and $A = \{(1, 3), (2, 1), (3, 1), (3, 2)\}$. Represent this network diagrammatically.

Solution

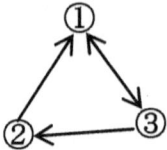

The above figure is an example of a digraph because each edge (line or arc) in a directed graph has at least an arrow at one end.

2. Represent the networks in below symbolically.

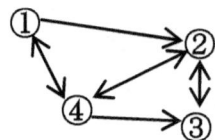

Solution
The symbolic representation of the graph is $D = (V, A)$ where
$V = \{1, 2, 3, 4, 5\}$ and
$A = \{(1, 2), (1,4), (2,3), (2,4), (3,2), (4, 1), (4, 2), (4, 3)\}$.

3. Draw the network structure for the graph $D = (V, A)$ where
$V = \{1, 2, 3, 4, 5, 6\}$ and
$A = \{(1,2),(2,1),(3,1),(3,2),(3,4), (3,5),(4,1), (4,5),(5,6)\}$.

Solution

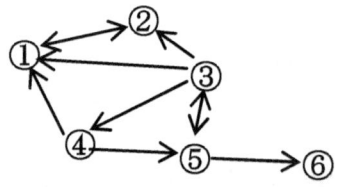

2. Undirected Graphs

An **undirected graph** is a graph for which the edges have no arrows. For such a graph, the edge connecting a and b is simply represented as $\{a, b\}$ or $\{b, a\}$. In other words, an undirected graph is a set of unordered pairs.

Example

Draw the network structure for the graph $G = (V, E)$ where $V = \{1, 2, 3\}$ and $E = \{\{1,2\}, \{1,3\}, \{2,3\}\}$.

Solution

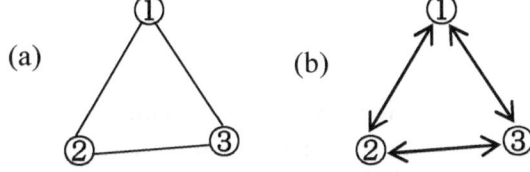

Figure (a) is an example of an undirected graph.

Notice that though the graphs in Figure (a) and (b) above are equivalent, it is needless putting the arrows as in (b).

3. Mixed Graph

A **mixed graph** is a graph with some edges directed and some edges undirected. This type of graph is written as an ordered triple $G = (V, E, A)$ where V is the set of vertices, E is the set of unordered edges and A is a set of ordered edges.

Example

Represent the networks below symbolically.

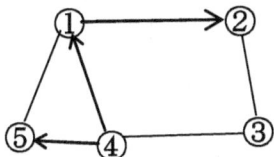

Solution
Symbolically, the graph of this network is $G = (V, E, A)$, where
$V = \{1, 2, 3, 4, 5\}$,
$E = \{\{1, 5\}, \{2, 3\}, \{3, 4\}\}$ and $A = \{(1, 2), (4, 1), (4, 5)\}$.
The figure above is an example of a mixed graph.

In real life situations, the mixed graph is what occurs predominantly. The directed and the undirected graphs are actually very special cases.

4. Complete Graph

A **complete graph** is one for which an edge connects each pair of vertices. This means that each vertex has an edge to every other vertex. The following is an example of a complete graph.

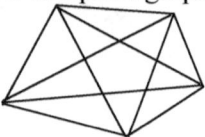

5. Weighted Graph

A **weighted graph** is one for which a number (weight) is assigned to each edge. Such weights may represent cost, length, capacity, importance etc. For instance, in electricity networks, the hospital line, the administrative lines and the commercial lines usually carry more weights than private lines. If there is low voltage and electricity has to be rationed, it is very unlikely to cut off such highly weighted lines.

 Example

Given that the numbers in the network show distances in km between towns in a municipality, find the shortest route from A to G, using the distances shown on the network below.

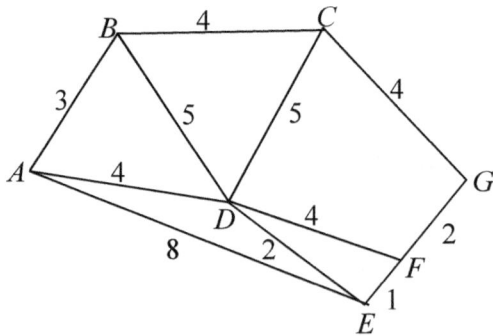

Solution
By adding the distances, the shortest distance is 9 km and is the road $ADEFG$.

6. Null Graph

A **null graph** is a graph whose edge set is empty. This means that all the vertices are isolated from each other. We denote a null graph on n vertices by Nn. The figure below is an example of a null graph on five vertices.

Module 16, Topic 11: Networks

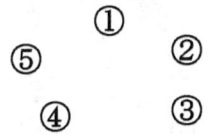

We can represent the graph in above symbolically by $G = (V, E)$ where $V = \{1, 2, 3, 4, 5\}$ and $E = \{\ \}$.

11.10 Network Trees

We can define a **network tree** in any of the following ways:
(i) A tree is a connected graph with no cycle.
(ii) A tree is a graph in which exactly one path connects any two vertices.
(iii) A tree is a connected graph in which

$$n(E) = n(V) - 1$$

The following are three different trees made from the same set of nodes. Check that each of these trees satisfies each of the above definitions.

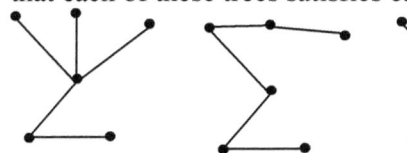

Examine the network in the figure below. Why do you think it is not a tree?

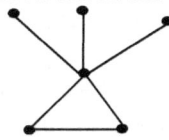

A **forest** is a collection of trees. This definition implies that a forest is a disjoint union of one or more trees with no cycles. The figure below is an example of a forest.

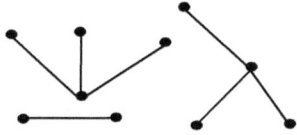

Exercise 11:5

1. The graph of a network structure is $D = (V, A)$, where $V = \{1, 2, 3, 4\}$ and $A = \{(1, 3), (2, 1), (3, 1), (3, 2), (1,4), (3,4)\}$. Represent this network diagrammatically.
2. Represent the following networks symbolically.

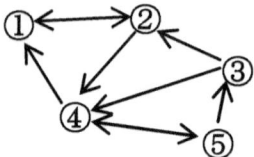

3. Draw the network structure for the graph $D = (V, A)$ where $V = \{1, 2, 3, 4\}$ and $A = \{(1,3),(2,1),(3,1),(3,2),(3,4), (4,1),(4,2)\}$.
4. Draw the network structure for the graph $G = (V, E)$ where $V = \{1, 2, 3, 4\}$ and $E = \{\{1,3\},\{2,1\},\{3,2\},\{3,4\}, \{4,1\},\{4,2\}\}$.
5. Represent the following networks symbolically.

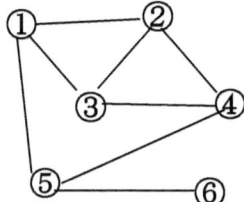

6. (a) Given that the numbers in the network below, show the time in hours taken to travel from one town to another in a certain region, find the shortest time required to travel from town A to G.
   (b) Given that a litre of fuel cost 550 FCFA and that the amount of fuel consumed is proportional to the time taken, determine the least cost of travelling from A to G, if hour requires half a litre of fuel.

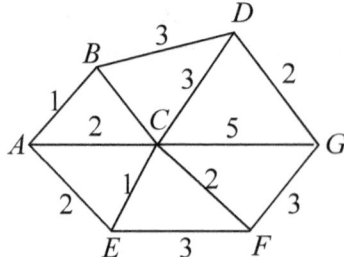

7. Draw five different trees with 6 vertices.
8. Determine which of the networks in Figure 42:46 (a) to (f) are trees, forests or neither giving reasons for answer.

Module 16, Topic 11: Networks

(a)

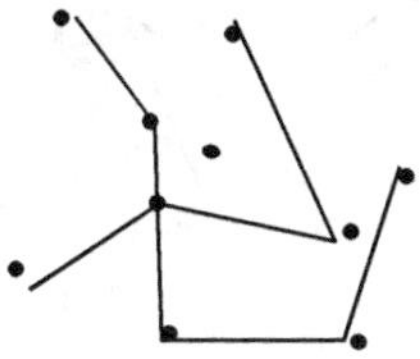

(b)

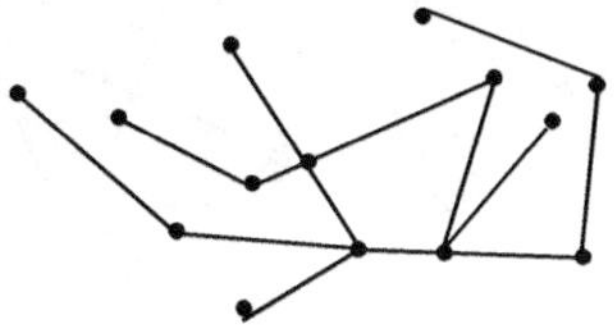

(c)

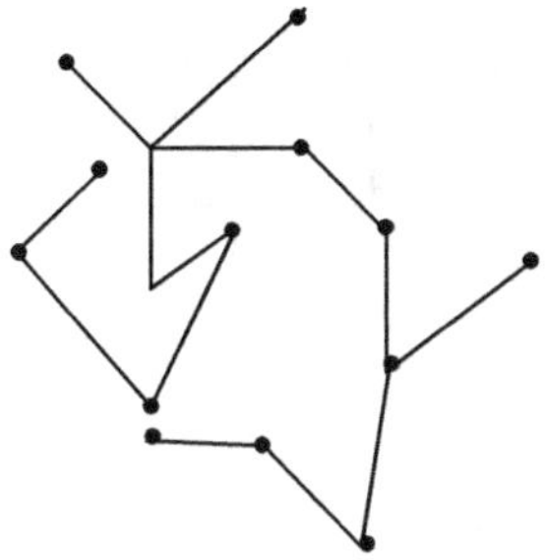

(d)

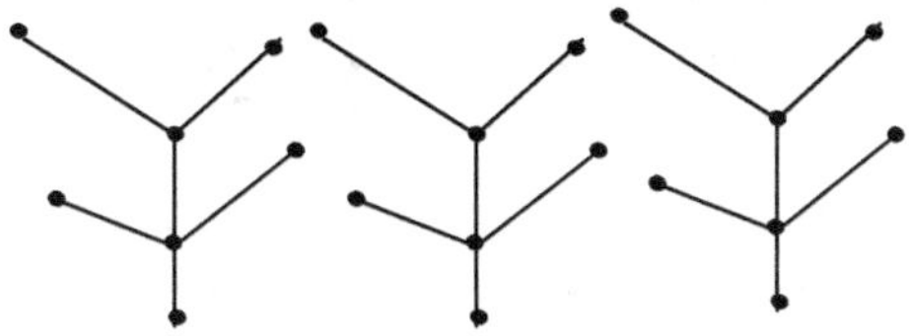

(e)

(f)

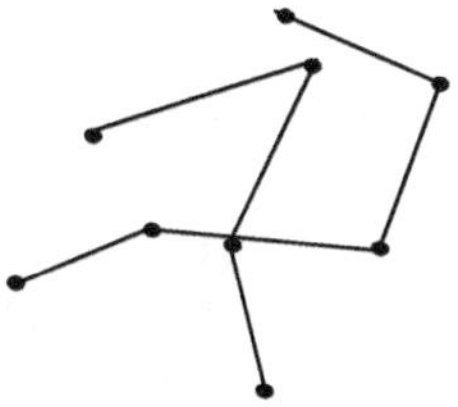

✎ Multiple Choice Exercise 11

1. Among the following, the network which is traversable is:

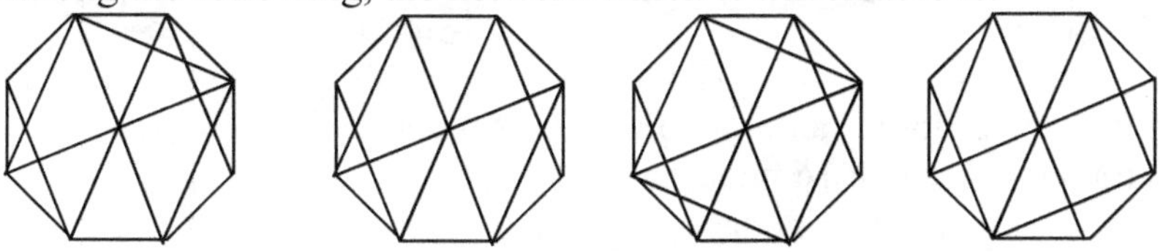

   [A]   [B]   [C]   [D]

2. Among the following, the network which is not traversable is:

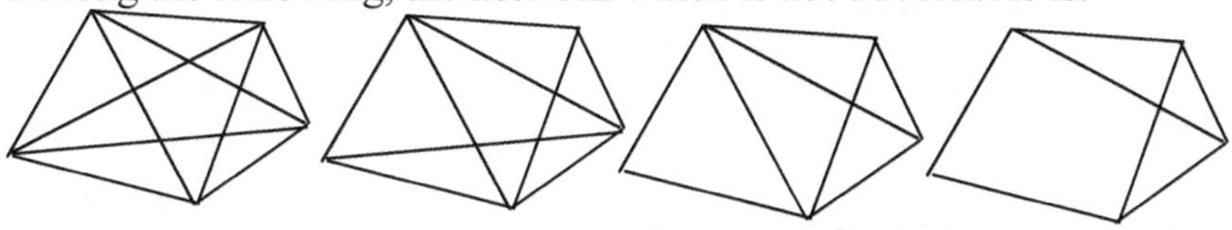

   [A]   [B]   [C]   [D]

3. Among the following, the network which is traversable is:

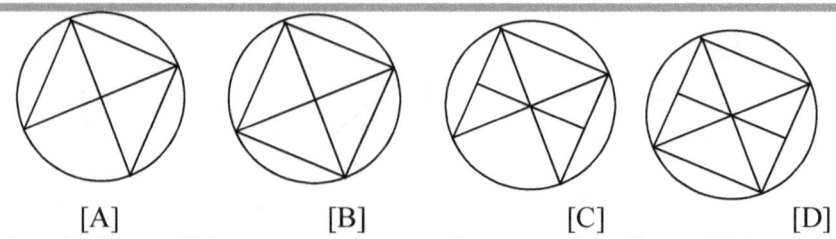

[A] [B] [C] [D]

4. One of the possible routes to traverse the network in figure (a) is:
   [A] (4, 2, 3, 5, 4, 3, 1, 4, 2)   [B] (3, 2, 3, 5, 4, 3, 1, 4, 2)
   [C] (1, 2, 3, 5, 4, 3, 1, 4, 2)   [D] (5, 2, 3, 1, 4, 3, 1, 4, 2)

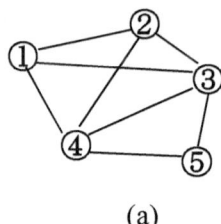 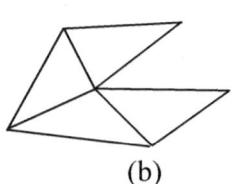

(a) (b)

5. The number of vertices, regions and arcs in the network in figure (b) are respectively:
   [A] 6, 9, 5   [B] 6, 5, 9   [C] 5, 9, 9   [D] 5, 6, 9

6. The smallest number of diagonals that we can draw on the faces of a cube to make the network between its vertices and edges traversable is:
   [A] 1   [B] 2   [C] 3   [D] 4

7. The vertices and edges of polyhedra are three-dimensional networks. Among the four regular polyhedra below, the one which is traversable is:

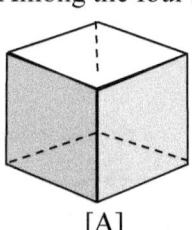 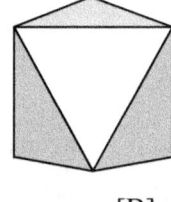 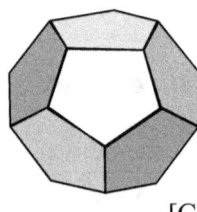 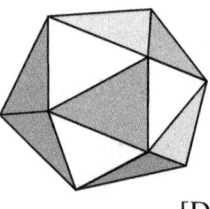

[A] [B] [C] [D]

8. The following floor plan of a house forms a network which is:
   [A] traversable, with 16 nodes and 6 arcs.
   [B] traversable, with 6 nodes and 16 arcs.
   [C] not traversable, with 16 nodes and 6 arcs.
   [D] not traversable, with 6 nodes and 16 arcs

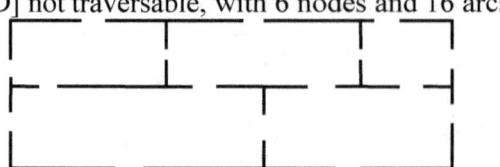

9. A house is to be traversable if one can enter all the rooms without passing through any door more than once. This is possible if:

[A] the network for the floor plan has no even vertices.
[B] the network for the floor plan has no even edges.
[C] the network for the floor plan has no odd vertices.
[D] the network for the floor plan has no odd edges.

10. The statement which is certainly true of networks is:
    [A] A network with exactly two odd vertices is not traversable.
    [B] A network with no odd vertices is traversable.
    [C] A network with more than two odd vertices is traversable.
    [D] A network with more than two even vertices is traversable.

11. The statement which is certainly true of networks is that the starting point in a traversable network with:
    [A] all even vertices is also the ending point.
    [B] all even vertices cannot be the ending point.
    [C] two odd vertices is also the ending point.
    [D] no odd vertices can be the ending points.

12. Given that $V = \{a, b, c, d, e\}$ and
    $E = \{\{a, b\}, \{a, c\}, \{b, c\}, \{b, d\}, \{c, d\}, \{a, e\}\}$.
    The number of regions in the network defined by $G = (V, E)$ is:
    [A] 3   [B] 4   [C] 5   [D] 6

13. Given that $V = \{1, 2, 3, 4, 5\}$ and $E = \{\{1, 2\}, \{1, 3\}, \{2, 3\}, \{2, 4\}, \{3, 4\}, \{1, 5\}\}$.
    The number of regions, nodes and arcs in the network defined by $G = (V, E)$ are respectively:
    [A] 3, 5 and 6   [B] 3, 6 and 5   [C] 5, 3 and 6   [D] 6, 3 and 5

14. The graph which represents the network $G = (V, E)$, $V = \{1,2,3\}$ and $E = \{\{1,2\},\{1,3\},\{2,3\}\}$ is:

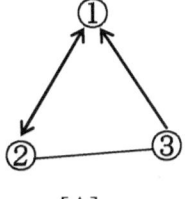 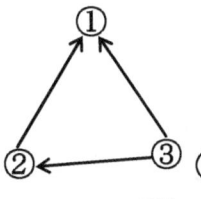 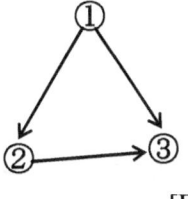

   [A]   [B]   [C]   [D]

15. A complete graph among the following is:

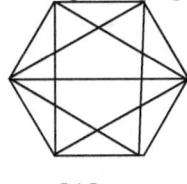  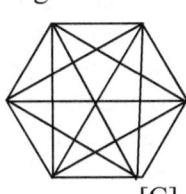 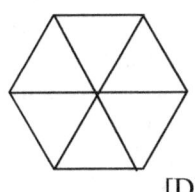

   [A]   [B]   [C]   [D]

16. In the graph of network below, the set of edges is:
    [A] $\{\{1,2\}, \{3, 2\}, \{3,4\}, \{4, 1\}\}$ [B] $\{(1,2), (3, 2), (3,4), (4, 1)\}$.
    [C] $\{(2,1), (2, 3), (4,3), (1, 4)\}$.   [D] $\{\{2,1\}, \{2, 3\}, \{4,3\}, \{1, 4\}\}$.

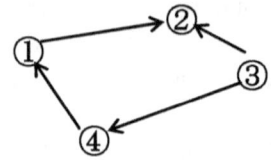

17. The network graph which is a tree among the following is:

18. The number of regions in a network with 4 vertices and 6 edges is:
    [A] 3   [B] 4   [C] 5   [D] 6
19. In the net work below, the set of odd vertices is:
    [A] {1, 2, 7, 8}   [B] {1, 3, 5, 7}   [C] {3, 4, 5, 6}   [D] {1, 3, 6, 8}

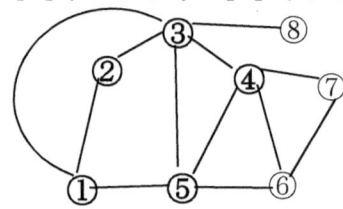

20. In the net work above, the set of even vertices is:
    [A] {1, 2, 7, 8}   [B] {2, 4, 5, 7}   [C] {3, 4, 5, 6}   [D] {2, 4, 6, 8}
21. In figure (a) below, the edge {1,4} is incident to node:
    [A] 2   [B] 1   [C] 3   [D] 5

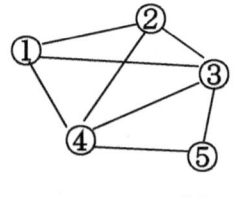

    (a)   (b)
22. In Figure (b) above, node 3 and 4 are said to be:
    [A] consecutive [B] coincident [C] adjacent   [D] incident
23. In the figure (b) above, edge {3,4} is said to be:
    [A] consecutive to node 3 and 4.   [B] coincident to node 3 and 4.
    [C] adjacent to node 3 and 4.   [D] incident to node 3 and 4.
24. In the figure (b) above, edge {3,4} and node 3 are said to be:
    [A] consecutive [B] inverted   [C] adjacent   [D] incident
25. In the figure below, the edges {1,2} and {1,3} are said to be:
    [A] consecutive [B] inverted   [C] adjacent   [D] incident

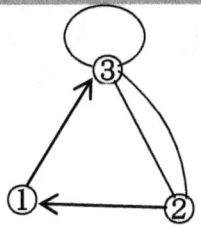

26. In the figure above, the edges {1,2} and {2,3} are said to be:
    [A] consecutive   [B] inverted   [C] adjacent   [D] incident
27. The network which is most likely to be a tree is the network which has:
    [A] 5 edges and 4 nodes   [B] 4 edges and 5 nodes
    [C] 6 edges and 4 nodes   [D] 4 edges and 6 nodes
28. The number of arcs in the network below is:
    [A] 2   [B] 3   [C] 4   [D] 5

Topic 12

FLOW DIAGRAMS

Objectives

At the end of this topic, the learner should be able to:

1. Appreciate the importance of the topic in computer technology and in real life.
2. Recognize and distinguish between various types of flow diagrams.
3. Distinguish between an instruction box and a decision box.
4. Interpret and draw flow charts or diagrams:
   (a) Representing or describing a process or the control processes of a system or program.
   (b) Showing the operations being carried out to solve simple problems.
   (c) Showing or representing the functioning of simple mechanical systems.
5. Understand the functioning of stores and use stores to store and retrieve data in flow charts.

Module 16, Topic 12: Flow Diagrams

12.1 The Concept of a Flow Diagram

Computers process large amounts of data. To do this the programmer first writes a list of instructions telling the computer the sequence of operations in a language that computers can understand. A **programming language** is the language used to write a list of instructions for computers. Using the instructions, a computer can solve in seconds a problem that might otherwise have taken weeks to solve.

Before writing a computer program, the programmer first writes the instructions in form of a **flow diagram**.

*A **flow diagram** is a schematic or graphic representation of the steps leading to the solution of a given problem.*

12.2 Types of Flow Diagrams

Flow diagrams are of different types and we use them in almost all disciplines such as mathematics, computer programming, business, physics, geography, sociology, biology, literature, engineering, production etc. Below is a brief outline of a few common flow diagrams.

1. A functional flow block diagram
This is a sketch showing the various parts of a system and their functions in relation to one another. The figure below is a functional flow block of a computer.

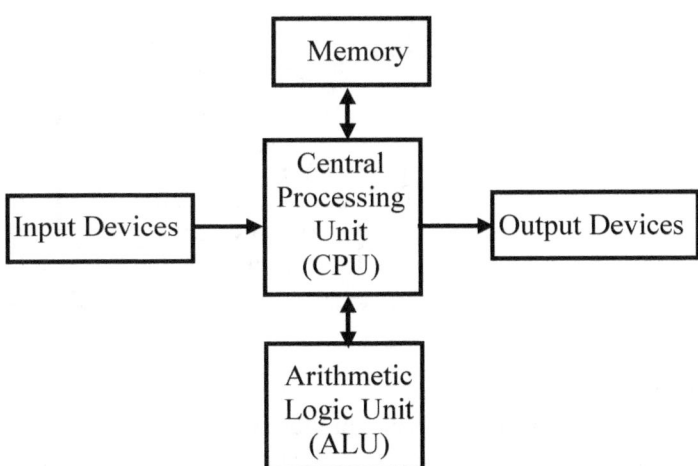

2. A process flow diagram
We usually use process flow diagrams in operations as a graphical representation of a process.

Example

Draw a process flow diagram to show the production and distribution of drinks by a brewery industry.

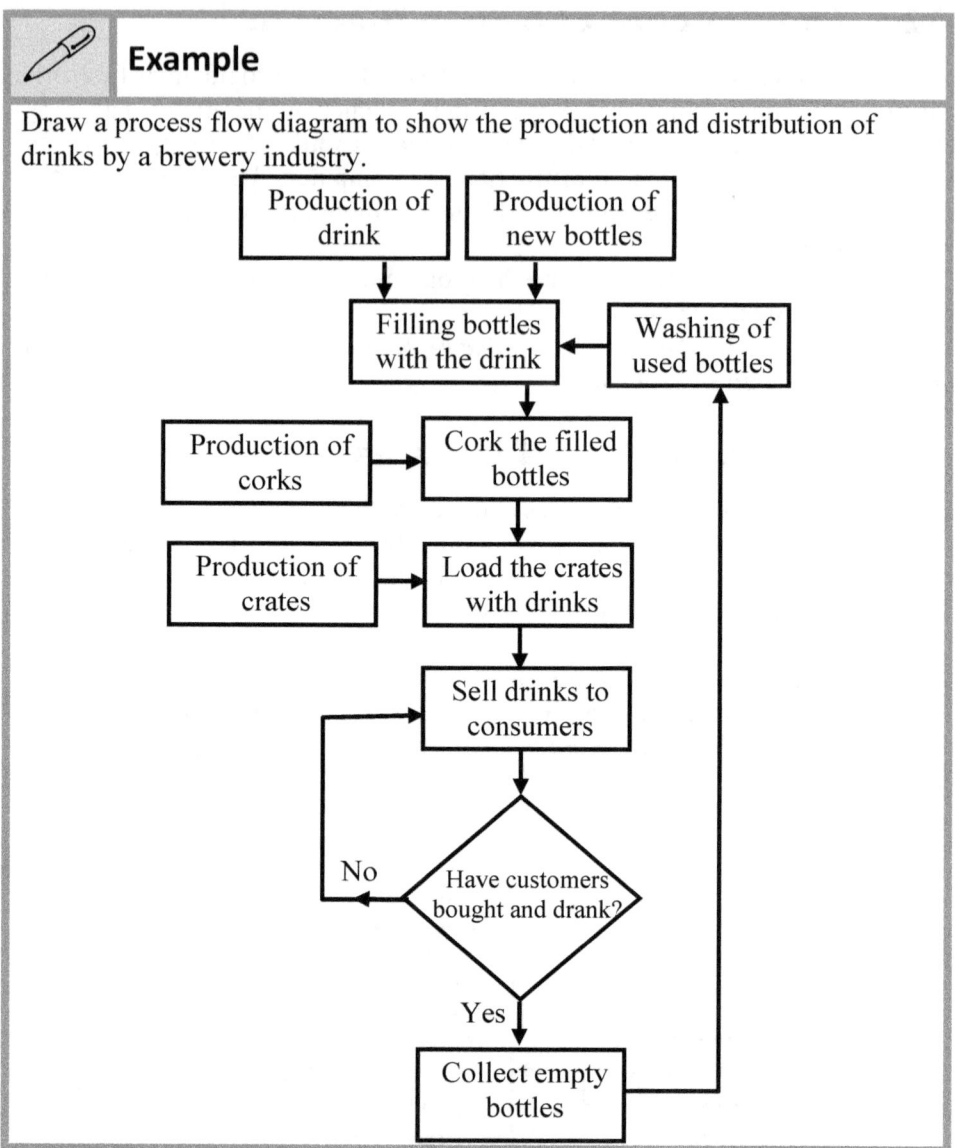

3. An alluvial diagram

An alluvial diagram is a graphical summary and highlight of the important structural changes in networks. Figure (a) and (b) are examples of alluvial diagrams.

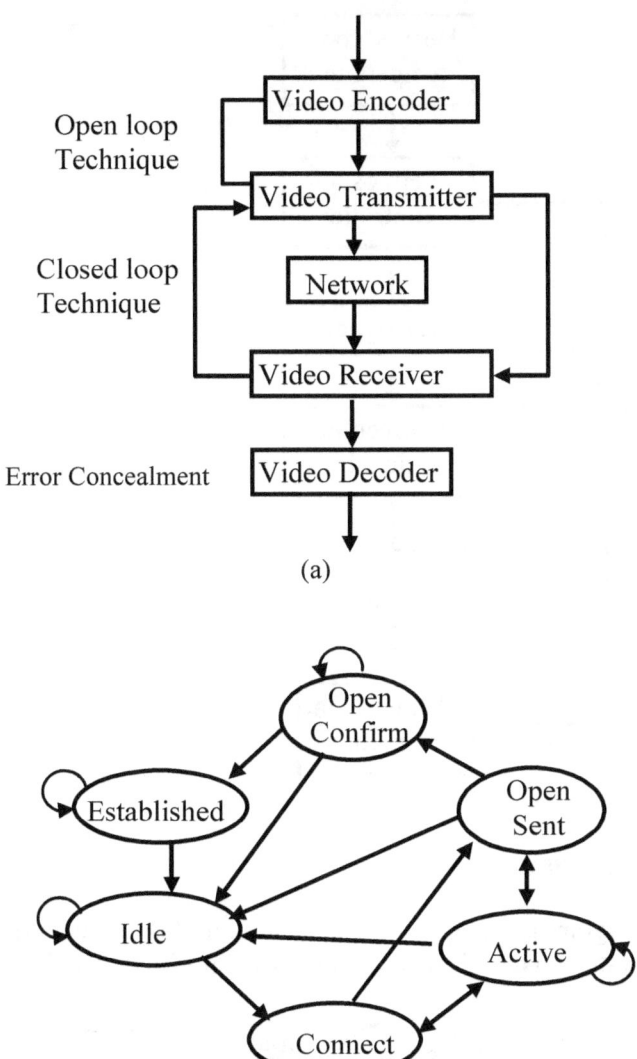

(a)

(b)

4. A control flow diagram
A control flow is a diagram describing the control flow of a business process, a program or any other process.
The following figure is a control flow diagram describing how an automated thermostat controls the temperature of a room.

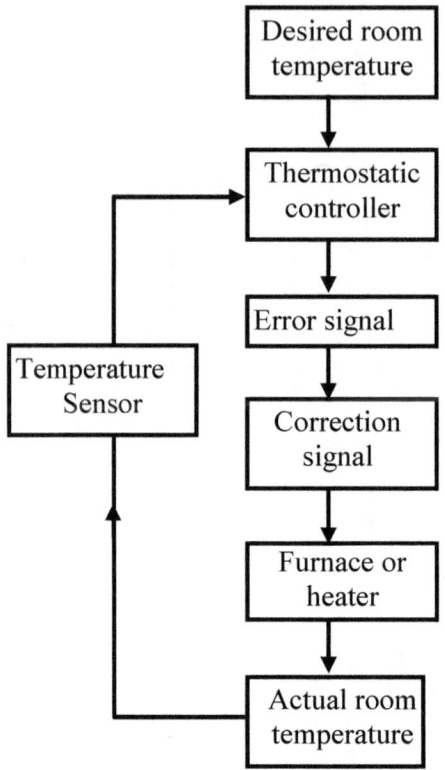

5. A data flow diagram

A data flow diagram is a graphical representation of the flow of data through an information system. The following figure is a simplified data flow diagram.

6. A flow map

A flow map is a mixture of maps and flow charts used in cartography, to show the movement of objects from one location to another. The following figure is a flow map of the voyage of Vasco da Gama from 1497 to 1498.

Module 16, Topic 12: Flow Diagrams

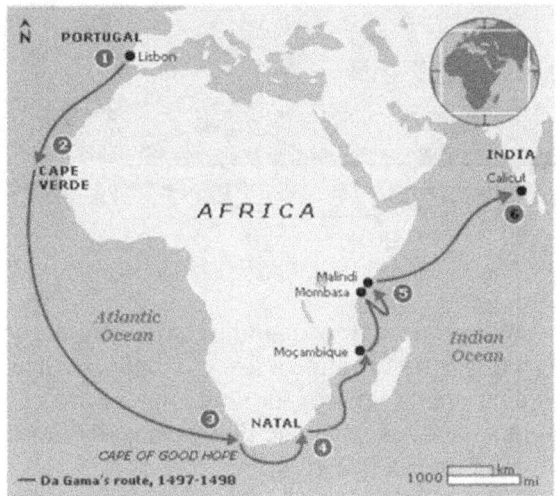

7. **A flowchart** is a graphical representation of a systematic solution to a given problem.

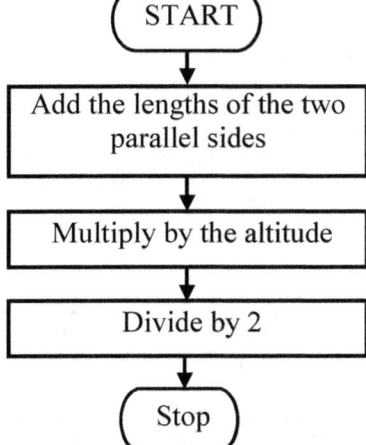

8. **A signal flow graph**
This is a graph used in mathematics to show the relations among the variables of a set of linear algebraic relations. A signal flow graph often look something like the figure below. Its treatment is beyond the scope of this book and for now, just be contented at being aware that such a graph exists.

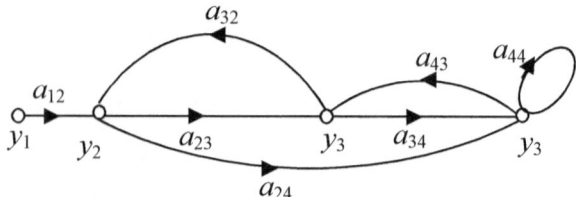

The full treatment of all the above flow diagrams is beyond the scope of this book. However, we shall briefly examine a few examples such as the flow chart and the process flow diagram. It suffices for now to be aware of the existence of the various flow diagrams and to appreciate that we usually draw and read flow diagrams from top to bottom or left to right.

12.3 Basic Flow Diagram Symbols

On a flow diagram, we use boxes connected by arrows pointing to the next box. Each box contains an instruction or question. There are many symbols, but in this topic, we will treat only the most common symbols shown in the table below.

	Symbol	Description and Function
1	START	We use this oval shape with the instruction "START" at the beginning of a flow diagram to indicate the starting point of the program.
2	STOP	We use this oval shape with the instruction "STOP" at the end of a flow diagram to indicate the ending point of the program.
3	▭	This rectangular shape called the **instruction block** is used to issue intermediate instructions
4	◇	We call this diamond shape used to ask precise questions with a "Yes" or "No" answer a **decision block**. A decision block usually has two exits, one for "Yes" and the other for "No".
5	↓	We use arrows connecting the instruction and decision boxes to show the direction of flow of the instructions.

12.4 Instruction Boxes

The following example illustrates the use of instruction boxes.

 Example

1. Draw a flow chart, which we can use to find the area of a trapezium.

   Solution

   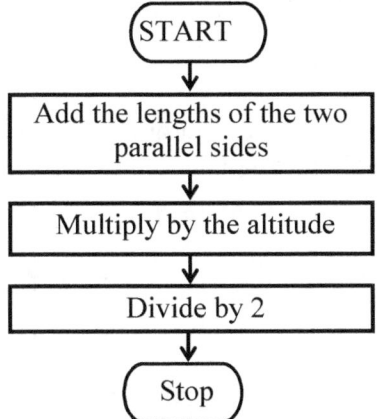

   We can easily carry out calculations using flow charts.

2. Use a flow Chart to evaluate $9 - 6 + 4 \times 6 \div 3$.

   Solution

   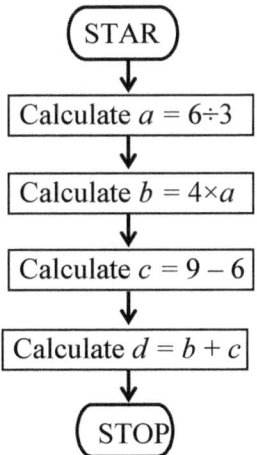

Exercise 12:1

1. Draw a flow chart for each of the following calculation.
   (a) $(4.7 \times 3 - 5.3) \div 2$ (b) $4.7 \times 3 - 5.3 \div 2$ (c) $4.7 \times (3 - 5.3 \div 2)$
   (d) $\sqrt{5.6 + 18.2 \div 7}$ (e) $\sqrt{5.6 + 18.2} \div 7$

2. Draw a flow chart to describe how to construct using ruler, pencil and compasses only:
   (a) the angle bisector between two rays OX and OY.
   (b) a line segment PQ, 6 cm long.
   (c) an equilateral triangle ABC of side 5 cm.
   (d) an angle BAC of $60°$.
   (e) a line parallel to a given line and passing through a given point P.

3. Write down the expression which the following flow charts are evaluating.

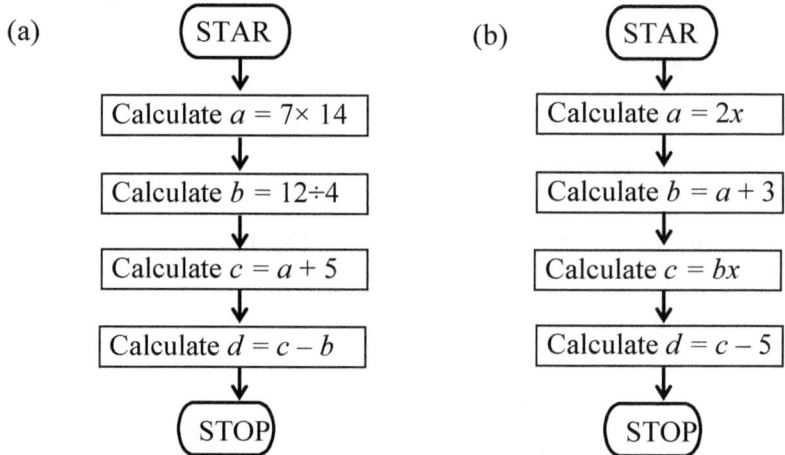

(c)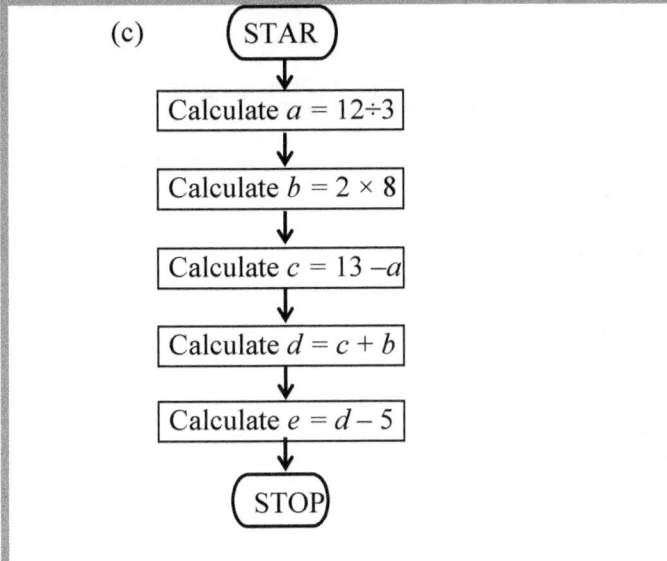

4. Draw a flow chart, which we can use to:
   (a) Find the L.C.M. of 12 and 42.   (b) Find the HCF of 12 and 42
   (c) Add three equal vectors that have the same direction and sense.

12.5 Decision Boxes

The following examples illustrate the use of decision boxes. There are usually two outlets from a decision box, one for yes and the other for no. Sometimes, an outlet from a decision box leads to an earlier stage of the flow diagram as shown in example 2 below. Sometimes, an outlet meets the other at a later stage of the flow diagram as shown in example 3 below. Whichever way, we call such an outlet a loop.

Example

1. Draw a flow diagram to show how you would close from school.

    Solution

    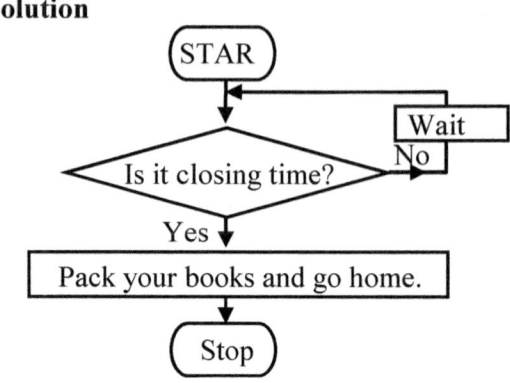

2. Draw a flow diagram, which you would use to draw a diagram with a pencil.

    Solution

    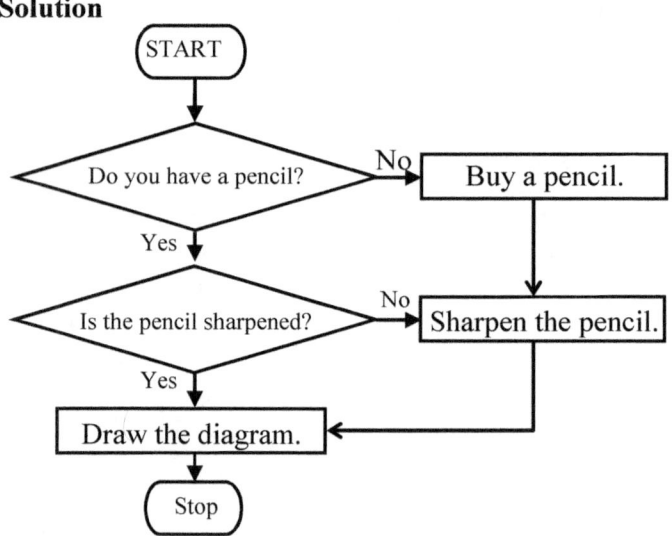

3. Five times a certain positive integer plus seven equals twenty-two. Find the number.

    Solution
    We may use an inverse operation method to solve the problem as follows.

Module 16, Topic 12: Flow Diagrams

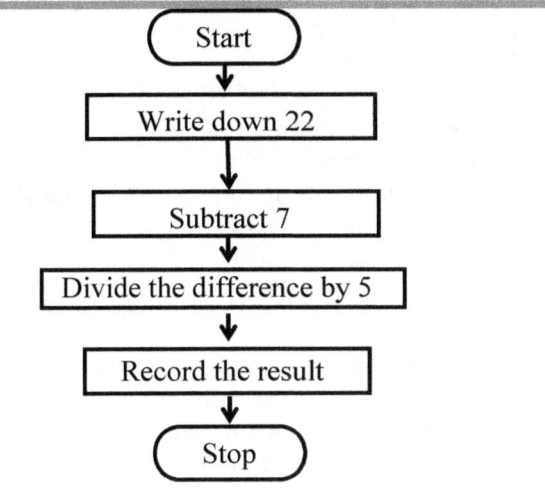

Alternatively, by a trial and error method, we can try the positive integers 1, 2, 3, ⋯ in turns.

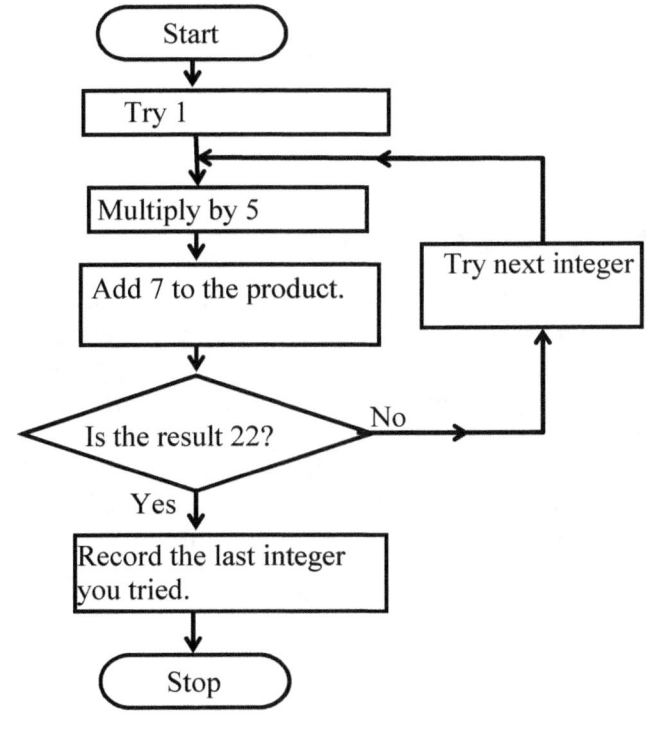

Exercise 12:2

Draw a flow diagram, which we can use to:
1. Find the arithmetic mean of the three numbers.
2. Find the number of elements in the union of two sets, A and B.
3. Factorise the algebraic expression $x^2 - 2x - 35$.
4. Change the subject in the formula $A = \pi r^2$.
5. Find the resultant of two vectors that are perpendicular to each other in direction and sense.
6. Find a number given that twice the number plus 3 is equal to 15.
7. Find a number given that, seven times the number, minus 9, is equal to 40.
8. Find a number given that the sum of the number, 2 and 5 is equal to 10.
9. Find a number given that, two and a half times the number plus 2.5, is equal to 7.5.
10. Find a number given that three times the number minus five is equal to that number plus five.
11. Find a number given that the number plus 2 is equal to twice the number minus 6.
12. Find the square root of 81 by listing factors.

12.6 Data and Stores

The arithmetic logic unit (ALU) of the computer holds numbers in stores. Each store has an address usually denoted each store by a single capital or small letter such as a, b, n, x, A, B etc. As in algebra, we take advantage of the simplicity and clarity of symbolic language in mathematical and scientific works. The number in a particular store often changes as the calculation progresses. For this reason, we use the symbolic language $n := 1$ in flow charts to represent the number in store n. We interpret $n := 1$ as "n takes the value 1". If we have to add 1 to the number in store n, we write as $n := n + 1$, which means, "n takes the new value $n + 1$".

Module 16, Topic 12: Flow Diagrams

Example

Five times a certain positive integer plus seven equals twenty-two. Use symbolic language to find the number.

Solution

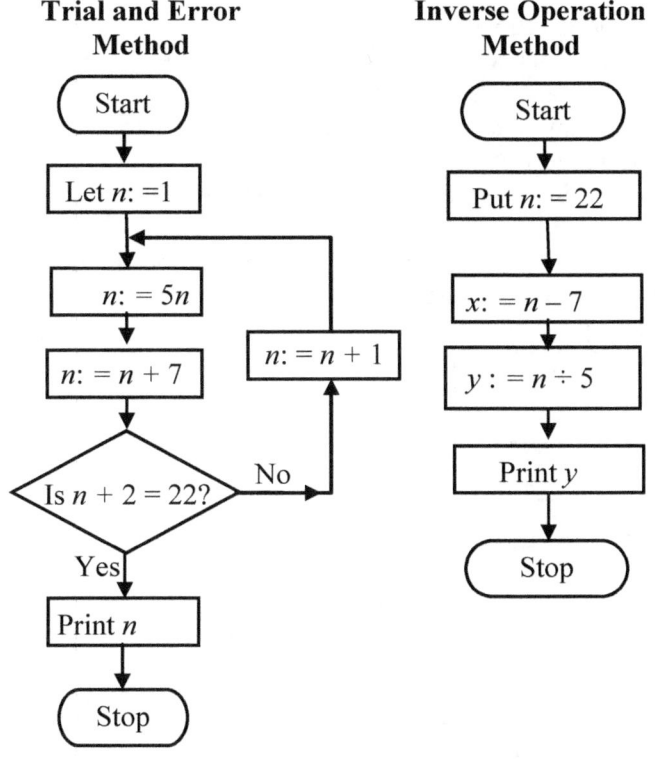

Exercise 12:3

1. Draw a flow diagram, which we can use to convert 874_{ten} to a number in base six.
2. A flow chart showing how to write a given number n ($n < 25$) as a product of prime factors contains:
    (i) The following instructions:
        Divide n by 2 and write 2 in the product.
        Divide the quotient by 2, and write 2 in the product.
        Divide the quotient by 3, and write 3 in the product.

Write the last quotient in the product.
(ii) The following questions:
Is the quotient even?
Is the quotient a multiple of 3?
(iii) Two loops
3. Draw the flow chart using all these facts and no more.
4. Draw a flow chart to describe the steps in the calculation of
$\sqrt{b^2 - 4ac}$ for, given values of a, b, c.
5. The Fibonacci sequence consists of terms, each one of which is the sum of the two previous ones.
If $u_1 = u_2 = 1$, draw a flow chart for calculating the later terms. Include the instruction record u_n and stop at u_{10}.
6. Repeat No. 4 taking $u_1 = 1$, $u_2 = 3$. Use your flow chart to calculate the first eight terms of this series.
In Problem 6(a) to (c), follow through the instructions in the flow chart in Figure (a) and record the results as requested.
7. (a) What pattern of numbers does each of figures (a) and (b) generate?

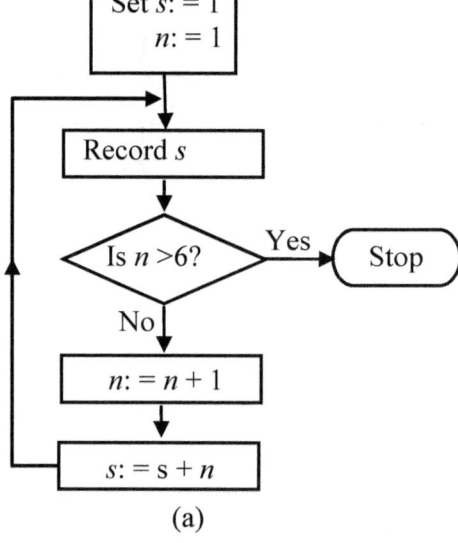

(a)

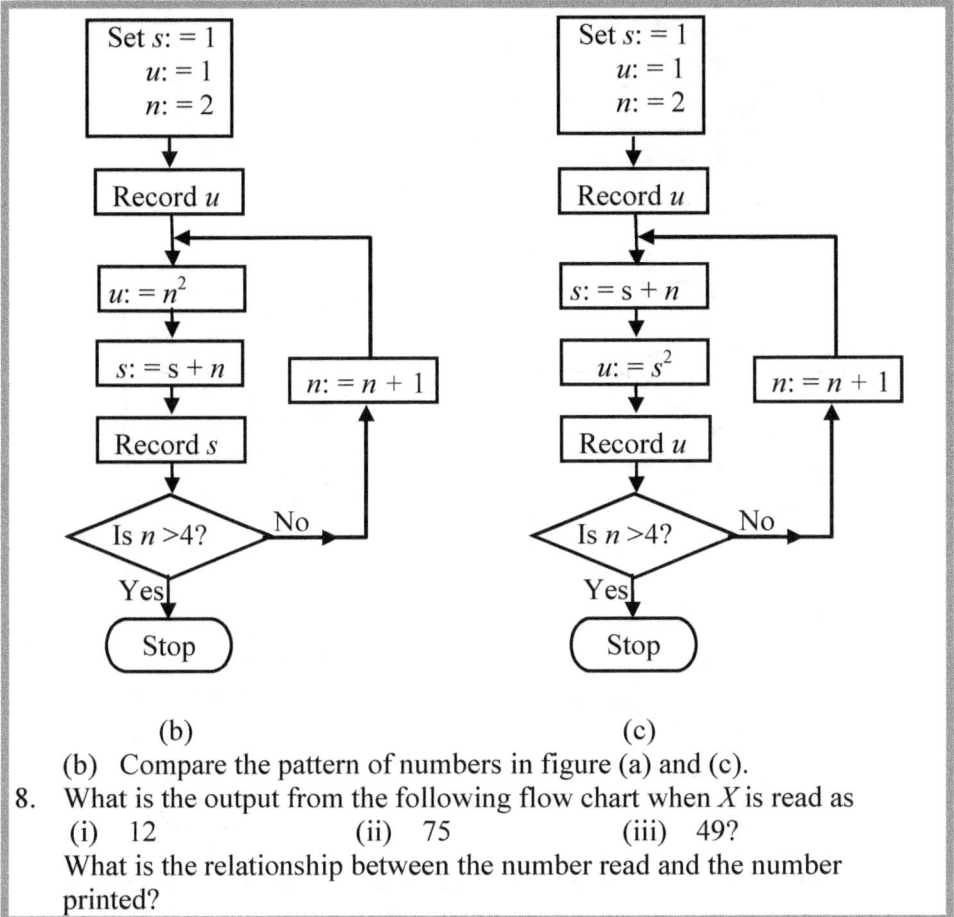

(b)   (c)
(b) Compare the pattern of numbers in figure (a) and (c).
8. What is the output from the following flow chart when X is read as
(i)   12   (ii)   75   (iii)   49?
What is the relationship between the number read and the number printed?

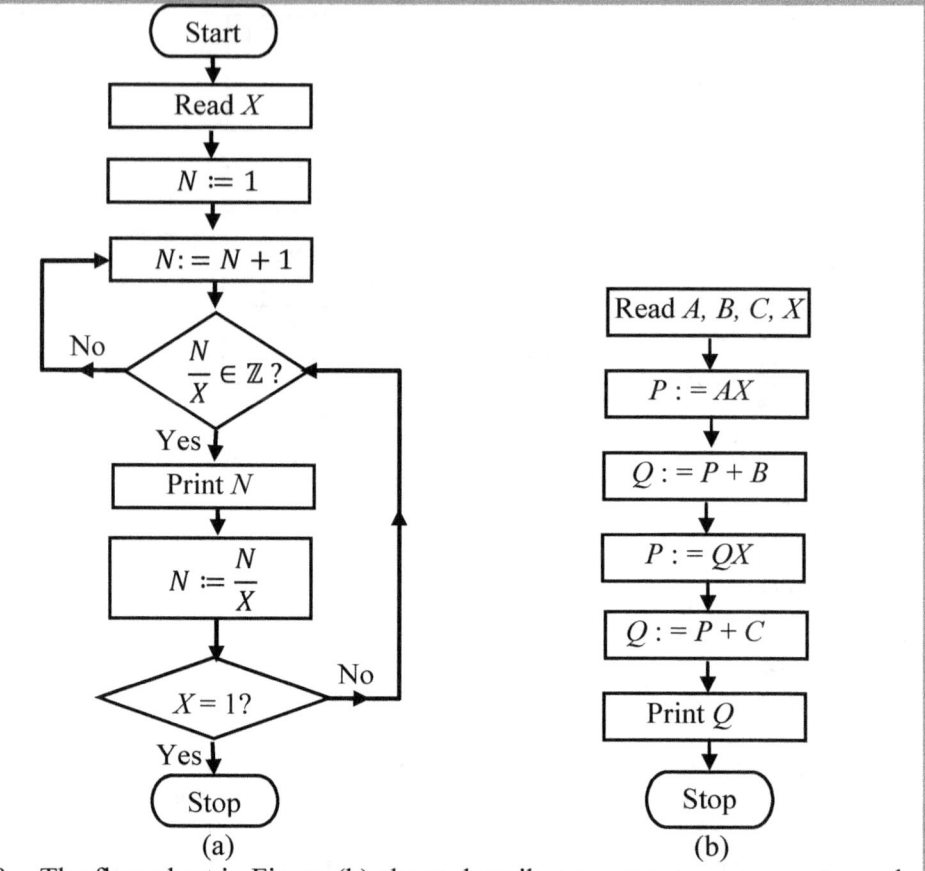

(a)   (b)

9. The flow chart in Figure (b) above describes a computer program to read four numbers A, B, C and X, and to print the result of an arithmetical calculation with them.
   (i) Find the number which will be printed when $A = 3$, $B = -4$, $C = -1$ and $X = 2$.
   (ii) Find the general formula that the program is designed to evaluate for any values of A, B, C and X.
   (iii) Modify the program so that the computer will printout a series of results for $X = 1, 2, 3, ..., 10$ (A, B and C remaining constant)

10. The program in the figure below is a design to print the highest common factor of two positive numbers x and y. Where $y > x$. Draw up a table of two columns showing the contents of stores A and B at each stage of the program when $x = 12$ and $y = 30$ and verify that the program does print the highest common factor.
    Show that the program fails when $x = 8$ and $y = 24$. Explain why it fails and show clearly how you would amend the program so that it will work correctly.

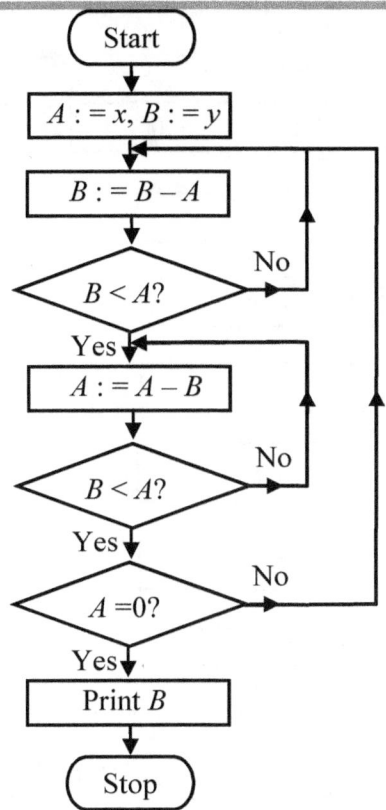

11. Work through the flow chart below, showing the content of every store after each instruction, and displaying your answer in tabular form, beginning with:

A	B	C
1	0	0
...	...	...

Underline each number that the computer will print. Do not carry out calculations to more than two decimal places.

If the original data store *A* had contained the number *X* instead of the number 1, find a formula involving *x* for the first number to be printed. The rows containing the printed values are given as guide:

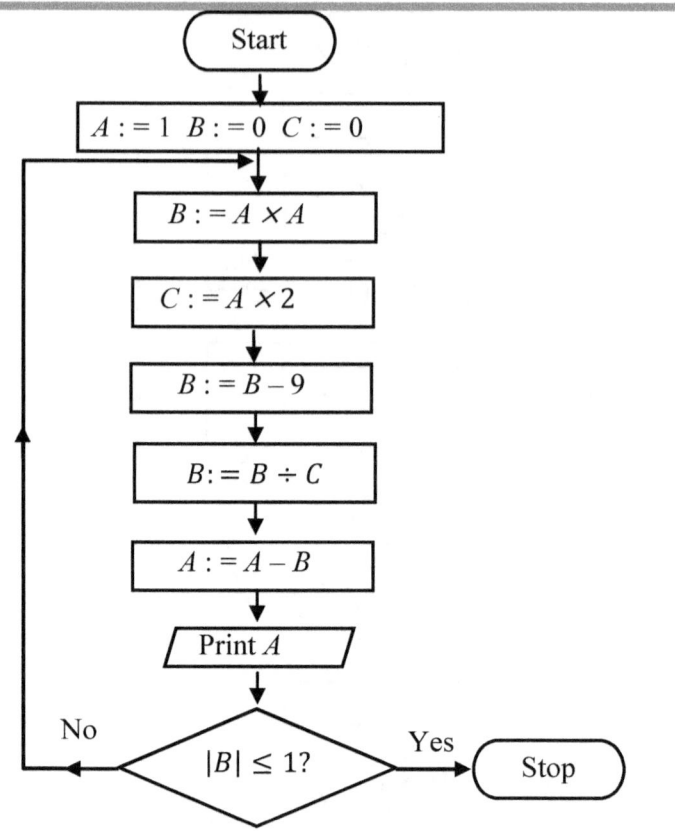

12. The following figure shows the flow chart for a simple computing device containing three stores A, B and C, each capable of storing one number only at a time.

 The instruction '$A := x$' is read 'A becomes x' and denotes that store A is to hold the number x in place of whatever number it held previously. The following flow diagram was intended to sum the series $1 + 3 + 5 + \ldots$ to 20 terms. It contains a mistake.

 (a) Say what it will actually print.
 (b) Which is the wrong instruction? Write a correct version

Module 16, Topic 12: Flow Diagrams

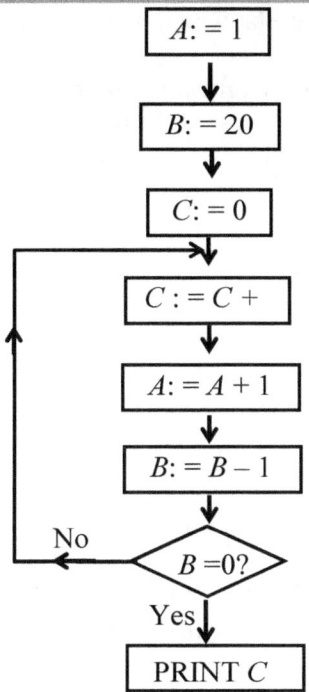

13. Draw up a flow chart, which we can use to add the terms of the series $12 + 15 + 18 + 21 + \ldots + 69$. Use three stores a, b and n and start with $a = b = 12, n = 1$.
14. Draw up a flow chart to show how we may sort a shuffled pack of cards into four suits.
15. We can convert temperature from the Celsius scale to the Fahrenheit and vice versa using the formula $F = \frac{9}{5}C + 32$ and an equivalent one with C as subject. Draw flow charts showing how you will obtain one formulae from one another and vice versa.
16. Make (a) r (b) v the subject of the formula $P = \frac{mv^2}{r}$. In each case, show the sequence of operations performed in flow chart form, and the sequence of inverse operations in a second flow chart.
17. Given that $r = \sqrt{(a^2 + b^2)}$, express b in terms of r and a.
    Assume that r, a and b are positive quantities. Show the sequence of operations performed in flow chart form. Show also in another flow chart, the reverse sequence of inverse operations.
18. If $x = \frac{\sqrt{b^2 - 4ac} - b}{2a}$, prove that $ax^2 + bx + c = 0$. Show in flow chart form the sequence of operations performed on this formula for x to obtain the final quadratic equation. Show also in another flow chart, the reverse sequence of inverse operations that leads from the general quadratic

equation $ax^2 + bx + c = 0$ to the formula, which gives the roots. Note that there should be two roots.

19. Before crossing the road, the Cameroon Highway Code recommends 'look left, look right, look left again and cross if the road is clear'. Complete the following flow diagram for this operation:

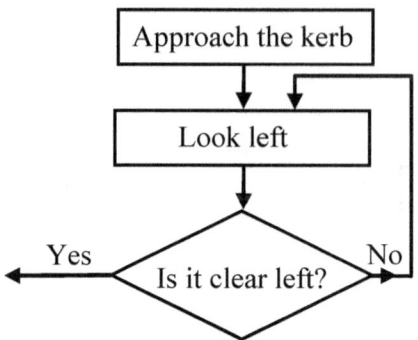

If the traffic flow in any road is such that at any instant the probability of there being traffic in one direction is $\frac{1}{2}$ and in the other is $\frac{2}{3}$, what is the probability of the road being completely clear?

Multiple Choice Exercise 12

1. On a flow chart, the shape used for a decision block is:

   [A]   [B]   [C]   [D]

2. The shape ▭ is used on a flow chart as:
   [A] a decision block [B] an instruction block
   [C] a question block [D] a description block

3. The type of flow diagram which highlights and summarizes the significant structural changes in networks is:
   [A] a data flow diagram [B] a process flow diagram
   [C] an alluvial diagram [D] a signal flow diagram

4. Given the line segment $XY = 8$ cm. The construction described by the flow chart below is:
   [A] the inscribed circles C_1 and C_2 [B] the locus of C_1 and C_2
   [C] the intersection of C_1 and C_2 [D] the mediator of XY

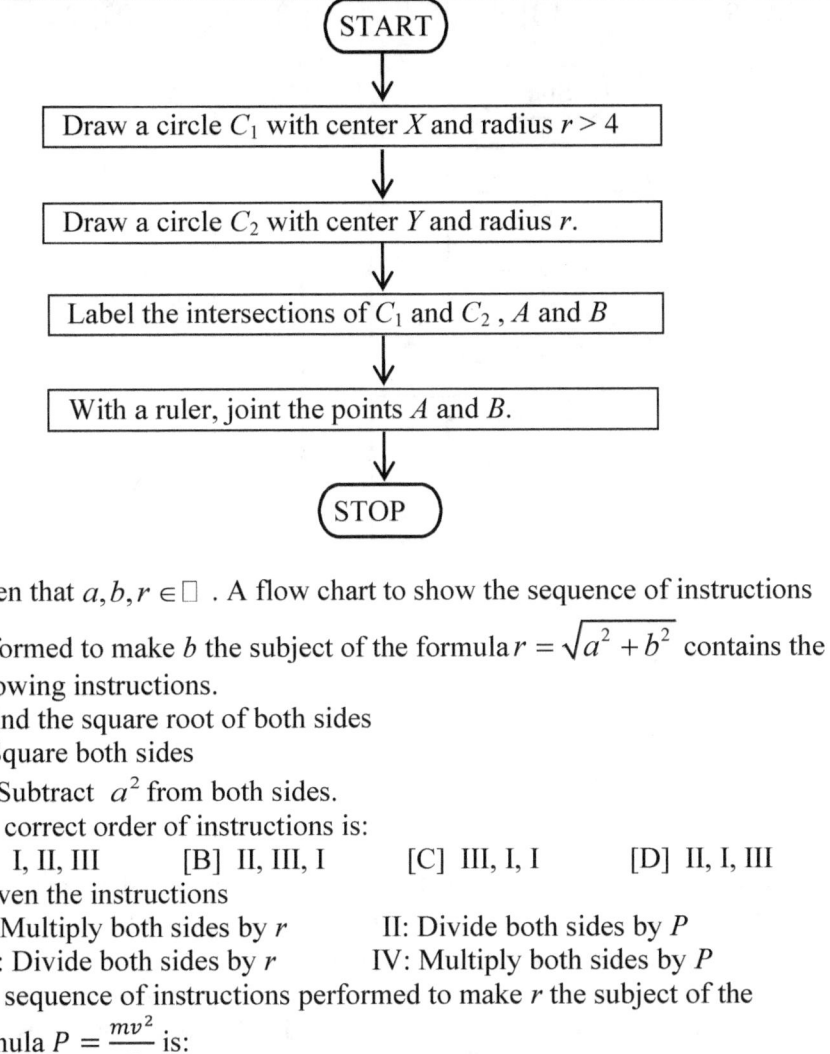

5. Given that $a, b, r \in \square$. A flow chart to show the sequence of instructions performed to make b the subject of the formula $r = \sqrt{a^2 + b^2}$ contains the following instructions.
   I: Find the square root of both sides
   II: Square both sides
   III: Subtract a^2 from both sides.
   The correct order of instructions is:
   [A] I, II, III [B] II, III, I [C] III, I, I [D] II, I, III

6. Given the instructions
   I: Multiply both sides by r II: Divide both sides by P
   III: Divide both sides by r IV: Multiply both sides by P
   The sequence of instructions performed to make r the subject of the formula $P = \frac{mv^2}{r}$ is:
   [A] I, II [B] I, III [C] I, IV [D] IV, III

7. In a flow chart to show the sequence of instructions performed to make C the subject of the formula $F = \frac{9}{5}C + 32$, the first two instructions are; multiply both sides by 5, divide both sides by 9 respectively. The third and final instruction should be:
   [A] Multiply both sides by $\frac{160}{9}$. [B] Divide both sides by $\frac{160}{9}$.
   [C] Add $\frac{160}{9}$ to both sides [D] Subtract $\frac{160}{9}$ from both sides

8. In the flow diagram in Figure 47:30, the number of loops is:
   [A] 9 [B] 4 [C] 2 [D] 3

9. In the flow diagram below, the number of stores is:

[A] 9   [B] 4   [C] 2   [D] 3

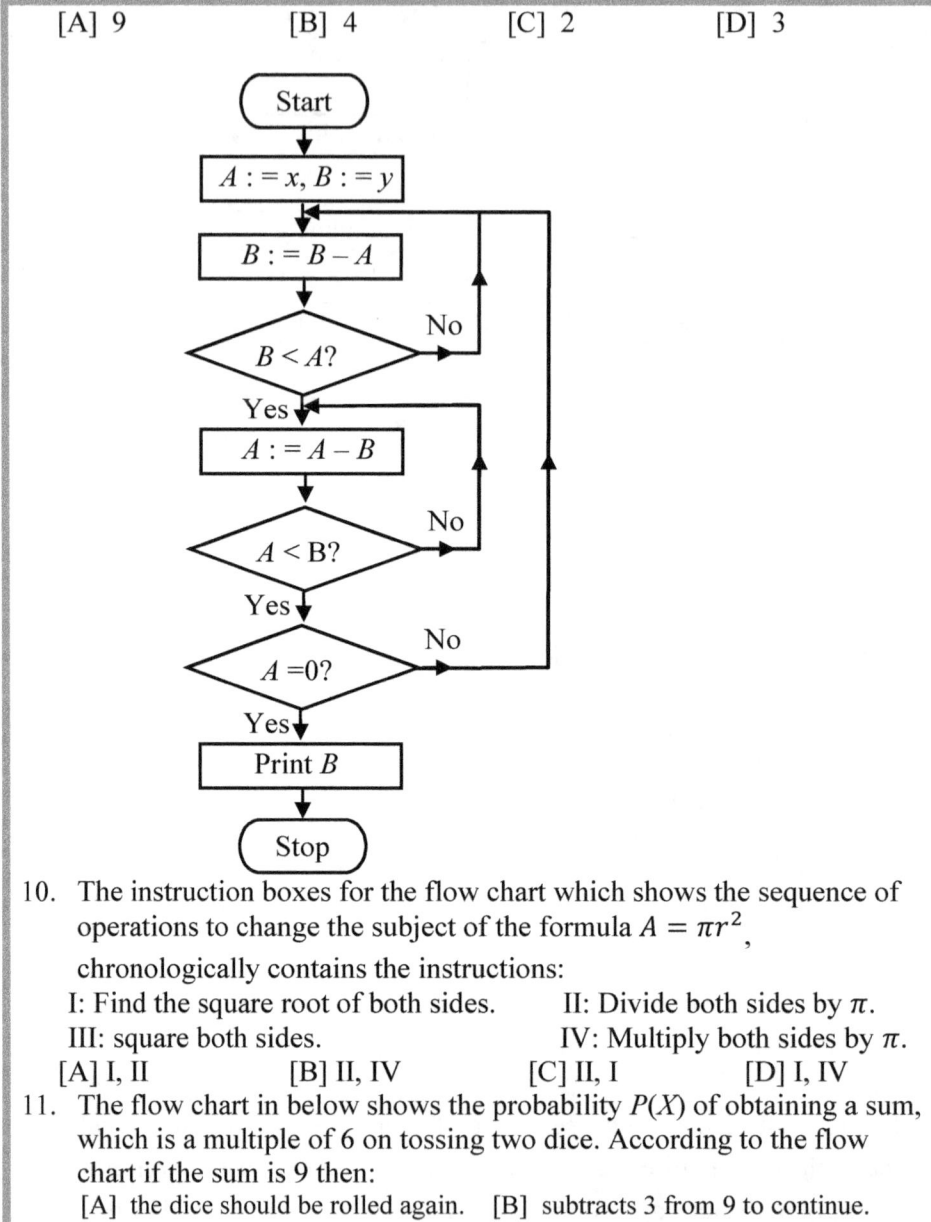

10. The instruction boxes for the flow chart which shows the sequence of operations to change the subject of the formula $A = \pi r^2$, chronologically contains the instructions:
I: Find the square root of both sides.   II: Divide both sides by π.
III: square both sides.   IV: Multiply both sides by π.
[A] I, II   [B] II, IV   [C] II, I   [D] I, IV

11. The flow chart in below shows the probability $P(X)$ of obtaining a sum, which is a multiple of 6 on tossing two dice. According to the flow chart if the sum is 9 then:
[A] the dice should be rolled again.   [B] subtracts 3 from 9 to continue.
[C] adds 3 to 9 to continue.   [D] the probability is $\frac{3}{4}$.

Module 16, Topic 12: Flow Diagrams

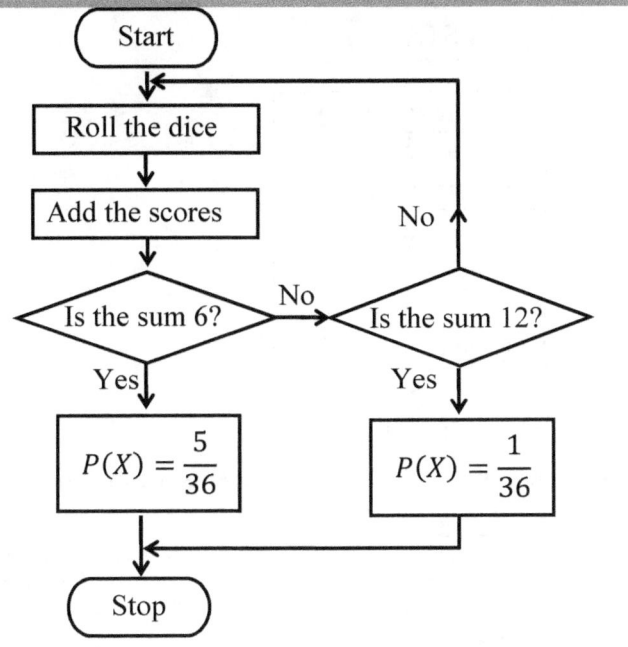

12. The expression which is being evaluated by the flow chart below is:
    [A] $3x^2 + 2x - 5$ [B] $2x^2 + 3x - 5$
    [C] $3a^2 + 2a - 5$ [D] $2a^2 + 3a - 5$

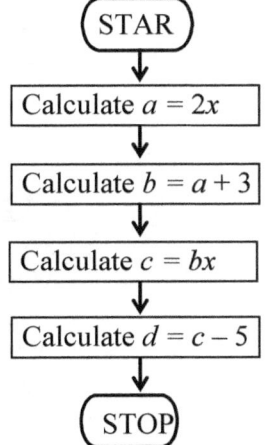

13. The expression which is being evaluated by the flow chart below is:
    [A] $13 - 12 \div 3 + 2 \times 8$ [B] $(13 - 12) \div 3 + 2 \times 8$
    [C] $13 - 12 \div (3 + 2) \times 8$ [D] $(13 - 12) \div (3 + 2) \times 8$

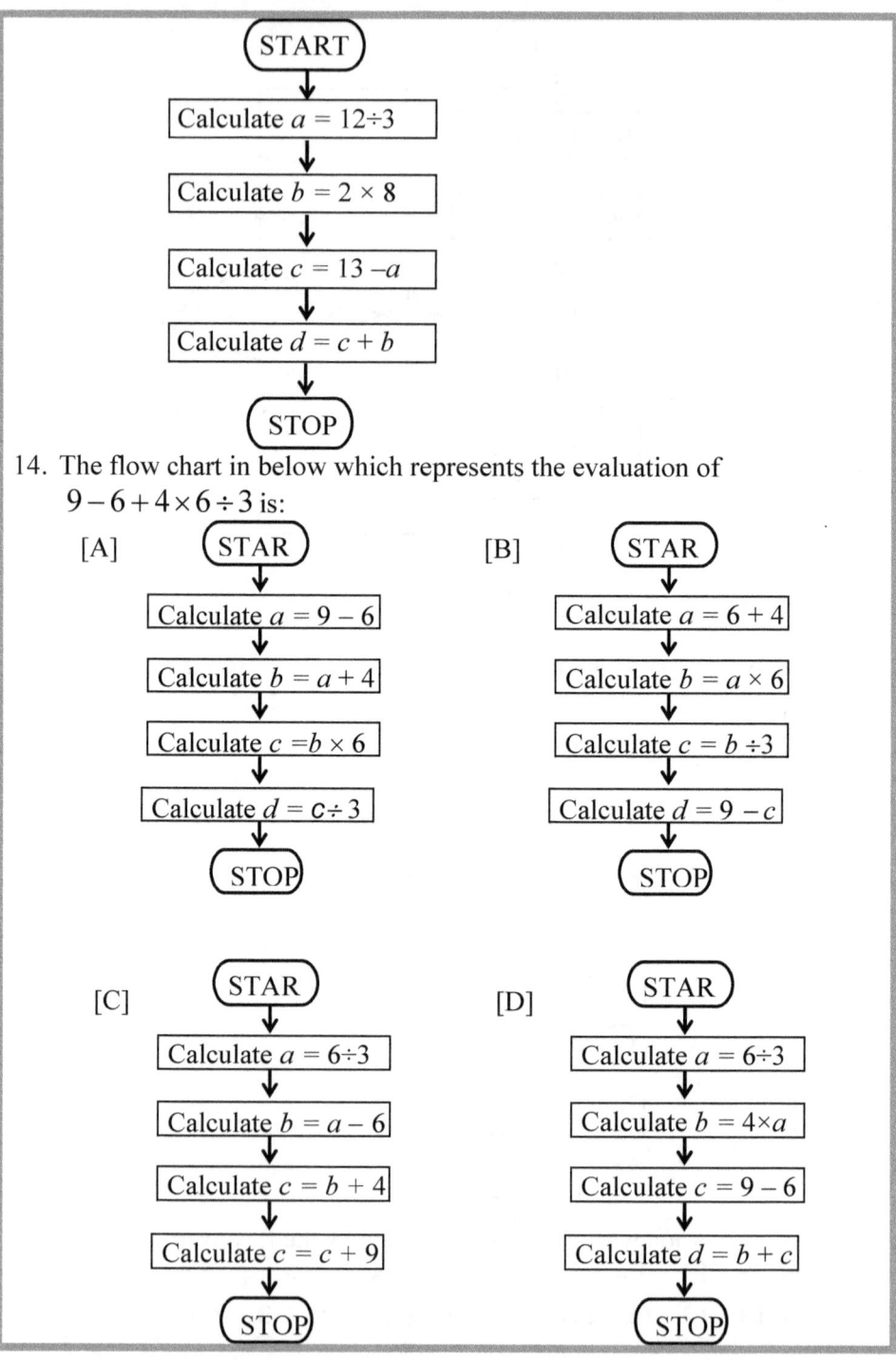

14. The flow chart in below which represents the evaluation of $9 - 6 + 4 \times 6 \div 3$ is:

Module 17

Algebra and Logic

Family of Situations
At the end of module 17; the student is expected to have acquired many more competencies within the **families of situations** *'Describing patterns and relationships between quantities using symbols'*.

Categories of Action
The categories of action for module 17 include:
1. Interpretation of algebraic models,
2. Determination of quantities from algebraic models,
3. Representation of quantities and relationships.

Credit
The module is expected to be covered within 9 weeks, teaching 4 periods of 50 minutes per week (or within 36 periods).

Topic 13

ALGEBRAIC PROCESSES

Objectives

At the end of this topic, the learner should be able to:

1. Expand and simplify expressions of the form $(a + b)(c + d), (a + b)(a - b), (a \pm b)^2$.
2. Factorize binomials and trinomials.
3. Factorize quadratic expressions.
4. Solve quadratic equations by factorization.
5. Solve quadratic equations by completing the square.
6. Solve quadratic equations by formula.
7. Develop quadratic equations from the roots.

13.1 Review and Revision

We did much work on algebra in topic 10 of book 3. However, owing to the importance of algebra the following reminders are very imperative.

1. $(a + b)(c + d) = ac + bc + ad + bd$

2. $(a+b)^2 = a^2 + 2ab + b^2$

3. $(a - b)^2 = a^2 - 2ab + b^2$

4. $(a+b)(a-b) = a^2 - b^2$.

5. To factorize a quadratic expression, take the following steps.
   (i) Multiply a by c to have ac
   (ii) Find the pair of integral factors p and q of ac whose sum or difference is b.
   (iii) Substitute the middle term bx with the sum or difference of px and qx in $ax^2 + bx + c$.
   (iv) Factorize the expression by grouping.

 Review and Revision Exercise

Expand the following.
1. $(p+3)(p-1)$ 2. $(x+3)^2$ 3. $(x-2)^2$
4. $(4+a)^2$ 5. $(3-y)^2$ 6. $(x-5)(x+5)$

Evaluate the following without using a calculator.
7. 101^2 8. 198^2 9. 442^2

Factorize the following.
10. $2q^3 - 14q^2 + 3q - 21$ 11. $p^2 - 5 - p^2q + 5q$
12. $4y^2 - 9x^2$ 13. $4u^2 - 25y^2$ 14. $36a^2 - 49b^2$

Evaluate the following without using a calculator.
15. $41^2 - 40^2$ 16. $124^2 - 120^2$ 17. $625 - 525^2$

Factorise the following.
18. $x^2 + x - 6$ 19. $10x^2 - x - 3$ 20. $5 - 16y + 12y^2$

Solve the following quadratic equations using the factorization method.
21. $x^2 - 2x - 8 = 0$ 22. $12x^2 + 18x - 15 = 0$ 23. $2x^2 - 9x - 5 = 0$

13.2 Quadratic Identities

 Investigative Activity

Consider the following equations.

(i) $4x^2 - 1 = (2x + 1)(2x - 1)$
(ii) $(x + 2)^2 = x^2 + 4x + 4$
(iii) $(x + 3)(2x - 1) = 2x^2 + 5x - 3$

1. Substitute any value of x on both sides of each of the above equations.
2. What can you say about the result of the right hand side and the left hand side on substituting any value of x?

Clearly, whatever value is substituted for x, the right hand side and the left hand side are equal. In other words the equations are true for all values of the variable x. An **identity** is a statement, which is true for all values of the variable. If the highest power of the variable in an identity is 2, the identity is called a **quadratic identity**. In an identity, the right hand side substitutes the left hand side and vice versa.

When it is clear that a statement is an identity, the symbol '≡', read 'identical to' is used instead of '='.

13.3 The Trinomial Perfect Square Test

From the expansions of the perfect square $(a \pm b)^2 = a^2 \pm 2ab + b^2$ above, we see that $(x \pm k)^2 = x^2 \pm 2kx + k^2$.

Comparing the right hand side of this equation with $x^2 + bx + c$, we see that,

$2k = b \Longrightarrow k = \frac{b}{2}$...................③   and $c = k^2$ ....④

Substituting ③ in ④, we have $c = \left(\frac{b}{2}\right)^2$.

Therefore, the necessary and sufficient condition for the quadratic expression $x^2 + bx + c$ to be a perfect square is that $c = \left(\frac{b}{2}\right)^2$. Therefore, to make

$x^2 + bx$ a perfect square, we add $\left(\frac{b}{2}\right)^2$ to $x^2 + bx$. Note that $\left(\frac{b}{2}\right)^2$ is the same as (half the coefficient of x)^2.

Completing the square is the process of adding the square of half the coefficient of x to a quadratic expression to make all or part of the resulting quadratic expression a perfect square.

 Example

Complete the square of $x(x-5)$.

Solution

$x(x-5) = x^2 - 5x$

Half the coefficient of x is $-\frac{5}{2}$.

Therefore the required perfect square is $x^2 - 5x + \left(-\frac{5}{2}\right)^2$ or $x^2 - 5x + \frac{25}{4}$

 Exercise 13:1

1. Complete the square in each of the following cases.
   - (a) $p^2 - 20p$
   - (b) $y^2 - 14y$
   - (c) $x^2 + 5x$
   - (d) $u^2 + \frac{4}{3}u$
   - (e) $x(x+18)$
   - (f) $x^2 - 4x$

2. Write the following perfect squares as a square of a binomial.
   - (a) $t^2 + 12t + 36$
   - (b) $x^2 - 18x + 81$
   - (c) $y^2 + 7y + \frac{49}{4}$
   - (d) $u^2 - 11u + \frac{121}{4}$
   - (e) $x^2 + 20x + 100$
   - (f) $m^2 - 30m + 225$

13.4 Quadratic Expressions in Two Variables

For a quadratic expression of the form $ax^2 + bxy + cy^2$, follow steps (i) and (ii) as in the preceding section. However, in (iii) substitute bxy with the sum or difference of pxy and qxy and then proceed to (iv).

Example

Factorize $3x^2 + xy - 10y^2$
Solution
$ac = -30$ and factors of -30 whose sum is $+1$ are -5 and $+6$.
$$3x^2 + xy - 10y^2 = 3x^2 - 5xy + 6xy - 10y^2$$
$$= x(3x - 5y) + 2y(3x - 5y)$$
$$= (x + 2y)(3x - 5y)$$

 Exercise 13:2

Factorize the following.
1. $x^2 + x - 6$
2. $p^2 + 12p + 11$
3. $y^2 - 7y + 12$
4. $x^2 + 6x - 16$
5. $x^2 - 2x - 15$
6. $10x^2 - x - 3$
7. $4a^2 - 3a - 10$
8. $5 - 16y + 12y^2$
9. $3 - x - 2x^2$
10. $x^2 + xy - 12y^2$
11. $x^2 + xy - 12y^2$
12. $2x^2 + 7xy - 15y^2$
13. $6k^2 + 17kx - 3x^2$
14. $12p^2 - 16pq + 5q^2$

13.5 Method of Completing the Square

Sometimes the quadratic expression is not factorable. In such a case, the method of completing the square discussed in Topic 10 is used.

 Example

1. Solve the equation $2x^2 - 5x - 5 = 0$, using the method of completing the square.

   Solution
   $$2x^2 - 5x - 5 = 0$$
   Add 5 to both sides.
   $$2x^2 - 5x = 5$$
   Divide each term by the coefficient of x^2.
   $$x^2 - \frac{5}{2}x = \frac{5}{2}$$
   Add the square of half the coefficient of x to both sides

Module 17, Topic 13: Algebraic Processes

$$x^2 + \left(-\frac{5}{2}\right)x + \left(-\frac{5}{4}\right)^2 = \frac{5}{2} + \left(-\frac{5}{4}\right)^2$$

The left hand side is now a perfect square and can now be factorized.

$$\therefore \left(x - \frac{5}{4}\right)^2 = \frac{5}{2} + \frac{25}{16}$$

Summing the right hand side and finding the square roots of both sides.

$$x - \frac{5}{4} = \pm \frac{\sqrt{65}}{4}$$

(Note that every positive number has two square roots, with the same absolute value but one positive and the other negative.)

Adding $\frac{5}{4}$ to both sides and rearranging gives:

$$x = \frac{5 \pm \sqrt{65}}{4}$$

$$\Rightarrow x = \frac{5 + \sqrt{65}}{4} \quad \text{or} \quad x = \frac{5 - \sqrt{65}}{4}$$

2. Solve the equation $x^2 + 2x - 7 = 0$, using the method of completing the square.

   Solution
   $$x^2 + 2x - 7 = 0$$
   $$x^2 + 2x = 7$$
   Completing the square
   $$x^2 + 2x + 1 = 8$$
   $$(x+1)^2 = 8$$
   $$x + 1 = \pm 2\sqrt{2}$$
   $$x = -1 \pm 2\sqrt{2}$$

3. Solve the equation $3x^2 + 2x - 7 = 0$, using the method of completing the square.

Solution
$$3x^2 + 2x - 7 = 0$$
Add 7 to both sides
$$3x^2 + 2x = 7$$
Divide both sides by 3
$$x^2 + \frac{2}{3}x = \frac{7}{3}$$
Add (half the coefficient of x)² to both sides.
$$x^2 + \frac{2}{3}x + \left(\frac{1}{3}\right)^2 = \frac{7}{3} + \left(\frac{1}{3}\right)^2$$
$$\left(x + \frac{1}{3}\right)^2 = \frac{22}{9}$$
$$x = \frac{1 \pm \sqrt{22}}{3} \Leftrightarrow x = \frac{1 + \sqrt{22}}{3} \text{ or } x = \frac{1 - \sqrt{22}}{3}$$

Exercise 13:3

Solve the following quadratic equations using the method of completing the square.
1. $x^2 + 12x - 45 = 0$ 2. $u^2 + 20u + 19 = 0$ 3. $p^2 + 7p - 8 = 0$
4. $y^2 - 18y + 65 = 0$ 5. $a^2 - 11a + 28 = 0$ 6. $x^2 - 30x - 99 = 0$
7. $x^2 + 14x - 32 = 0$ 8. $x^2 + 18x - 19 = 0$

13.6 The Quadratic Formula

When the general form of the quadratic equation is solved using the method of completing the square, this leads to the quadratic formula derived below.

13.7 Derivation of the quadratic formula

Consider the general form of the quadratic equation
$$ax^2 + bx + c = 0$$
Subtract c from both sides.
$$ax^2 + bx = -c$$
Divide both sides by a.

Module 17, Topic 13: Algebraic Processes

$$x^2 + \frac{b}{a}x = -\frac{c}{a}$$

To make the left hand side a perfect square add $\left(\dfrac{b}{2a}\right)^2$ to both sides

$$x^2 + \frac{b}{a}x + \left(\frac{b}{2a}\right)^2 = \left(\frac{b}{2a}\right)^2 - \frac{c}{a}$$

Factorizing the left hand side and rearranging the right hand side,

$$\left(x + \frac{b}{2a}\right)^2 = \frac{b^2 - 4ac}{4a^2}$$

Finding the square roots of both sides and rearranging,

$$x + \frac{b}{2a} = \pm \frac{\sqrt{b^2 - 4ac}}{2a}$$

Subtracting $\dfrac{b}{2a}$ from both sides and rearranging,

$$x = \frac{-b \pm \sqrt{b^2 - 4ac}}{2a}$$

i.e. $x = \dfrac{-b + \sqrt{b^2 - 4ac}}{2a}$ or $x = \dfrac{-b - \sqrt{b^2 - 4ac}}{2a}$

Therefore, if $ax^2 + bx + c = 0$, where a, b and c are constants and $a \neq 0$, then
$$x = \frac{-b \pm \sqrt{b^2 - 4ac}}{2a}.$$

To find the roots of any quadratic equation, use this formula, which we call the **quadratic formula**.

Take note that the division bar goes across!

i.e. $x \neq -b \pm \dfrac{\sqrt{b^2 - 4ac}}{2a}$

Remarks!

In the quadratic formula $x = \dfrac{-b \pm \sqrt{b^2 - 4ac}}{2a}$;

1. The sum of the roots is equal to $-\frac{b}{a}$. This is very useful in checking the solutions to quadratic equations.
2. If $b^2 - 4ac$ called the discriminant is equal to zero, the expression $ax^2 + bx + c$ is a perfect square.

 Example

Use the quadratic formula to solve the equation $2x^2 - 5x - 5 = 0$.

Solution

$$x = \frac{-b \pm \sqrt{b^2 - 4ac}}{2a}, \quad a = 2, b = -5, c = -5$$

$$x = \frac{-(-5) \pm \sqrt{(-5)^2 - 4(2)(-5)}}{2(2)} = \frac{5 \pm \sqrt{65}}{4}$$

$$x = \frac{5 - \sqrt{65}}{4} \quad \text{or} \quad x = \frac{5 + \sqrt{65}}{4}$$

 Exercise 13:4

Solve the following quadratic equations using the formula method, leaving your answer in surd form.

1. $2x^2 - 5x - 4 = 0$
2. $x^2 + 7x + 5 = 0$
3. $2x^2 + 5x + 1 = 0$
4. $x^2 + 6x - 10 = 0$
5. $3x^2 - 4x - 2 = 0$
6. $6x^2 - 10x + 3 = 0$
7. $5x^2 - 10x + 4 = 0$
8. $9x^2 + x - 2 = 0$

13.8 Non-Standard Quadratic Equations

Sometimes a quadratic equation is not in the standard form. In this case, first rearrange it to be in standard form before solving.

Example

Solve the equation $x - 6 = \sqrt{x}$.

Solution
Squaring both sides
$$x^2 - 12x + 36 = x$$
$$\Rightarrow x^2 - 13x + 36 = 0$$
$$(x-9)(x-4) = 0$$
$$\Rightarrow x = 9 \text{ or } x = 4$$

Testing in the equation, we see that $x = 9$ and $x \neq 4$. If the equation were $x - 6 = \pm\sqrt{x}$, $x = 4$ could have been a solution.

Exercise 13:5

Rearrange the following and solve by factorization.

1. $u^2 + 6u = -8$
2. $a^2 = 5a + 24$
3. $x(3x+2) = 7$
4. $5 = m(2m+3)$
5. $p + 8 = \dfrac{20}{p}$
6. $\dfrac{7}{x} = 9 - 2x$
7. $2 + \dfrac{5}{x} = \dfrac{12}{x^2}$
8. $(a-8)(2a-3) = 34$

13.9 Incomplete Quadratic Equations

Incomplete quadratic equations are quadratic equations of the form $ax^2 + bx = 0$ or $ax^2 + c = 0$ which lack the constant term or the term in x. The methods of solving incomplete quadratic equations are differently from the standard quadratic equation form as shown below.

1. When $c = 0$, the quadratic equation $ax^2 + bx + c = 0$ becomes $ax^2 + bx = 0$.
$$x(ax + b) = 0$$
Therefore, $x = 0$ or $ax + b = 0$

Hence, $ax^2 + bx = 0 \Leftrightarrow x = 0$ or $x = -\dfrac{b}{a}$.

It is unwise to divide both sides by any multiple of the unknown for this leads to the loss of the solution $x = 0$.

 Example

Solve the quadratic equation $3x^2 + 4x = 0$.

Solution

$$3x^2 + 4x = 0 \Rightarrow x(3x + 4) = 0$$
$$\Leftrightarrow x = 0 \text{ or } x = -\dfrac{4}{3}$$

2. When $b = 0$, the quadratic equation $ax^2 + bx + c = 0$ becomes
$$ax^2 + c = 0 \Leftrightarrow x^2 = -\dfrac{c}{a}.$$

This equation will have real solutions only if $\dfrac{c}{a} \leq 0$.

Hence, if $ax^2 + c = 0$ and $\dfrac{c}{a} \leq 0$ then $x = \pm\sqrt{-\dfrac{c}{a}}$.

 Example

Solve the equation $4x^2 - 1 = 0$.
Solution

$$4x^2 - 1 = 0 \Leftrightarrow x^2 = \dfrac{1}{4} \Leftrightarrow x = \pm\dfrac{1}{2}$$

Alternatively, using the difference of two squares identity,

$$4x^2 - 1 = (2x)^2 - 1^2 = (2x+1)(2x-1)$$

$$4x^2 - 1 = 0 \Leftrightarrow (2x+1)(2x-1) = 0 \Leftrightarrow x = \pm\dfrac{1}{2}$$

Note that any positive number has two square roots. Do not forget one of the square roots. For instance, $(-2)^2 = 4$ and $2^2 = 4$ so the square roots of 4 are 2 or -2.

 Exercise 13:6

Solve the following equations.
1. $x^2 - 64 = 0$ 2. $4y^2 = 28y$ 3. $4a^2 = 1$ 4. $9p^2 = p$
5. $x^2 - 4x = 0$ 6. $y^2 - 9 = 0$ 7. $3b^2 = 18$ 8. $\dfrac{8p}{25} = \dfrac{16}{p}$

13.10 Worded Problems that lead to Quadratic Equations

Many real life problems such as the macaroni problem below involve quadratic equations and until the solving begins, it will not be easy to predict what the problem may entail. This is illustrated by the following macaroni problem which we saw in module 14 topic 11 of book 3. You are required to solve the problem again.

 Group Activity

The cost of a carton of macaroni is normally 4000 Francs. Due to a 20 Francs discount per sachet, a woman buys 10 more sachets at the same amount. You need to organize a party which will be attended by 240 people and you estimate that one person will eat half a sachet of macaroni. How much will you budget for this item and how many sachets of macaroni will you expect to be bought?

 Example

A man wants to buy a piece of land. The owner tells him that the area of the piece of land is 80 square metres and that the length is 2 metres longer than the width. The man hires you to work out the length and width for him. Go ahead and do your job.

Solution
Let the width be x m, then the length will be $(x + 2)$ m.

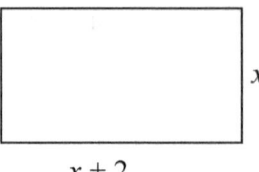

$$x(x+2) = 80 \Leftrightarrow x^2 + 2x = 80$$
$$\Leftrightarrow x^2 + 2x - 80 = 0 \text{ or } (x-8)(x+10) = 0$$
$$\Leftrightarrow x = 8 \text{ or } x = -10$$

Since $x > 0$, so the solution $x = -10$ should be discarded. Clearly if the width is 8 m, the length will be 10 m, since $8 \times 10 = 80$.

 Exercise 13:7

Translate and solve the following equations.
1. The sum of the squares of two consecutive integers is 13. Find the numbers.
2. The square of a number is 12 more than the number. Find the number and its square.
3. At the beginning of a school year, a man bought for his two children $(x - 5)$ exercise books at $(3x + 75)$ FCFA each. Given that, he spent 2025 FCFA altogether, find the number of exercise books he bought and the exact price of each.
4. A woman bought a piece of land in Bamenda. While in Yaounde, she decided to sell it. The buyer was interested in knowing the dimensions of the piece of land. What the woman could remember was that its area is 247.5 square metres and the length is 1.5 m more than the width. Help the woman out of this problem.
5. The formula $T_n = \dfrac{n(n+1)}{2}$ is a formula for finding any triangular number. Find the value of n for which $T_n = 78$.
6. The product of two positive integers is 27. One is three more than twice the other is. Find the numbers.
7. The capacity of a hall is 144 persons. The number of rows is 7 more the number of seats in each row. Determine the number of rows of seats in the hall.
8. The area and perimeter of a rectangular farm are 24 square metres and 20 metres respectively. Find the length and width of the farm.
9. The height of a triangular flowerbed is 4 m less than the base. The area of the flowerbed is 48 m^2. Calculate the height and the length of the base of the flowerbed.
10. The product of a positive integer and a number three less than the integer is

equal to the integer increased by 32. What is the integer?
11. Two positive integers differ by 8. The sum of the square of the larger and the smaller is 124. Find the numbers.
12. During a procession, the signboard bearer and the choir leader are alone on their own rows. The flag bearer and shield bearer occupy the next row. The arrangement of the rest of the choir members is such that the number of choristers on each row exceeds the number of rows by three. There are altogether 112 choristers. Find the total number of rows on this procession.

13.11 Simultaneous Equations, One Linear-One Quadratic

We can solve simultaneous equations, one linear one quadratic using the method of substitution. To do this, we make one of the unknowns a subject, using the linear equation and then substitute into the quadratic equation.

Example

1. Solve the simultaneous equations $y = 2x - 3$ and $y = x^2 + 5x - 7$.

   Solution
   $y = 2x - 3$ ....①
   $y = x^2 + 5x - 7$ ....②
   Substitute ① in ②.
   $x^2 + 5x - 7 = 2x - 3$
   $\quad\quad x^2 + 3x - 4 = 0$
   $\quad\quad (x + 4)(x - 1) = 0$
   $\therefore\ x = -4$ or $x = 1$
   Substitute in ①,
   When $x = -4$, $y = -11$ or when $x = 1$, $y = -1$.
   $\therefore x = -4$ and $y = -11$ or $x = 1$ and $y = -1$.

2. Solve the simultaneous equations
   $2x^2 + 7x - y^2 = 21$ and $3x + 2y = 8$

   Solution
   $\quad\quad 3x + 2y = 8$ ....①
   $2x^2 + 7x - y^2 = 21$ ....②

From ①: $y = \dfrac{8-3x}{2}$ ....③

Substitute ③ in ②

$$2x^2 + 7x - \left(\dfrac{8-3x}{2}\right)^2 = 21$$

$$2x^2 + 7x - \left(\dfrac{64-48x+9x^2}{4}\right) = 21$$

Multiply all through by 4,
$$8x^2 + 28x - (64 - 48x + 9x^2) = 84$$
$$x^2 - 76x + 148 = 0$$
$$(x - 74)(x - 2) = 0$$
$$\therefore x = 2 \text{ or } x = 74$$

Substitude in ③:

When $x = 2$, $y = \dfrac{8-3(2)}{2} \Rightarrow y = 1$

When $x = 74$, $y = \dfrac{8-3(74)}{2} \Rightarrow y = -107$

$\therefore x = 2$ and $y = 1$ or $x = 74$ and $y = -107$.

 Exercise 13:8

Solve the following simultaneous equations.
1. $x - 2y = 7$, $x^2 + 4y^2 = 37$
2. $y + 7x = 3x^2$, $y - 11x + 16 = 0$
3. $x + y = 7$, $xy = 12$
4. $2x - y = 2$, $4x^2 + y^2 = 2$
5. $3x - 4y + 9 = 0$, $2x^2 + y^2 = 11$
6. $2x - 3y + 11 = 0$, $2x^2 - xy = 36$
7. $5x + 3y + 7 = 0$, $3y^2 = x^2 - 4y + 3$
8. $x(x+y) = 3$, $y - x = 1$
9. $\dfrac{x}{2} + \dfrac{y}{3} = 2$, $\dfrac{x^2}{4} + \dfrac{y^2}{9} = 2$
10. $x + 2y = 5$, $2x^2 + y^2 = 6$

Module 17, Topic 13: Algebraic Processes

 Multiple Choice Exercise 13

1. An identity among the following is:
   [A] $(x+1)(2x+3) = 2x^2 + 5x + 3$ [B] $(3p-1)(2p+1) = 3p^2 - 5p - 1$
   [C] $2y + 7 = 3y - 5$ [D] $x^2 + 2x + 1 = 6$

2. The statement, which is not an identity, is:
   [A] $(x+2)(x-1) = x^2 + x - 2$ [B] $(5x-1)(x+1) = 5x^2 + 4x - 1$
   [C] $(3x+1)(2x-1) = 6x^2 - x - 1$ [D] $(3x-1)(2x+1) = 3x^2 - 5x - 1$

3. The quadratic equation whose roots are $x = -2$ and $x = 7$ is:
   [A] $x^2 + 2x - 7 = 0$ [B] $x^2 - 2x + 7 = 0$
   [C] $x^2 - 5x - 14 = 0$ [D] $x^2 + 5x - 14 = 0$

4. A quadratic equation has roots $-\frac{2}{3}$ and $-\frac{1}{4}$. The required equation is:
   [A] $x^2 - \frac{11}{12}x + 2 = 0$ [B] $12x^2 - 3x + 2 = 0$
   [C] $12x^2 - 11x + 2 = 0$ [D] $12x^2 - 11x - 2 = 0$

5. The only quadratic equation with roots $-\frac{1}{2}$ and 2 is:
   [A] $3x^2 - 3x + 2 = 0$ [B] $3x^2 + 3x + 2 = 0$
   [C] $2x^2 - 3x + 2 = 0$ [D] $2x^2 - 3x + 2 = 0$

6. The quadratic equation whose roots are 3 and $\frac{2}{3}$ is:
   [A] $3x^2 - 11x + 6 = 0$ [B] $x^2 - 11x + 6 = 0$
   [C] $3x^2 - 11x + 2 = 0$ [D] $x^2 - 11x - 2 = 0$

7. Given that the roots of a quadratic equation are $\frac{1}{4}$ and 3 then the quadratic equation is:
   [A] $4x^2 - 13x + 3 = 0$ [B] $4x^2 - 13x - 3 = 0$
   [C] $4x^2 + 13x - 3 = 0$ [D] $4x^2 + 13x + 3 = 0$

8. The equation whose roots are 4 and -5 is:
   [A] $x^2 - x - 20 = 0$ [B] $x^2 + x + 20 = 0$
   [C] $x^2 - x + 20 = 0$ [D] $x^2 + x - 20 = 0$

9. The equation whose roots are $\frac{2}{3}$ and $-\frac{1}{4}$ is:
   [A] $12x^2 + 11x + 2 = 0$ [B] $12x^2 - 11x + 2 = 0$
   [C] $12x^2 - 5x - 2 = 0$ [D] $12x^2 - 11x - 2 = 0$

10. The roots of a quadratic equation in x are $-m$ and $2n$. The equation is:
    [A] $x^2 + x(m - 2n) - 2mn = 0$ [B] $x^2 - x(m - 2n) - 2mn = 0$
    [C] $x^2 - x(m - 2n) + 2mn = 0$ [D] $x^2 + x(m - 2n) + 2mn = 0$

11. The equation whose roots are -8 and 5 is:
    [A] $x^2 + 3x + 40 = 0$ [B] $x^2 - 3x - 40 = 0$
    [C] $x^2 + 3x - 40 = 0$ [D] $x^2 - 3x + 40 = 0$

12. The sum of the roots of the equation $2x^2 + 3x - 9 = 0$ is:
    [A] -18 [B] -6 [C] $\frac{9}{2}$ [D] $-\frac{3}{2}$

13. The two values of x that satisfy the equation $5x^2 - 4x - 1 = 0$ are:
    [A] $1, -\dfrac{1}{5}$ [B] $-1, -\dfrac{1}{5}$ [C] $-1, \dfrac{1}{5}$ [D] $1, \dfrac{1}{5}$

14. The solutions of the equation $3 + 5x - 2x^2 = 0$ are:
    [A] $-\dfrac{1}{2}, -3$ [B] $2, 3$ [C] $-2, 3$ [D] $-\dfrac{1}{2}, 3$

15. Given that $10 - 3x - x^2 = 0$. The values of x are:
    [A] $x = 2$ or -5 [B] $x = -2$ or 5 [C] $x = -1$ or 10 [D] $x = 2$ or 5

16. The equation $3a + 10 = a^2$ gives rise to the roots:
    [A] $a = 5$ or 2 [B] $a = -5$ or 2 [C] $a = -5$ or -2 [D] $a = 5$ or -2

17. One of the roots of the equation $6x^2 = 5 - 7x$ is:
    [A] $-\dfrac{1}{2}$ [B] $-\dfrac{1}{3}$ [C] $\dfrac{1}{2}$ [D] $2\dfrac{1}{2}$

18. The values of y, which satisfy the equation $3y^2 = 3y$ are:
    [A] $y = -3$ or 9 [B] $y = 0$ or 9 [C] $y = -3$ or 3 [D] $y = 3$ or 9

19. The equation $7y^2 = 27y$ has roots:
    [A] $y = 3$ and $y = 7$ [B] $y = 0$ and $y = 7$
    [C] $y = 0$ and $y = \dfrac{3}{7}$ [D] $y = 0$ and $y = 9$

20. A root of the equation $x^2 + 6x = 0$ is:
    [A] 0 [B] 6 [C] 2 [D] 3

21. If $x^2 - k^2 = 0$ where k is an integer, the truth about the roots of the equation is:
    [A] The two roots are equal.
    [B] The sum of the two roots is zero.
    [C] The difference of the two roots is $2x$.
    [D] The sum of the two roots is zero and the difference of the two roots is $2x$.

22. The value of k, which makes the expression $m^2 - 8m + k$ a perfect square, is:
    [A] 2 [B] 4 [C] 8 [D] 16

23. For $x^2 - 6x$ to be a perfect square:
    [A] add 9 [B] add 36 [C] 1 [D] add 3

24. The number to the expression $x^2 - 8x$ to make a perfect square is:
    [A] 36 [B] 9 [C] 25 [D] 16

25. The nature of the roots of the quadratic equation $x^2 - 3x + 2 = 0$ is that the roots are:
    [A] Real and equal [B] Real and distinct
    [C] Imaginary [D] Imaginary and equal

Topic 14

POLYNOMIALS

Objectives

At the end of this topic, the learner should be able to:

1. State the degree of a polynomial.
2. Develop identities for the sum and difference of cubes
3. State and use the remainder and factor theorems.

14.1 Concept of Remainder and Factors

> **? Brainstorming Exercise**
>
> 1. Given that a and b are real number, what conclusion can you draw if a divides b exactly without any remainder?
> 2. Given that $f(x)$ and $g(x)$ are polynomial expressions, what conclusion can you draw if $f(x)$ divides $g(x)$ exactly without any remainder?
> 3. Given that $f(x) = (x-1)(x+3)g(x)$. State the relationship between $f(x)$ and each of $(x-1)$ or $(x+3)$.
> 4. Given that $f(x) = (x-1)(x+3)g(x)$ and that $f(x) = 0$. What are the possible values of x?

In module 1, we learnt that, provided there is no remainder; the divisor and the quotient are factors of the dividend. Thus,

$$\underbrace{\overset{\text{dividend}}{21}}_{\text{multiple}} \div \underbrace{\overset{\text{divisor}}{7} = \overset{\text{quotient}}{3}}_{\text{factors}}$$

$$\overset{\text{dividend}}{25} \div \overset{\text{divisor}}{7} = \overset{\text{quotient}}{3} \text{ Remainder } 4$$

$\Rightarrow 25 = 7 \times 3 + 4$

If $f(x) = (x-1)(x+3)g(x)$, then $f(x) = 0 \Longleftrightarrow x = 1, x = -3$.

We call the values $x = 1$ and $x = -3$, which make the polynomial $f(x)$ equal to zero, **zero values** or the **roots** of the equation $f(x) = 0$, and call $(x-1)$ and $(x+3)$ the factors of the expression $f(x)$.

Module 17, Topic 14: Polynomials

14.2 Long Division of Polynomials

We can do long division in algebra using the same rules as in arithmetic.

 Example

1. Divide $x^3 - 2x^2 - x - 2$ by $x + 1$ stating clearly the remainder if any.
   Solutions

   $$\begin{array}{r}
   x^2 - 3x + 2 \\
   (x+1)\overline{)\, x^3 - 2x^2 - x - 2}\\
   \underline{-(x^3 + x^2)}\\
   -3x^2 - x - 2\\
   \underline{-(3x^2 - 3x)}\\
   2x - 2\\
   \underline{-(2x + 2)}\\
   -4
   \end{array}$$

   $\therefore \dfrac{x^3 - 2x^2 - x - 2}{x + 1} = x^2 - 3x + 2$ remainder -4.

   Therefore, the remainder is -4 which is equal to $f(-1)$.

2. Divide $x^3 + 2x^2 - x - 2$ by $x + 1$ stating clearly the remainder if any.
   Solutions

   $$\begin{array}{r}
   x^2 + x - 2 \\
   (x+1)\overline{)\, x^3 + 2x^2 - x - 2}\\
   \underline{-(x^3 + x^2)}\\
   x^2 - x - 2\\
   \underline{-(x^2 + x)}\\
   -2x - 2\\
   \underline{-(2x - 2)}\\
   0
   \end{array}$$

   $\therefore \dfrac{x^3 + 2x^2 - x - 2}{(x + 1)} = x^2 + x - 2$ remainder 0.

   Since the remainder on dividing $x^3 + 2x^2 - x - 2$ by $x + 1$ is 0, we can deduce that $x + 1$ is a factor of $x^3 + 2x^2 - x - 2$.

14.3 Variable Substitution in Polynomials

Given a polynomial function $f(x) = \cdots$, we can find the value of $f(a)$ of the polynomial for any given value a of x.

Example

1. Given that $f(x) = x^3 - 2x^2 - x - 2$, find $f(-1)$.

   Solution
   $f(-1) = (-1)^3 - 2(-1)^2 - (-1) - 2 = -1 - 2 + 1 - 2 \Rightarrow f(-1) = -4$

2. Given that $f(x) = x^3 + 2x^2 - x - 2$, find $f(-1)$.

   Solutions
   $f(-1) = (-1)^3 + 2(-1)^2 - (-1) - 2 = -1 + 2 + 1 - 2 \Rightarrow f(-1) = 0$

Investigative Activity

Given that $f(x) = x^3 + 3x - 4$.
1. Find the remainder when $f(x)$ is divided by $x - 4$.
2. Find $f(4)$?
3. What relationship exist between the remainder when $f(x)$ is divided by $x - 4$ and $f(4)$?
4. Compare your result in the example in section 14.2, number 1 and section 14.2, number 1. What conclusion do you draw concerning $f(a)$ and the remainder when $f(x)$ is divided by $(x - a)$?
5. Compare your result in the example in section 14.2, number 2 and section 14.2, number 2.
6. What conclusion do you draw about $(x - a)$ if $f(a) = 0$?
7. What conclusion do you draw about $f(a)$ if $f(x)$ if the remainder when $f(x)$ is divided by $(x - a)$ is zero?

From above we see that we can express any polynomial $f(x)$ in the form
$$f(x) = (ax + b)Q(x) + R$$

Module 17, Topic 14: Polynomials

We call $Q(x)$ the **quotient**, $ax + b$ the **divisor** and R the **remainder**.

The above investigations reveal the following very important polynomial theorems- the **remainder theorem** and the **factor theorem**.

14.4 Remainder theorem
The **remainder theorem** states that,

When we divide a polynomial $f(x)$ by $ax + b$, the remainder is $f\left(-\frac{b}{a}\right)$.

Note that the remainder is always of a lower degree than the divisor.

14.5 Factor theorem
We can extend the remainder theorem to the **factor theorem**, which states that,

If the remainder when $f(x)$ is divided by $ax + b$ is zero, then $ax + b$ is a factor of $f(x)$.
Conversely, if $f\left(-\frac{b}{a}\right) = 0$, then $ax + b$ is a factor of $f(x)$.

Example

Given that $f(x) = 6x^3 - 5x^2 - 17x + 7$.

(a) Find $f\left(-\frac{3}{2}\right)$ (b) Divide $f(x)$ by $2x+3$

Solution

(a) $f\left(-\frac{3}{2}\right) = 6\left(-\frac{3}{2}\right)^3 - 5\left(-\frac{3}{2}\right)^2 - 17\left(-\frac{3}{2}\right) + 7.$

$= -\frac{81}{4} - \frac{45}{4} + \frac{51}{2} + 7 = 1$

(b)
$$
\begin{array}{r}
3x^2 - 7x + 2 \\
(2x+3) \overline{\smash{\big)}\, 6x^3 - 5x^2 - 17x + 7} \\
\underline{-(6x^3 + 9x^2)} \\
-14x^2 - 17x + 7 \\
\underline{-(-14x^2 - 21x)} \\
4x + 7 \\
\underline{-(4x + 6)} \\
1
\end{array}
$$

$\therefore \dfrac{6x^3 - 5x^2 - 17x + 7}{2x + 3} = 3x^2 - 7x + 2$ remainder 1.

Again the remainder is $f\left(-\dfrac{3}{2}\right)$.

14.6 Trial and Error Method for Finding Roots of Polynomials

Bearing in mind the factor theorem, we often obtain the factors of $f(x)$ by trial and error method as follows.

 Example

Determine the factors of $f(x) = 2x^3 + x^2 - 7x - 6$.

Solution
By trial and error method,
$f(1) = 2(1)^3 + (1)^2 - 7(1) - 6 = 2 + 1 - 7 - 6 \neq 0$
$f(-1) = 2(-1)^3 + (-1)^2 - 7(-1) - 6 = -2 + 1 + 7 - 6 = 0$
Therefore, $(x + 1)$ is a factor of $f(x)$.
$f(2) = 2(2)^3 + (2)^2 - 7(2) - 6 = 16 + 4 - 14 - 6 = 0$
Therefore, $(x - 2)$ is a factor of $f(x)$.
$\Rightarrow f(x) = (x + 1)(x - 2)(ax + 3)$

Since the coefficient of x^3 and the independent terms are 2 and −6 respectively, then $a = 2$.
$\Rightarrow f(x) = (x + 1)(x - 2)(2x + 3)$

Module 17, Topic 14: Polynomials

 Exercise 14.1

1. Determine the factors of the following polynomials.
   (a) $x^3 - 2x^2 - x + 2$ (b) $2x^3 + x^2 - 2x - 1$
2. Solve the equations
   (a) $2x^3 - x^2 - 3x + 2 = 0$ (b) $6x^3 - x^2 - 6x + 1 = 0$
3. Find the remainder when $x^3 + 5x^2 + 7x - 6$ is divided by $x + 2$.
4. Given that $x + 3$ is a factor of $f(x) = x^3 + 6x^2 + kx + 6$, find the value of k. Hence, factorize the expression completely.
5. Given that $x - 2$ is a factor the polynomial $f(x) = ax^2 + bx + c$ and that the remainders when $f(x)$ is divided by $x - 2$ and $x + 1$ respectively are -4 and 6. Find the values of a, b and c.
6. Find the equation whose roots are $-1, -2$ and 3.
7. The expression $ax^2 + 7x + b$ leave remainders 8 and 10 when divided by $x + 1$ and $x + 2$ respectively. Find the values of a and b.
8. If $x-2$ is a factor of $x^2 + kx + 6$, find the value of k. Using this value of k find the roots of the equation $x^2 + kx + 6 = 0$.
9. Find the value of k for which $x^3 + kx + 6 = f(x)$ has a root -3. Hence, factorize $f(x)$ completely.
10. Given that $f(x) = x^3 + kx^2 - x - 2$ and that $x + 1$ is a factor of $f(x)$. Find the value of k and hence factorize $f(x)$ completely.
11. Find the remainder when $x^3 + 5x^2 - 7x - 6$ is divided by $x + 1$. Hence, determine the number, we must add to $x^3 + 5x^2 - 7x - 6$, to make the result divisible by $x + 1$.

14.7 Sum and Difference of Two Cubes

Consider the expansion of $(a - b)^3$ below.
$$(a - b)^3 = (a - b)(a - b)^2$$
$$= (a - b)(a^2 - 2ab + b^2)$$
$$= a^3 - 2a^2b + ab^2 - a^2b + 2ab^2 - b^3$$
$$= a^3 - b^3 - 3a^2b + 3ab^2$$
$$= a^3 - b^3 - 3ab(a - b)$$

Adding $3ab(a - b)$ to both sides,

$$a^3 - b^3 = (a - b)^3 + 3ab(a - b)$$
$$= (a - b)[(a - b)^2 + 3ab]$$
$$= (a - b)(a^2 + ab + b^2)$$
$$\Rightarrow a^3 - b^3 = (a - b)(a^2 + ab + b^2)$$

Now consider the expansion of $(a + b)^3$.
$$(a + b)^3 = (a + b)(a + b)^2$$

$$= (a+b)(a^2 + 2ab + b^2)$$
$$= a^3 + 2a^2b + ab^2 + a^2b + 2ab^2 + b^3$$
$$= a^3 + b^3 + 3a^2b + 3ab^2$$
$$= a^3 + b^3 + 3ab(a+b)$$

Subtracting $3ab(a+b)$ from both sides,

$$a^3 + b^3 = (a+b)^3 - 3ab(a+b)$$
$$= (a+b)[(a+b)^2 - 3ab]$$
$$= (a+b)(a^2 - ab + b^2)$$
$$\Rightarrow a^3 + b^3 = (a+b)(a^2 - ab + b^2)$$

From the above we see that the sum and difference of two cubes can be factored according to the formulae

$$a^3 - b^3 = (a-b)(a^2 + ab + b^2)$$

$$a^3 + b^3 = (a+b)(a^2 - ab + b^2)$$

Exercise 14.2

Factorize the following.
1. $x^3 + 64$ 2. $x^3 - 27$ 3. $x^3 - 8y^3$ 4. $8x^3 + 125y^3$ 5. $64x^3 - 27y^3$

Multiple Choice Exercise 14

1. The factor theorem states that if the remainder when we divide a polynomial $f(x)$ by $ax + b$ is 0 then:

   [A] $f\left(\dfrac{a}{b}\right) = 0$ [B] $f\left(-\dfrac{a}{b}\right) = 0$ [C] $f\left(-\dfrac{b}{a}\right) = 0$ [D] $f\left(\dfrac{b}{a}\right) = 0$

2. The remainder theorem states that when we divide a polynomial $f(x)$ by $ax + b$ the remainder is:

   [A] $f\left(\dfrac{a}{b}\right)$ [B] $f\left(-\dfrac{a}{b}\right)$ [C] $f\left(-\dfrac{b}{a}\right)$ [D] $f\left(\dfrac{b}{a}\right)$

3. When we divide a polynomial $f(x)$ by $ax - 1$ the remainder is:

   [A] $f(a)$ [B] $f(-a)$ [C] $f\left(-\dfrac{1}{a}\right)$ [D] $f\left(\dfrac{1}{a}\right)$

Module 17, Topic 14: Polynomials

4. The remainder when we divide $x^3 + 2x^2 + 1$ by -1 is:
   [A] x [B] 1   [C] 2   [D] -1

5. $f(x) = -x^2 + kx - 6$ has a factor $(x+2)$. The value of k is:
   [A] -3 [B] -7 [C] -5 [D] 7

6. Given that $f(x) = 2x^2 + 3x - 2$. It is true to say that:
   [A] $(x-2)$ is a factor of $f(x)$.
   [B] $(x+2)$ is a factor of $f(x)$.
   [C] $f(x)$ leaves a remainder -2 when divided by $(x+2)$.
   [D] $(2x+1)$ is a factor of $f(x)$.

7. The remainder when $x^3 - 2x^2 - x - 2$ is divided by $x+1$ is:
   [A] -2 [B] 4   [C] -4 [D] 2

8. If $f(x) = x^3 - 3x - 4$. The remainder when $f(x)$ is divided by $x-4$ is:
   [A] -56 [B] 48   [C] -80 [D] 72

9. The remainder when $x^3 + 3x - 4$ is divided by $x+1$ is:
   [A] -2 [B] -8 [C] 0   [D] -7

10. Given that $f(x) = 2x^3 - 3x^2 - 3x + 11$. The remainder when $f(x)$ is divided by $2x + 1$ is:
    [A] 1.5   [B] 2.5   [C] -1.5 [D] -2.5

11. The remainder when $3x^3 - x^2 - 2x + 13$ is divided by $x+2$ is:
    [A] 4   [B] -3 [C] -4 [D] 3

12. Given that when divided by $x+1$ and $x+2$ the expression $ax^2 + bx + 3$ leaves remainders 6 and 9 respectively. The values of a and b are:
    [A] $a = -2, b = 1$ [B] $a = -3, b = 0$ [C] $a = 0, b = -3$ [D] $a = 2, b = -1$

13. The remainder when $x^3 + 3x^2 - 5x - 6$ is divided by $x + 2$ is:
    [A] -12 [B] 8   [C] 4   [D] 12

14. $x - 1$ and $x - 2$ are both factors of $x^3 + ax^2 + bx - 6$ when:
    [A] $a = -6, b = 11$ [B] $a = 6, b = 11$ [C] $a = 6, b = -11$ [D] $a = -6, b = -11$

15. If $(x+2)$ is a factor of $x^3 + kx^2 - 2x + 4$. The value of k is:
    [A] 2   [B] -2 [C] 1   [D] 0

16. $x^3 - 3x^2 + 6x - 2$ has remainder 2 when divided by:
    [A] $x - 1$ [B] $x + 1$ [C] $x + 2$ [D] $2x - 1$

17. $x^3 - 3x^2 + 2x - 6$ has a factor:
    [A] $x - 4$ [B] $x - 2$ [C] $x - 3$ [D] $x + 3$

18. Given that $f(x) = 3x^3 + 4x^2 - 3x - 4$. One of the factors of $f(x)$ is:
    [A] $3x - 4$ [B] $4x + 7$ [C] $x - 1$ [D] $x - 4$

19. A factor of $x^3 + 2x^2 - 5x - 6$ is:
    [A] $x + 2$ [B] $x - 1$ [C] $x + 1$ [D] $x - 2$

20. A factor of $x^3 + 3x^2 - 4x - 12$ is:
    [A] $x - 4$ [B] $x + 4$ [C] $x - 3$ [D] $x + 3$

21. $x - 2$ is a factor of:
    [A] $x^3 - 3x^2 - 4x + 12$ [B] $x^3 + 3x^2 + 4x + 12$
    [C] $x^3 - 3x^2 + 4x + 12$ [D] $x^3 + 3x^2 - 4x + 12$

Topic 15

INEQUALITIES AND INEQUATIONS

Objectives

At the end of this topic, the learner should be able to:

1. Identify and denote intervals.
2. Interpret real life problems involving unequal situations using (at least, at most etc)
3. Use < or >.
4. Represent intervals on the number line.
5. Solve inequalities in one unknown and represent their solutions on the number line.
6. Define and identify absolute inequalities.
7. Solve quadratic inequlities.

Module 17, Topic 15: Inequalities and Inequations

Inequalities were introduced in Module 9, topic 16. It will be useful to revise that section before continuing.

15.1 Representation of inequalities

We can represent the solutions of inequations on a number line in the same way as intervals.

In sections 5.6, 5.7 and 5.8 of book two, we treated intervals to a very reasonable depth and at this point the learner is advised to revise that section. At this point we shall examine in detail "unbounded intervals" and will use these intervals to represent the solution of inequalities from section 15.5 to 15.9

Unbounded Intervals
An unbounded interval is an interval that extends indefinitely in one or both directions.

Right-unbounded Open Intervals

$x > a$ or (a, ∞)

$x > 2$ or $(2, \infty)$

The boundary point a is not included and all points to the right of a are included.

Left-unbounded Open Intervals

$x < a$ or $(-\infty, a)$

$x < 2$ or $(-\infty, 2)$

Right-unbounded Closed Interval

$x > a$ or (a, ∞)

$x > 2$ or $(2, \infty)$

The boundary point a is included and all points to the left of a are included.

Intervals may also be unbounded on both ends.

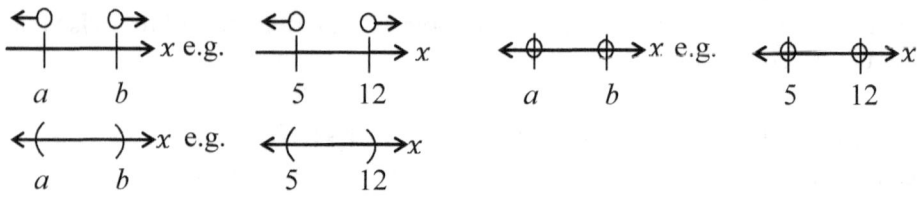

The boundary points a and b are not included and the points between a and b are not included.

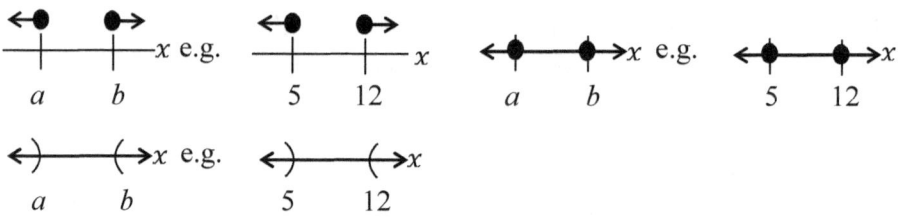

The boundary points a and b are included and points between a and b are not included.

15.2 Conditional Inequalities or Inequations

Conditional inequalities are inequalities that are sometimes true and sometimes false depending on the range of values of the variable. Another name for conditional inequalities is inequations. Inequations are analogous to equations because we can solve them to obtain the range of values of the variable. The mean difference between inequalities and inequations is that inequalities are always true (e.g. $-5 < 8$, $13 > 4$) while inequations are sometimes true and sometime false (e.g. $x^2 - 2x - 3 \geq 0$), $x + 2 < 5$

15.3 Building up Inequations

Inequations are built in the same way in which equations are built but for the fact that the inequality sign is used instead of the equal to sign.

 Example

1. The product of 2 and a number is at least 8.
2. When we add 5 to a number, the result is greater than 9.
3. When we increase an amount of money 70 FCFA the total amount is at most 200 FCFA.
4. Four girls shared a number of oranges and each had at least 8 oranges.

Module 17, Topic 15: Inequalities and Inequations

Solutions
1. Let the number be n. Then, $2n \geq 8$.
2. Let the number be x. Then, $x + 5 > 9$.
3. Let the amount of money be m. Then, $m + 70 \leq 200$.
4. Let the basket contain b oranges. Then, $\frac{b}{4} \geq 8$.

 Exercise 15:2

Build inequations from the following situations
1. When we increase a number by 8 the resulting sum is equal to or less than 14.
2. On dividing 40 by a certain number, the resulting quotient is greater than 4.
3. Thrice the sum of a number and 7 is not more than 27.
4. The product of a number and 4 is at least 20.
5. 60 % of a number decreased by 10 is less than 12.
6. On subtracting half of a number from three times the number, the difference is less than the number increased by 6.
7. Twice the sum of a number and 7 is at most 12 more than the number.
8. Three times the sum of a number and 5 is less than four times the number increased by 2.
9. Thirty minus five times a certain number is at least 4.
10. Three boys earn at most 600,000 FCFA. Ambe earns 20,000 FCFA less than Ambe earns Anye and Ndoh earns twice as much as Ambe.

15.4 Solving Inequations

Inequations are solved in the same way as equations but for the fact that when multiplying or dividing both sides by a negative number, the inequality sign changes from < to > or from ≤ to ≥ and vice versa.

15.5 Linear Inequalities

 Example

Solve the following inequalities and represent the solution on a number line.
1. $2x + 5 \leq 13$
2. $\frac{2}{3}n + 5 > 11$

Solution

1. $2x + 5 \leq 13$
   $2x \leq 8$
   $\Rightarrow x \leq 4$

2. $\frac{2}{3}n + 5 > 11$
   $\Rightarrow \frac{2}{3}n > 6$
   $2n > 18$
   $n > 9$

15.6 Compound Inequalities

A **compound inequality** is an inequality, which consist of two inequalities joined by '*and*' or '*or*'. The following are some examples of compound inequalities.

- $3 \leq 6 + 3x \leq 9$ read '$6x + 3$ is greater than or equal to 3 <u>and</u> less than or equal to 9.
- $3x - 4 < x \leq 5x + 12$ read 'x is greater than $3x - 4$ <u>and</u> less than or equal to $5x + 12$.

We can solve such inequalities by either performing the same operation to each expression or separating them into two entities before solving them.

Example

1. Solve the following inequalities and represent your result on a number line.
   (a) $3 \leq 6 + 3x \leq 9$
   (b) $3x - 4 < x \leq 5x + 12$

   Solution
   (a) $3 \leq 6 + 3x \leq 9$
   Subtract 6 from each expression.
   $-3 \leq 3x \leq 3$
   Divide through by 3.
   $-1 \leq x \leq 1$

   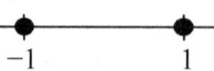

   (b) $3x - 4 < x \leq 5x + 12$.
   This can better be solved by separating $3x - 4 < x$ and $x \leq 5x + 12$.

Module 17, Topic 15: Inequalities and Inequations

$3x - 4 < x$	$x \leq 5x + 12$
Subtract x from both sides	Subtract $5x$ from both sides
$2x - 4 < 0$	$-4x \leq 12$
Add 4 to both sides	Divide both sides by -4 and
$2x < 4$	reverse the inequality sign.
Divide both sides by 2.	$x \geq -3$
$x < 2$	

The results can now be combined to have $-3 \leq x < 2$.

2. Solve the inequality $2x + 3 < -1$ or $3x - 5 > -2$ and represent your solution on a real number line.

   Solution

   $2x + 3 < -1$ or $3x - 5 > -2$
   $2x < -4$ or $3x > 3$
   $x < -2$ or $x > 1$

15.7 Absolute Value Inequalities

An **absolute value inequality** is an inequality, which involves absolute values. For example, $|3x + 2| \leq 5$.

To solve an absolute value inequality, appreciate that $|n| < k$ then, $-k < n < k$.

Example

1. Solve the inequality $|3x + 2| \leq 5$ and represent your solution on a real number line.

   Solution

   We can write $|3x + 2| \leq 5$ as $-5 \leq 3x + 2 \leq 5$

   $\Rightarrow \quad -7 \leq 3x \leq 3$

   $\quad\quad -\dfrac{7}{3} \leq x \leq 1$

Exercise 15:3

Solve the following inequalities and represent your solution on a real number line.
1. $2x < 4$
2. $2 + x > 11$
3. $16 \leq 2x + 8$
4. $2y + 4 \geq 12$
5. $7x - 5 \leq 2x + 11$
6. $2(n+5) > 18$
7. $\dfrac{2x-5}{3} < 25$
8. $5x > 40 - 3x$
9. $4y + 5 \leq 5y - 30$
10. $3t + 10 < 2t + 20$
11. $5u < 2u + 27$
12. $2x \geq 90 - 7x$
13. $10t - 11 > 8t$
14. $18 - 5a \leq a$
15. $13y \leq 15 + 3y$
16. $100 + 3\dfrac{1}{2}x > 23\dfrac{1}{2}x$
17. $9u \geq 16v - 105$
18. $5p + 3 - 2p < p + 8$
19. $7m + 10 - 2m > 16m - 12$
20. $20 \geq 3x - 1$
21. $0.6x - 0.4 \geq 0.4x + 0.6$
22. $3x - 2 \leq 2 - x$
23. $15 \geq 2 - 4x \geq 2$
24. $|2w - 7| \leq 9$
25. $x - 1 < 3x + 1 < x + 5$
26. $|2x - 1| < 3$
27. $|4x - 2| \leq 6$

15.8 Quadratic Inequations

A quadratic inequation is an inequation, which involves a quadratic expression.
Consider the inequation $(x-2)(x+3) < 0$.
$(x-2)(x+3) = 0$ is the related quadratic equation and its roots are 2 and -3. 2 and -3 are called the **critical values** of the inequality $(x-2)(x+3)<0$. We can plot these values on a number line as follows.
These boundary points define three intervals
$x < -3, -3 < x < 2$ and $x > 2$ of which at least one must satisfy the inequality. By testing a point within each interval, the range of values, which satisfies the inequation, can easily be determined.

Testing the point -4 in the interval $x < -3$,
Left hand side $= (-4 - 2)(-4 + 3) = (-6)(-1) = +6$ and $+6 \not< 0$

Testing the point 0 in the interval $-3 < x < 2$,
Left hand side $= (0 - 2)(0 + 3) = (-2)(+3) = -6$ and $-6 < 0$

Testing the point 4 in the interval $x > 2$,
left hand side $= (4 - 2)(4 + 3) = (+2)(+7) = +14$ and $+14 \not< 0$

Module 17, Topic 15: Inequalities and Inequations

So clearly the interval which is satisfied by the inequality $-3 < x < 2$ is, shaded in the number line below.

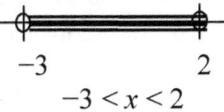

$$-3 < x < 2$$

The range of values which satisfy the inequality always lie either within the limits of the interval or outside the limits, but not both. So testing a point within or without suffices. If the interval within satisfies the interval outside does not and vice versa. We often take advantage of the origin since computing with zero can be very easy.

Example

1. Find the solution set of the inequation $x^2 - 5x + 6 < 0$ and represent it on a number line.

   Solution
   $$x^2 - 5x + 6 < 6 \Rightarrow (x-2)(x-3) < 0$$

   $$2 < x < 3$$

   Critical values are 2 and 3
   Solution set = $\{x: 2 < x < 3, x \in \mathbb{R}\}$

2. Find the set of real values of x, which satisfy the inequality $x^2 + 2x - 3 \geq 0$. Represent your solution on a number line.

   Solution
   $$x^2 + 2x - 3 \geq 0 \Rightarrow (x+3)(x-1) \geq 0$$

   Critical values are -3 and 1. Testing a value the range outside is satisfied.

   $$x \leq -3 \text{ and } x \geq 1$$
   Solution set = $\{x: x \leq -3 \text{ or } x \geq 1, x \in \mathbb{R}\}$

15.9 Absolute Inequalities

An **absolute inequality** is an inequality, which is always true for all real values of the variables involved. In fact, an absolute inequality is analogous to an identity. Examples of absolute inequalities are:

$x^2 \geq 0, x \in \mathbb{R}$; $(x-1)^2 \geq 0, x \in \mathbb{R}$; $(x \pm y)^2 \geq 0, x, y \in \mathbb{R}$.

Notice that in these examples, provided the variables are real, the value of the quantity on the left hand side is always positive no matter the value of the variables-positive or negative. We should not confuse the concept of an absolute inequality with the absolute value inequality, which is true only for values of the variable within a certain range.

Exercise 15:4

Solve the following inequalities and represent your solution on the number line.

1. $u^2 + 6u < -8$
2. $a^2 > 5a + 24$
3. $x(3x+20) \geq 7$
4. $5 \leq m(2m+3)$
5. $p + 8 > \dfrac{20}{p}$
6. $\dfrac{7}{x} \geq 9 - 2x$
7. $2 + \dfrac{5}{x} < \dfrac{12}{x^2}$
8. $(a-8)(2a-3) \leq 34$

15.10 Nature of roots of a Quadratic Equation

In the quadratic formula above, we call $b^2 - 4ac$ the **discriminant** of the expression $ax^2 + bx + c$ denoted by Δ.
Thus, $\Delta = b^2 - 4ac$.

The nature of Δ can give rise to three different types of roots as follows:
1. If $\Delta > 0$, the equation has two real and distinct roots.
2. If $\Delta = 0$, the equation has repeated or equal real roots.
3. If $\Delta < 0$, the equation has no real roots i.e. the roots are complex or imaginary.

It is necessary to check the discriminant of a quadratic equation always to see if the equation has real roots or if it is solvable.

Module 17, Topic 15: *Inequalities and Inequations*

 Exercise 15:5

Determine the nature of the roots of each of the following quadratic equations.

1. $x^2 + 2x + 3 = 0$
2. $u^2 - 4u - 4 = 0$
3. $4x^2 - 4x = -9$
4. $2p^2 + 5p = -8$
5. $3y^2 - 12y = -6$
6. $9x + 12 + \dfrac{4}{x} = 0$
7. $2x^2 - 7x = -13$
8. $2x^2 + 8x = 17$
9. $1 - 4x + 4x^2 = 0$
10. $4x^2 - 9x + 9 = 0$
11. $2x^2 - 7x = -13$
12. $x^2 + 1 = -2x$

15.11 Simultaneous Linear Equations (Graphical Method)

To solve simultaneous linear equations using the graphical method, plot the graphs of the two equations on the same Cartesian axes. The coordinates of the point of intersection of the lines gives the solution of the equation.

 Example

Solve the simultaneous equations $y = 2x$ and $y + x = 3$, using the graphical method.

Solution

$y = 2x$

x	0	1	2
y	0	2	4

$y + x = 3$

x	0	2	4
y	3	1	-1

The figure below shows the graph.
From the figure below, the two lines meet at the point (1,2). Therefore, the solution of the simultaneous equations is $= 1, y = 2$.

285

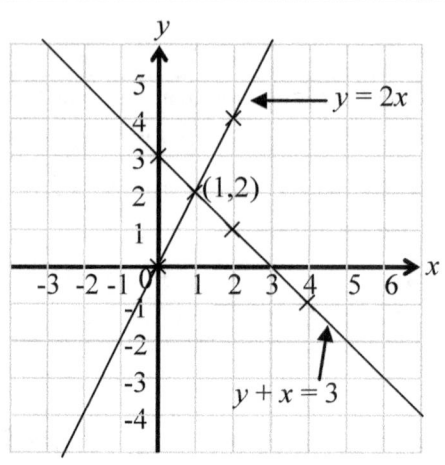

Exercise 15:6

Find graphically the solutions of the simultaneous equations.

1. $3x - 3y = 5$, $3x + 2y = 14$
2. $7x + 3y = 6$, $5y - 9x = 10$
3. $6x + 5y = 11$, $5x - 2y = 3$
4. $5x - y = 9$, $y - 2x = 3$
5. $7x + 2y = -3$, $3x - 5y = 28$
6. $4x + 3y = 1$, $6x - 5y = -8$
7. $3x - 7 = y$, $4x - 5y = 2$
8. $y = 4x$, $3x + y = 21$
9. $5x - 2y = 9$, $x - 5y = 7$
10. $3x + 4 = 0$

Hint: The solution will be the point where the line $y = 3x + 4$ cuts the x-axis.

15.12 Half planes

A straight line divides a plane into three set of points, namely:
(1) The set of points on the line
(2) The set of points on either side of the line

The planes mentioned in (2) above are called **half planes**.

Example

Represent the inequation $2x + y \geq 3$ on a graph.

Solution

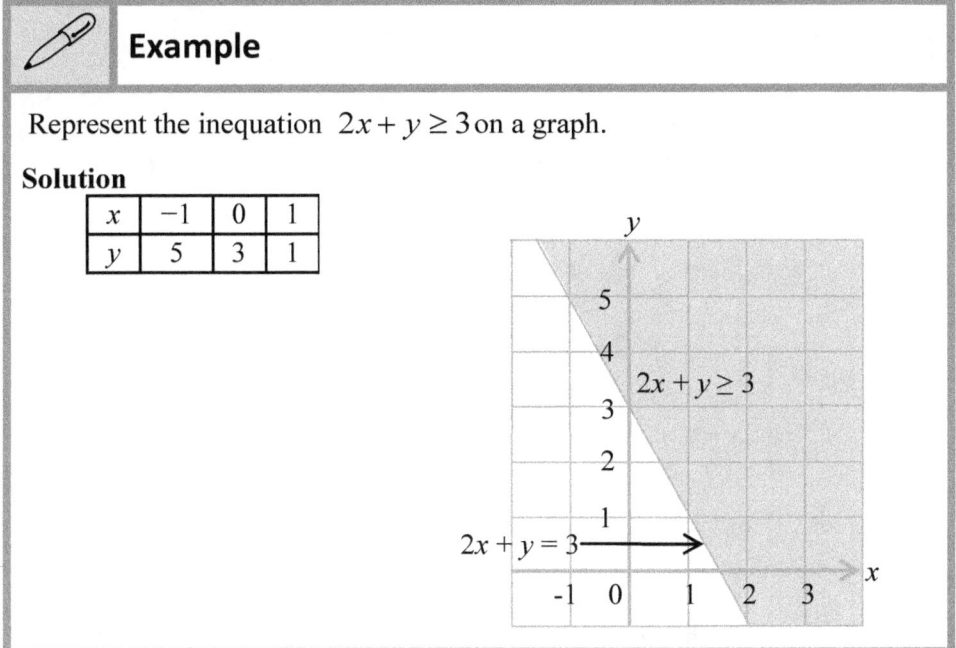

Since the points on the line $2x + y = 3$ are included, we draw the line as continuous. By testing any point (a, b) on any side of the line, we can determine the region which satisfies the inequation. The shaded region in the diagram satisfies the inequation. No point on the other side of the line whatsoever will satisfy the inequation. Testing using the point $(0,0)$ is very convenient since calculation with 0 is very easy to perform. There is no need testing using more than one point, for if an inequality is true for a point, it must be true for all other points in that region and false for all points on the other region and vice versa.

Testing using the point (0,0), Left hand side = 2(0) + 0 = 0
Therefore, the point does not satisfy the inequality since 0 < 3.
Testing using the point (2, 2), Left hand side = 2(2) + 2 = 6
Therefore, the point (2,2) satisfies the inequation since 6 > 3.

Example

Represent the inequation $2x + y > 3$ on a graph.

Solution

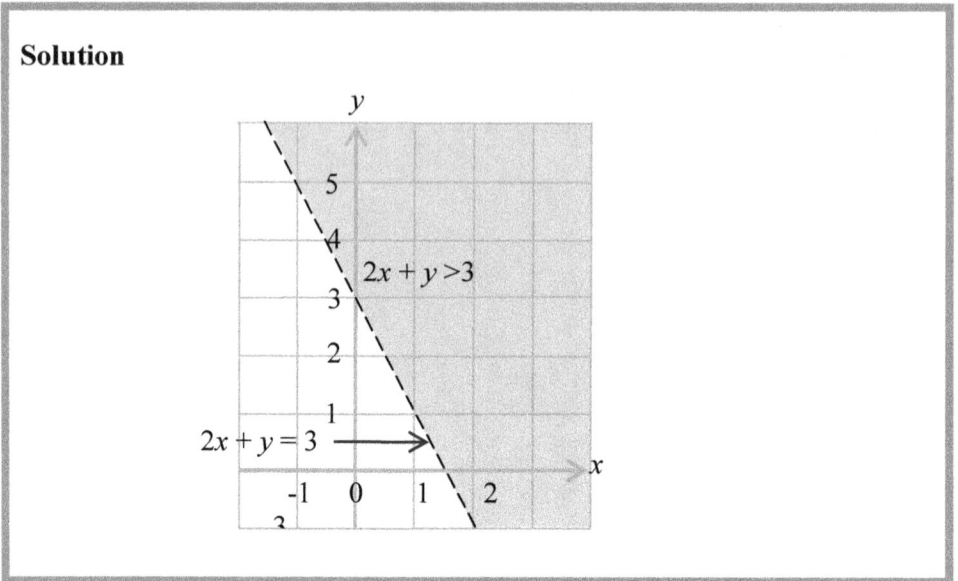

Notice that, we draw the line $2x + y = 3$ as a broken line because it is not included in the solution of the inequation.

15.13 Vertical and Horizontal Boundary Lines

Consider the inequations involving the lines $x = a$ and $y = b$, where a and b are constants. The line $x = a$ will be vertical and the line $y = b$ will be horizontal.

Example

Represent the following inequations on separate Cartesian planes.
(a) $x > 2$ (b) $x < 0$ (c) $y \leq 3$ (d) $y > 0$

Solution

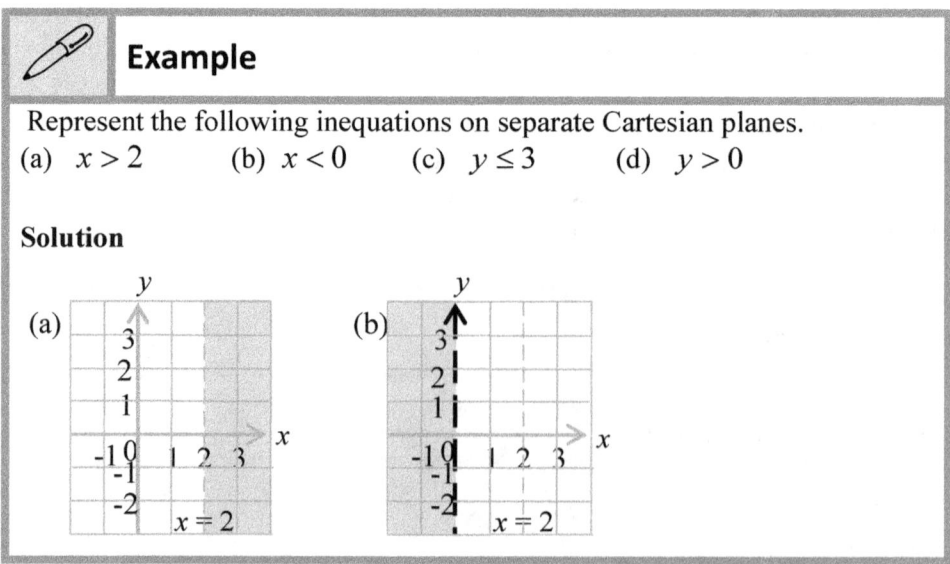

Module 17, Topic 15: Inequalities and Inequations

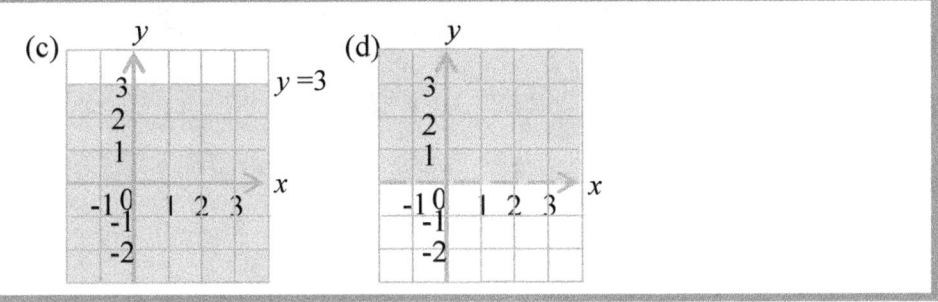

15.14 Simultaneous Linear Inequations

Recall that graphically, the point of intersection of the two lines representing the equations gives the solution of simultaneous linear equations. Equally, the region described by the intersection of two or more inequations gives the solution of the inequations.

 Example

Shade the region in which the inequations $x + y < 7$, $x - y \geq 3$ and $y \geq -4$ are all satisfied.

Solution

$x + y < 7$

x	0	2	7
y	7	5	0

$x - y \geq 3$

x	0	2	3
y	-3	-1	0

The graph is shown below.

$R = \{(x,y): x - y \geq 3\} \cap \{(x,y): x + y < 7\} \cap \{(x,y): y \geq -4\}$

Since the line $x - y = 3$ satisfies the inequation $x - y \geq 3$, we draw it as a continuous line. The line on the other hand does not satisfy the inequation $x + y < 7$; hence, we draw it as a broken line.

Competency Based Mathematics for Secondary Schools. Book 4

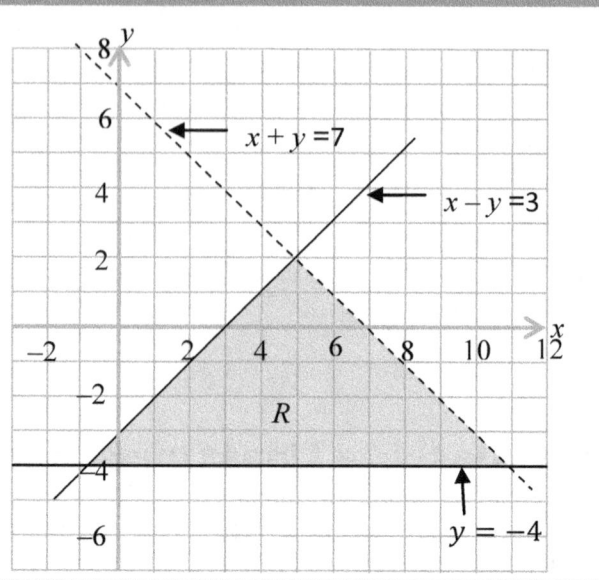

 Exercise 15:7

1. By drawing suitable straight lines on a graph, show by shading the region R for which all the three $x+y \leq 8$, $y < x$, $y < 3$ inequalities are satisfied.
2. Shade the unwanted regions in each of the following.
   (a) $y \geq 0$, $x \geq 0$, $x+y \leq 9$ (b) $x \geq 0$, $y \geq x$, $2x+y \leq 10$
   (c) $y < 6$, $x < 5$, $x+y \geq 5$ (d) $y \geq \frac{1}{2}x$, $x+y \leq 10$, $y \leq 3x$
   (e) $x < 4$, $y < x+3$, $2y+x > 4$ (f) $y < 4x$, $x+y < 10$, $3y > x$
   (g) $x \geq 0$, $x+y < 11$, $2y > x+5$ (h) $x \geq 0$, $x+2y \geq 12$, $x+y \geq 12$
   (i) $y \geq 0$, $x+y \leq 12$, $3x+y \geq 12$

3. Describe the region bounded by the shaded regions in each of the following.

Module 17, Topic 15: Inequalities and Inequations

(a)

(b)

(c)

(d)

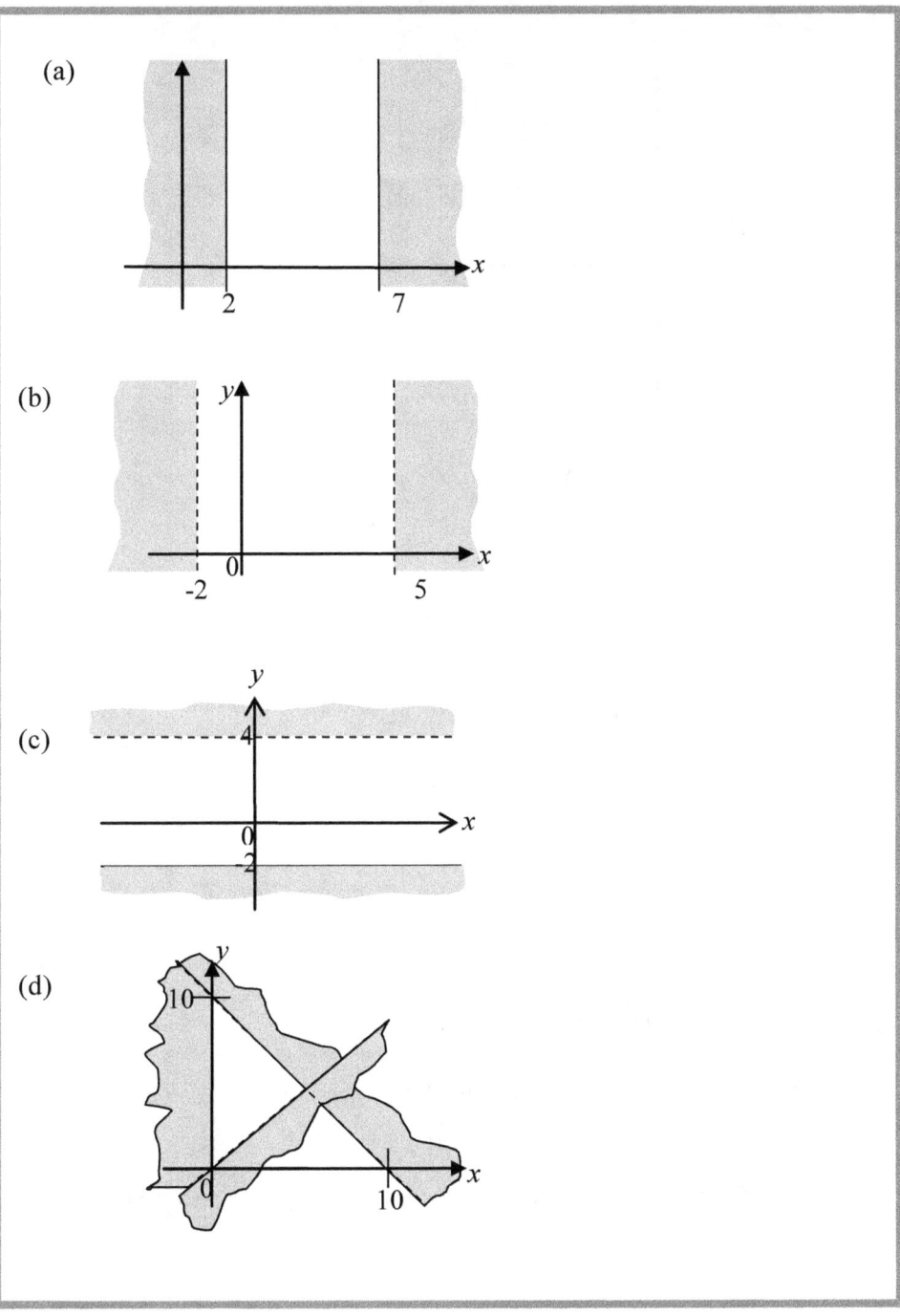

(e)

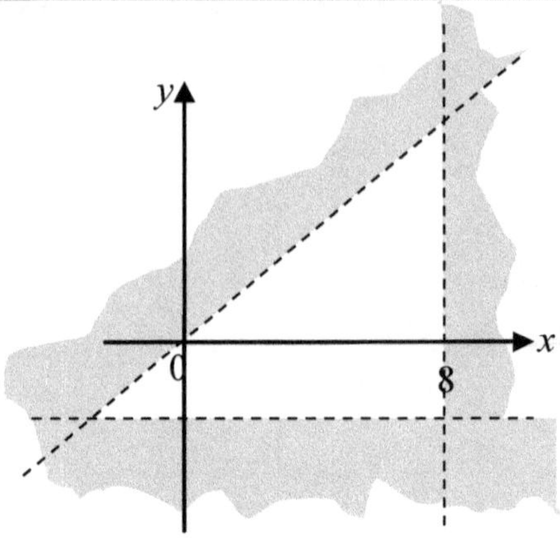

(f)

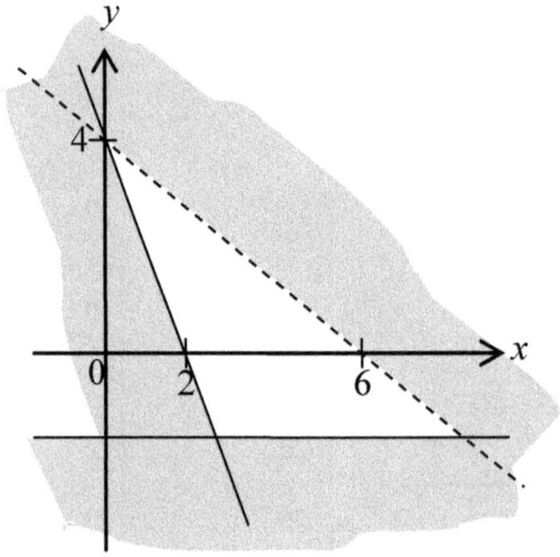

Multiple Choice Exercise 15

1. The range of values of x for which $\frac{1}{2}(4x+2)-(x-5) \leq \frac{1}{4}(3x-1)$ is:

   [A] $x \geq 25$ [B] $x \leq 25$ [C] $x \geq -25$ [D] $x \leq -25$

2. In the figure below, the solution to the inequality $\frac{x}{3} - \frac{(x-3)}{2} < 1$ is represented by the line:

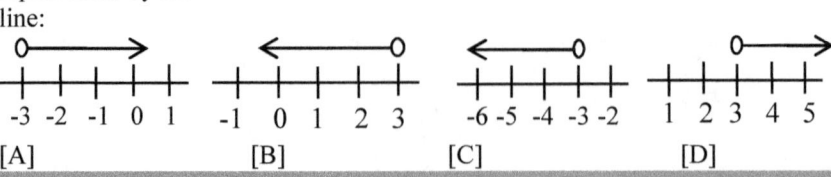

Module 17, Topic 15: Inequalities and Inequations

3. If x varies over the set of real numbers, the inequality illustrated in the figure below is:
   [A] $-3 < x \leq 2$ [B] $-3 \leq x < 2$ [C] $-3 < x < 2$ [D] $-3 \leq x \leq 2$

   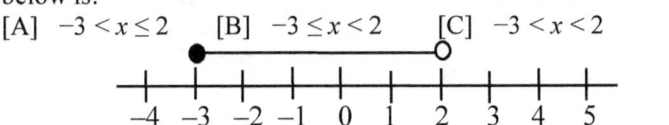

4. If x is a real number the inequality represented in the figure below is:
   [A] $\{x: -5 < x \leq 3\}$ [B] $\{x: -5 \leq x \leq 3\}$
   [C] $\{x: -5 \leq x < 3\}$ [D] $\{x: -5 < x < 3\}$

   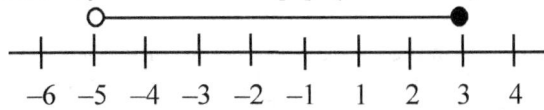

5. If x is a real number, the inequality more illustrated in the number line in the figure below is:
   [A] $x < 4$ [B] $x > -2$ [C] $-2 < x \leq 4$ [D] $-2 \leq x < 4$

   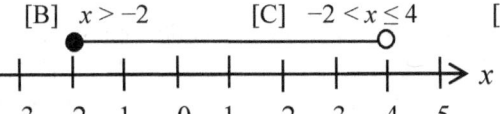

6. If x is a real number, the inequality more illustrated in the number line in the figure below is:
   [A] $x < 4$ [B] $x > -2$ [C] $-2 < x \leq 4$ [D] $-2 \leq x < 4$

   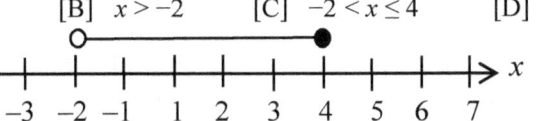

7. If x varies over the set of real numbers, the inequality illustrated in the figure below is:
   [A] $-2 \leq x < 3$ [B] $-2 < x \leq 3$ [C] $-2 < x < 3$ [D] $-2 \leq x \leq 3$

8. The pairs of inequalities is represented on the number line in the figure below are:
   [A] $x < -2$ and $x \geq 1$ [B] $x \leq -2$ and $x > 1$
   [C] $x \leq -2$ and $x < 1$ [D] $x < -2$ and $x > 1$

   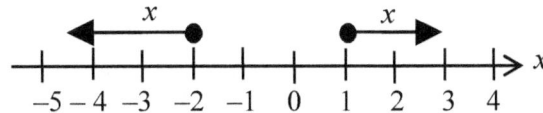

9. The number lines in the figure below which represents the inequality $2 \leq x < 9$ is:

[A] 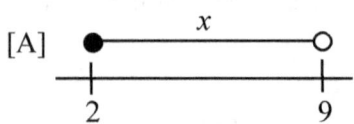   [B]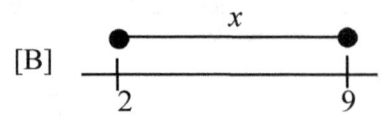

[C] 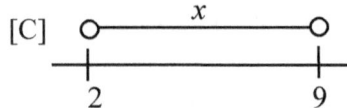   [D]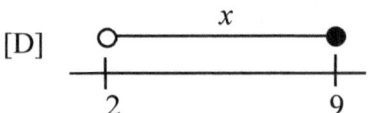

10. The number line in the figure below represents:
    [A] $0 < x < -7$ [B] $-7 < x < 1$ [C] $-7 < x \leq -1$ [D] $-7 \leq x < -1$

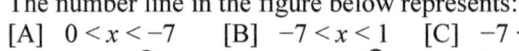

11. Given that, $x \in \mathbb{R}$. An absolute inequality among the following is:
    [A] $|3x + 2| \leq 5$ [B] $(x-1)^2 \geq (x+1)^2$ [C] $(x-1)^2 \geq 0$ [D] $-2 < x < 0$

12. The inequalities represented on the number line in the figure below are:
    [A] $-2 < x \leq 0$ or $x > 3$ [B] $-2 \leq x < 0$ or $x \geq 3$
    [C] $-2 \leq x < 0$ or $x > 3$ [D] $-2 \leq x \leq 0$ or $x > 3$

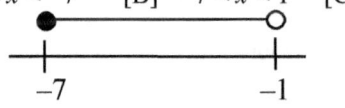

13. Given that, $x \in \mathbb{R}$. An absolute value inequality among the following is:
    [A] $|7x - 16| \geq 0$ [B] $(x+3)^2 \geq (x-3)^2$
    [C] $(x-3)^2 \geq 0$ [D] $-4 < x + 2 < 10$

14. A conditional inequality is an inequality which is:
    [A] always true   [B] sometimes true
    [C] always false   [D] sometimes false

15. The root of the equation represented by the graph below is:
    [A] 4   [B] 7   [C] −4   [D] −7

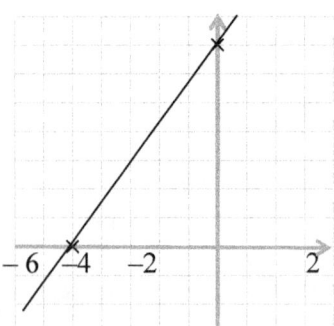

16. The graph representing inconsistent lines is:

294

[A] 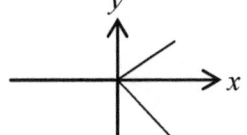 [B]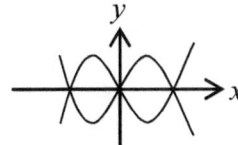

[C] 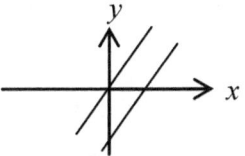 [D]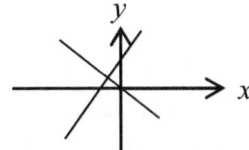

17. The table of values for $y = x - 6$ is:

[A]
	−5	−8	−7
	1	−14	−13

[B]
x	−5	−8	−7
y	−11	−2	−13

[C]
	−5	−8	−7
	−11	−14	−13

[D]
x	−5	−8	−7
y	1	−2	−1

18. The equation, which corresponds to the table of values, is:
   [A] $y = 4 + 5x$ [B] $y = 3 + 6x$ [C] $y = 5 + 4x$ [D] $y = 6 + 3x$

Input (x)	1	2	3	4	5
Output (y)	9	12	15	18	21

19. In the figure below, the shaded portion shows the boundary of the half plane defined by the inequality:
   [A] $4x + 3y > 6$
   [B] $4x + 3y < 6$
   [C] $4x + 3y \geq 6$
   [D] $4x + 3y \leq 6$

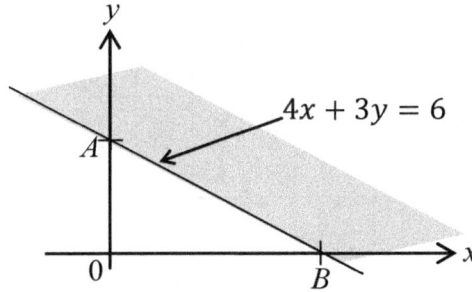

20. In the figure above, the co-ordinates of point B is:
   [A] $\left(0, 1\frac{1}{2}\right)$ [B] $(0, 2)$ [C] $(2, 0)$ [D] $\left(1\frac{1}{2}, 0\right)$

21. The inequality illustrated in the sketch graph in the figure below is:
   [A] $y \geq -2x + 3$ [B] $y \geq -3x + 3$ [C] $y \geq 3x + 3$ [D] $y \geq 3x + 2$

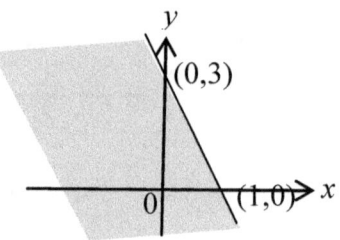

22. In the figure below, the region P, Q, R or T which satisfies the inequalities $0 < y < 1$, $y < x + 2$ and $x < 0$ is:

   [A] P [B] Q [C] R [D] S

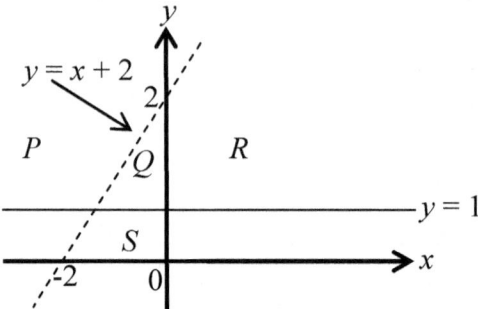

23. In the figure below, we can represent the region defined by triangle OPQ by the inequalities:

   [A] $y \geq x, y \leq 0, y + x \geq 7$ [B] $y \leq x, y \geq 0, y + x \leq 7$
   [C] $y \geq x, y \geq 0, y + x \leq 7$ [D] $y \leq x, y \leq 0, y + x \leq 7$

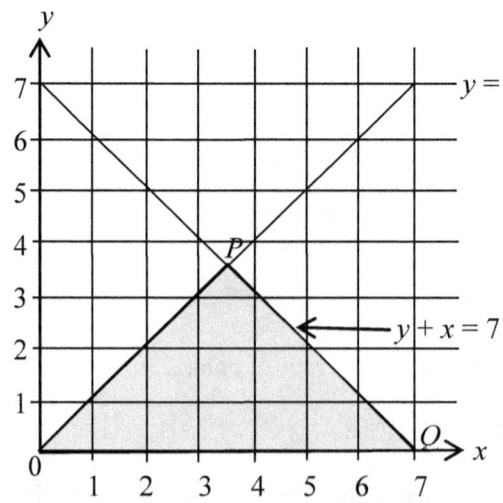

24. In the figure below, the equations of the lines AC, AB, and BC are:

[A] $x = 2, y = 3, y + x = 7$ [B] $x = 3, y = 2, y + x = 7$
[C] $x + 2 = 0, y = 3, y + x = 7$ [D] $x = 2, y + 3 = 3, y + x = 7$

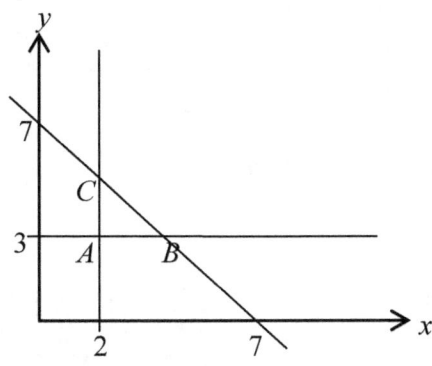

25. In the figure above, the three inequalities, which define the triangle ABC, are:
    [A] $x \geq 2, y \leq 3, y + x \leq 7$ [B] $x \geq 3, y \geq 2, y + x \leq 7$
    [C] $x \geq 2, y \geq 3, y + x \leq 7$ [D] $x \geq 2, y + 3 \geq 0, y + x \geq 7$

26. We can best describe the shaded region in the figure below by:
    [A] $x \leq 3, y \leq 4$ and $4x + 3y \leq 12$ [B] $x \leq 3, y < 4$ and $4x + 3y < 12$
    [C] $x \geq 3, y \geq 4$ and $4x + 3y > 12$ [D] $x \leq 3, y < 4$ and $4x + 3y > 12$

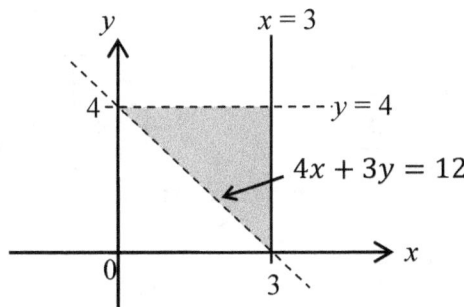

27. The graph below which represents the inequality, $x+1 \leq 0$ is:

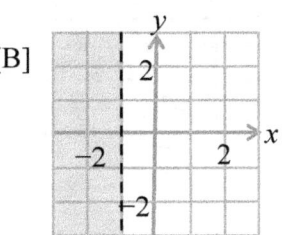

[C] [image of graph] [D] [image of graph]

28. The correct graphical representation of the inequalities $y > x + 2$ and $y \geq 1 + 2x$ is:

[A] 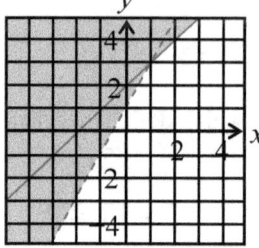 [B]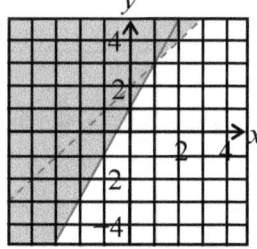

[C] 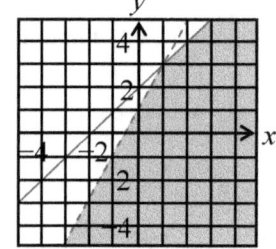 [D]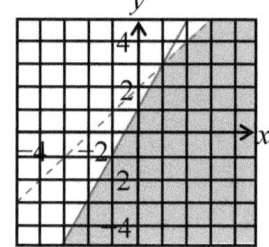

29. The graph in the figure below which represents the inequality $y \geq x - 1$ is:

[A] 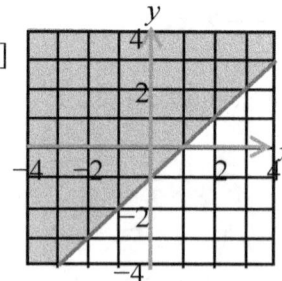 [B]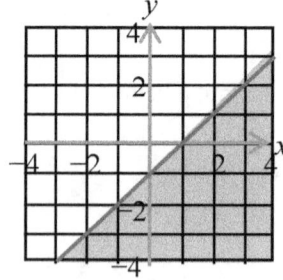

Module 17, Topic 15: Inequalities and Inequations

[C] 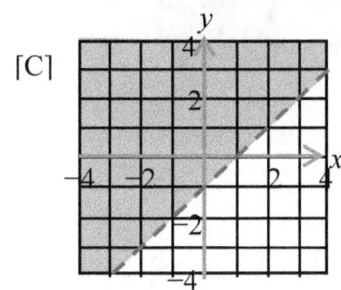 [D]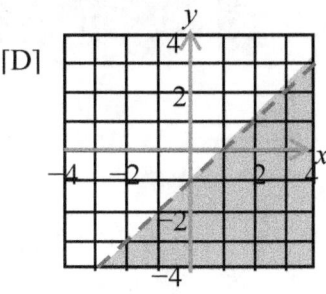

30. The linear inequality, which represents the graph below is:
    [A] $y \leq -x + 1$ [B] $y \geq -x + 1$ [C] $y \leq x + 1$ [D] $y \geq x + 1$

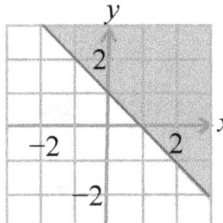

Topic 16

SEQUENCES AND SERIES

Objectives

At the end of this topic, the learner should be able to:

1. Identify number patterns.
2. Deduce a general rule for a simple number pattern and sequence.
3. Find the subsequent terms of sequence.
4. Recognize APs and GPs as special sequences.
5. Find a common difference for an AP.
6. Find a common ratio for a GP.
7. Find the nth term of an AP and a GP.
8. Find the arithmetic mean and geometric mean.
9. State and apply the formula for the sum of the first n terms of an AP and that of a GP.

Module 17, Topic 16: Sequences

16.1 Number Sequence

In section 2.5 of book 2, we saw that we can write a number sequence using a given rule or state the rule given a number sequence. In this topic, we shall learn how to obtain a sequence from a given formula or define a given sequence using a given formula.

16.2 The nth Term of a Sequence

Sometimes instead of listing terms of sequences as above, a formula is usually given which defines the general term U_n of the sequence. If instead of repeating terms separating them with commas, we repeat terms separating them with (+) signs, we call such an ordered arrangement of numbers a **series**. For instance, 1+2+3+4+5+6+... is a series. Using the given general term of a sequence or series, we can obtain every other term. Sometimes we denote the first term by a instead of U_1.

 Example

1. The nth term U_n of a series is given by $U_n = 2n^2 + 1$, where $n \in \mathbb{N}$. List the first 5 terms of the series.

   Solution
   Therefore the first five terms of the series are, $3 + 9 + 19 + 33 + 51 + \cdots$

   Write down the sixth term U6 of a sequence whose n^{th} term is given by $U_n = n(n+1)$, where $n \in \mathbb{N}$.

   Solution
   $U_n = n(n+1) \Rightarrow U_6 = 6(6+1) = 42$

2. Find the nth term of each of the following sequences.
   (a) 1, 4, 7, 10, 13, 16 ...   (b) 2, 5, 10, 17 ...
   Solution

   (a) 1, 4, 7, 10, 13, 16 ...

   $1 + 3(0) = 1$
   $1 + 3(1) = 4$
   $1 + 3(2) = 7$
   $1 + 3(3) = 10$
   $1 + 3(4) = 13$
   $1 + 3(5) = 16$
   $\therefore U_n = 1 + 3(n-1) = 3n - 2$

   (b) 2, 5, 10, 17 ...
   $1 + 1^2 = 2$
   $1 + 2^2 = 5$
   $1 + 3^2 = 10$
   $1 + 4^2 = 17$
   $\therefore U_n = 1 + n^2$

16.3 Sum of the First n Terms of a Sequence

Just as there can be a formula for the nth term U_n of a sequence, there can be a formula for the sum S_n, of the first n terms of a sequence, where $n \in \mathbb{N}$.

Example

The sum S_n of the first n terms, of a sequence is given by $S_n = 3n^2 - 2n$, where $n \in \mathbb{N}$. Find (a) S_1, S_2 and S_{10}. (b) U_1, U_2 and U_{10}.

Solution

$S_n = 3n^2 - 2n$

(a) $S_1 = 3(1)^2 - 2(1) = 1$
$S_2 = 3(2)^2 - 2(2) = 8$
$S_{10} = 3(10)^2 - 2(10) = 280$

(b) $U_1 = S_1 = 1$
$U_2 = S_2 - S_1 = 8 - 1 = 7$
$U_{10} = S_{10} - S_9$
But $S_9 = 3(9)^2 - 2(9) = 225$
$U_{10} = 280 - 225 = 55$

Exercise 16:2

1. Write down the nth term of the following sequences:
   (a) $3, 6, 12, 24, \ldots$
   (b) $5, 8, 11, 14, \ldots$
   (c) $1, 8, 27, 64, \ldots\ldots$
   (d) $10, 40, 160, 640 \ldots$
   (e) $1, \frac{1}{3}, \frac{1}{4}, \frac{1}{8}, \ldots$
   (f) $\frac{1}{2}, \frac{2}{3}, \frac{3}{4}, \frac{4}{5}, \ldots$

2. Write down the first three terms in the sequences whose nth terms are given below:
   (a) $u_n = n(n+4)$
   (b) $u_n = n(n+1)$
   (c) $u_n = \frac{n}{10} + \frac{1}{n}$
   (d) $u_n = n^2 - 5n$
   (e) $u_n = 3(2)^n$

3. The sum of the first n terms of a sequence is $\frac{1}{3}n(n+1)(n+2)$. Find the first four terms and the nth term, where $n \in \mathbb{N}$.

4. The sum of the first n terms of a sequence is $n(n + 3)$, where $n \in \mathbb{N}$. Find the first three terms and the nth term.

5. The sum of the first n terms of a sequence is $\frac{1}{6}n(n+1)(2n+1)$, where $n \in \mathbb{N}$. Find S_1, S_2, S_3, S_4 and S_5. Hence, find the first five terms. Write down the nth term of the sequence.

6. The sum of the first n terms of a series is $\frac{1}{2}n(n+1)$, where $n \in \mathbb{N}$, Find the first three terms and the nth term.
7. Give, with reasons, the next two terms in each of the following sequences:
   (a) 1,3,7,13,21, ...
   (b) $1, \frac{1}{2}, \frac{1}{6}, \frac{1}{24}, \frac{1}{120}, ...$
8. The nth term U_n of a certain sequence is given as $U_n = (4-n)^2 - 1$, where $n \in \mathbb{N}$. Write down the first five terms of the sequence.
9. The sum of the first n terms of a sequence is given by $n.2^n$, where n is a positive integer. Find the third term of the sequence.
10. The sum of the first n terms of a sequence of number is given by $3n^2 + 2n$, where $n \in \mathbb{N}$.
    (a) Find the first 3 terms of the sequence.
    (b) Find the sum of first 12 terms of the sequence.
11. The first term of a sequence of numbers is –10 and each succeeding term of the sequence is obtained by adding 4 to the previous term.
    (a) Write down the first 5 terms of the sequence.
    (b) Given that the sum of the first n terms of the sequence is $2n^2 + 12n$, where $n \in \mathbb{N}$, calculate the sum of the first 11 terms.

16.4 Arithmetic and Geometric Progressions

The arithmetic and geometric progressions are two very important types of sequences in daily life especially in fields like banking and Economics. These progressions are so rampant that they require some special focus. This section will give many Economic students, an opportunity to appreciate better, the so-called Malthusian theory published sometimes in 1798. What did Malthus, Thomas Roberts really mean by population was growing at a geometric progression while food production was growing at an arithmetic progression?

16.5 Arithmetic Progression (A.P.)

The **common difference** of an arithmetic progression (A.P.), denoted by d is the difference between a term and the preceding one and is constant. In other words, to obtain each term we add the common difference d to the preceding one. Some examples of arithmetic progression are, 14,17,20,23 ... and 4,2,0,–2,–4 ...

> **Example**
>
> 1. In the following A.P's, write down the first term a and the common difference d.
>   (a) 14, 17, 20, 23, ...   (b) 4, 2, 0, −2, −4, ...   (c) 1, 2, 3, ...
>   (d) 4, 7, 10, ...   (e) $1\frac{1}{2}, -1\frac{1}{2}, -4\frac{1}{4}, ...$
>
>   **Solution**
>   (a) $a = 14, d = 3$ (b) $a = 4, d = -2$ (c) $a = 1, d = 1$
>   (d) $a = 4, d = 3$ (e) $a = 1\frac{1}{2}, d = -3$
>
> 2. Insert four numbers between 7 and 67 so that the sequence is an A.P.
>
>   **Solution**
>   Let the common difference of the A.P. be d then consecutive terms of the A.P. are $7, 7+d, 7+2d, 7+3d, 7+4d, 67$.
> $$\Rightarrow d = 67 - (7+4d)$$
> $$d = 67 - 7 - 4d$$
> $$5d = 60 \Rightarrow d = 12$$
>   Therefore the four numbers are
> $7 + 12, 7 + 2(12), 7 + 3(12), 7 + 4(12)$ i.e.
> $19, 31, 43$ and 55 in that order.

16.6 The nth term of an A.P.

Consider an A.P. whose first term is a and whose common difference is d. Then we can write this A.P. as $a, a + d, a + 2d, a + 3d, ...$
$$\Rightarrow U_1 = a + (1-1)d = a$$
$$U_2 = a + (2-1)d = a + d$$
$$U_3 = a + (3-1)d = a + 2d$$
$$U_4 = a + (4-1)d = a + 3d$$
And so on
$$\therefore U_n = a + (n-1)d \text{ where } n \in \mathbb{N}.$$

> Hence the general term of an A.P. is $U_n = a + (n-1)d$.

Module 17, Topic 16: Sequences

 Example

1. Find the 200th term of the A.P. 4, 2, 0, – 2, – 4 ...

   Solution
   $U_n = a + (n-1)d$
   $a = 4$, $d = -2$ and $n = 200$
   $U_{200} = 4 + (200 - 1)(-2) = 4 + 199(-2) = -394$

2. The nth term of the A.P. $\frac{1}{2}, \frac{3}{2}, \frac{5}{2}, \frac{7}{2}, ...$ is $\frac{649}{2}$. Find the value of n.

   Solution
   $U_n = a + (n-1)d$
   $a = \frac{1}{2}$, $d = 1$ and $U_n = \frac{649}{2}$
   $\frac{1}{2} + (n-1)(1) = \frac{649}{2}$
   $\qquad n - 1 = 324$
   $\qquad\quad n = 325$

16.7 Sum of the first *n* terms of an A.P.

 Example

1. Find the sum of the first 5 terms of the *A.P.* 3, 7, 11, 15, 19....

   Solution
   $S_5 = 3 + 7 + 11 + 15 + 19 = 55$

   Alternatively;
   $S_5 = 3 + 7 + 11 + 15 + 19 = 55$

   $\quad S_5 = 3 + 7 + 11 + 15 + 19$ ....①
   Or $S_5 = 19 + 15 + 11 + 7 + 3$ ....②

①+②:
$$2S_5 = 22+22+22+22+22$$
$$2S_5 = 110$$
Divide both sides by 2,
$$S_5 = 55$$
This could have been done more easily by adding the first and last terms, multiplying by the number of terms, then dividing by 2.
When the numbers of terms is large, the alternative method is a better method for finding the sum of the first n terms of an A.P.

2. Find the sum of the first 75 terms of the A.P. 3,7,11,15,19 ....

Solution
To list all these terms before adding them is cumbersome. So let us use the alternative method.
$$a = 3 = U_1, \ d = 4$$
$$U_n = a+(n-1)d$$
$$U_{75} = 3+(75-1)4 = 299$$
$$2S_{75} = 75(U_1 + U_{75})$$
$$\Rightarrow 2S_{75} = 75(3+299)$$
$$\Rightarrow 2S_{75} = 22650$$
$$\therefore S_{75} = 11325$$

To generalize the above, suppose the first term of an A.P. is a and the common ratio is d, then the nth term of the A.P. is
$$U_n = a+(n-1)d$$
$$2S_n = n(U_1+U_n) \Rightarrow 2S_n = n\{2a+(n-1)d\}$$
$$\Rightarrow S_n = \frac{n}{2}\{2a+(n-1)d\}$$

Therefore if $n \in \mathbb{N}$, then the sum of the first n terms of an A.P. is given by;

$$S_n = \frac{n}{2}\{2a+(n-1)d\} \text{ or } S_n = \frac{n}{2}(a+l)$$

Example

Find the sum of the first 20 terms of the A.P. 3, 7, 11, 15...

Solution

$a = 3, d = 4, n = 20$

$$S_n = \frac{n}{2}\{2a + (n-1)d\}$$

$$S_{20} = \frac{n}{2}\{2(3) + (20-1)4\} = 10\{6 + 19(4)\} = 820$$

16.8 The Arithmetic Mean

If p, q, r are consecutive terms of an A.P. then, the common difference of the A.P. will be

$$d = q - p = r - q$$
$$\Rightarrow 2q = p + r$$
$$q = \frac{p+r}{2}$$

We call q the arithmetic mean of p and r. Another name for arithmetic mean is the average denoted by $\bar{x}$. Hence, the formula for finding the arithmetic mean of two numbers x_1 and x_2 is

$$\bar{x} = \frac{x_1 + x_2}{2}$$

We define the arithmetic mean of a set of n numbers $x_1, x_2, x_3, \ldots, x_n$ as the quotient of the sum of the numbers and the number of numbers. i.e.

$$\bar{x} = \frac{x_1 + x_2 + x_3 + \cdots + x_n}{n}$$

We can insert n numbers $x_1, x_2, x_3, \ldots, x_n$ between two numbers a and b, so that the sequence is an arithmetic progression.

Example

Find the arithmetic mean of 13 and 17.

Solution
$$\bar{x} = \frac{a+b}{2} = \frac{13+17}{2} = 15$$

 Exercise 16:3

1. Find the 24th term of the sequence $7, 13, 19, ...$.
2. The common difference of an arithmetic progression is 2 and the sum of the first 28 terms is 0. Find the first term and the 28th term.
3. Given that, the nth term of the sequence $6, 11, 16, 21 ...$ is 121. Find the value of n.
4. Find the first four terms of the arithmetic progression with first term 5 and the seventh term is -11.
5. The first and last terms of an arithmetic progression are -3 and 25. The sum of all the terms is 1837. Find (a) The number of terms;
   (b) The common difference, and (c) The middle term.
6. Insert four arithmetic means between the numbers 9 and 29. [Hint: The four arithmetic means are numbers such that the sequence is an A.P.]
7. Find the sum of the first ten terms of the sequence $3, 7, 11...$. What is the sum of the next ten terms?
8. The sum of the first n terms of the sequence $7, 11, 15,...$ is 250. Find the value of n.
9. The sum of five consecutive numbers of an arithmetic progression is 50, and the sum of their squares is 590. Find the numbers.
10. The first term of an A.P. is 4 and the common difference is 7. How many terms are required in order that the sum may exceed 500?
11. Insert three arithmetic means between a and b. In an A.P., the first term is a and the common difference is $2a$. Show that the sum to $2n$ terms is always equal to the sum of the n terms.

16.9 Geometric Progression (G.P.)

In a Geometric progression, the ratio of a term and the preceding one called the **common ratio** r is constant. In other words, each term is obtained by multiplying the preceding one by the common ratio r. Some examples of geometric progressions are,

(a) $1, 4, 16,...$ (b) $4, 2, \frac{1}{2}...$

Module 17, Topic 16: Sequences

 Example

In the following G.P's, write down the first term a and the common ratio r.

(a) 1, 4, 16,... (b) 4, 2, $\frac{1}{2}$... (c) 6, 18, 54,...

(d) 7, 14, 28,... (e) $\frac{1}{9}, \frac{1}{3}, 1, 3, \cdots$

Solution

(a) $a = 1, r = 4$ (b) $a = 4, r = \frac{1}{2}$ (c) $a = 6, r = 3$

(d) $a = 7, r = 2$ (e) $a = \frac{1}{9}, r = 3$

16.10 The nth term of a G.P.

Consider a G.P. whose first term is a and whose common ratio is r. Then we can write this G.P. as

$$a, ar, ar^2, \ldots$$

$$\Rightarrow U_1 = ar^{1-1} = a, \quad U_2 = ar^{2-1} = ar, \quad U_3 = ar^{3-1} = ar^2$$

And so on

$$\therefore U_n = ar^{n-1}$$

Hence, the general formula for the nth term of a G.P. is $U_n = ar^{n-1}$, where $n \in \mathbb{N}$.

 Example

1. Find the 30th term of the G.P. $4, 2, 1, \frac{1}{2}, \ldots$, leaving your answer in index form.

   Solution
   $$U_n = ar^{n-1}$$
   $$a = 4, r = \frac{1}{2} \text{ and } n = 30 \Rightarrow U_{30} = 4\left(\frac{1}{2}\right)^{30-1} = 4\left(\frac{1}{2}\right)^{29} = 4(2)^{-29} = 2^{-27}$$

2. The nth term of the G.P. 2, 4, 8, 16… is 2048. Find the value of n.
   Solution
   $$U_n = ar^{n-1}$$
   $a = 2, r = 2$ and $U_n = 2048 \Rightarrow 2(2)^{n-1} = 2048$
   $$2^n = 2^{11} \Rightarrow n = 11$$

16.11 The Sum of n terms of a G.P.

 Example

Find the sum of the first 5 terms of the G.P. 2, 6, 18, 54, 162 ….

Solution
$S_5 = 2 + 6 + 18 + 54 + 162 = 242$

Alternatively;
$S_5 = 2 + 6 + 18 + 54 + 162$ …………………①
The common ratio $r = 3$
Multiply both sides by the common ratio 3
$3S_5 = 6 + 18 + 54 + 162 + 486$ ……………②
②− ①:
$\quad 2S_5 = 486 - 2$
$\Rightarrow 2S_5 = 484$
Divide both sides by 2,
$\quad S_5 = 242$

Notice the way the terms cancel out on subtracting ① from ②, leaving only the last term of ② and the first term of ①.
When the number of terms is large, the alternative method is a better method for finding the sum of the first n terms of a G.P.

Example

Find the sum of the first 18 terms of the G.P. $\frac{1}{4}, \frac{1}{2}, 1, 2, 4, 8, \ldots$

Solution
To list all these terms before adding them is cumbersome. So let us use the alternative method.

$$a = \frac{1}{4}, \ r = 2, \ n = 18, \ U_n = ar^{n-1}$$

$$U_{18} = \frac{1}{4}(2^{18-1}) = 2^{15}$$

$$S_{18} = \frac{1}{4} + \frac{1}{2} + 1 + 2 + 4 + \ldots + 2^{15} \ldots \ldots \ldots \textcircled{1}$$

Multiply both sides by $r = 2$

$$2S_{18} = \frac{1}{2} + 1 + 2 + 4 + \cdots + 2^{16}$$

$$2S_{18} - S_{18} = 2^{16} - \frac{1}{4}$$

$$S_{18} = \frac{4(2^{16}) - 1}{4} = 65535\frac{3}{4}$$

To generalize the above, suppose the first term of a G.P. is a and the common ratio is r, then the nth term of the G.P. is

$$U_n = ar^{n-1}$$

$$S_n = a + ar + ar^2 + \ldots + ar^{n-1} + \ldots \ldots \ldots \textcircled{1}$$

Multiply both sides by r.

$$rS_n = ar + ar^2 + \ldots + ar^n + \ldots \ldots \ldots \textcircled{2}$$

$\textcircled{2} - \textcircled{1}$:

$$rS_n - S_n = ar^n - a$$
$$rS_n - S_n = ar^n - a$$
$$S_n(r-1) = a(r^n - 1)$$

$$\Rightarrow S_n = \frac{a(r^n - 1)}{(r-1)}, |r| > 1 \text{ or } S_n = \frac{a(1 - r^n)}{(1 - r)}, |r| < 1$$

Depending on the individual or problem, either of the forms of this formula may be used.

It is not necessary to memorize the two since one can easily be obtained from the other and whichever is used the final result is the same.

Therefore, the sum of the first n terms of a G.P. is given by

$$S_n = \frac{a(r^n - 1)}{(r - 1)} \quad \text{or} \quad S_n = \frac{a(1 - r^n)}{(1 - r)}$$

 Example

Find the sum of the first 20 terms of the G.P. 3, 6, 12, 24, ...

Solution

$$S_n = \frac{a(r^n - 1)}{(r - 1)}$$

$a = 3, r = 2$ and $n = 20$

$$\Rightarrow S_{20} = \frac{3(2^{20} - 1)}{2 - 1} = 3(1048576 - 1) = 3145725$$

16.12 Sum to infinity, S_∞

Consider a G.P. with $|r| < 1$. As the number of terms, n increases (i.e. $n \to \infty$) the value of r^n decreases (i.e. $r^n \to 0$) and the sum S_n of the G.P. approaches a limiting value called **sum to infinity** denoted by S_∞.

$$S_\infty = \frac{a}{(1 - r)}, \quad |r| < 1$$

 Example

Find the sum to infinity of the G.P. whose first term is 81 and whose common ratio is $\frac{1}{3}$.

> **Solution**
> $S_\infty = \frac{a}{1-r}$, $a = 81, r = \frac{1}{3}$
> $S_\infty = \frac{81}{1-\frac{1}{3}} = \frac{81}{\frac{2}{3}} = \frac{243}{2}$

16.13 Geometric Mean

Let a, b, c be three consecutive terms of a G.P. Then by the definition of a G.P., the common ratio r is given by

$$r = \frac{b}{a} \text{ or } r = \frac{c}{b} \Leftrightarrow \frac{b}{a} = \frac{c}{b}$$
$$b^2 = ac$$
$$b = \pm\sqrt{ac}$$

b is said to be the **geometric mean** of a and c, usually denoted by GM.

$$GM = \pm\sqrt{ac}$$

Generally, the geometric mean of a set of n numbers $x_1, x_2, x_3, \ldots, x_n$ is defined as the nth root of the product of the n numbers. i.e.

$$GM = \pm\sqrt[n]{x_1 x_2 \ldots x_n}$$

We can insert n numbers between two numbers a and b, so that the sequence is a G.P.

Example

1. Find the geometric mean of 6 and $13\frac{1}{2}$.

   Solution

   $$GM = \pm\sqrt{ac} = \pm\sqrt{6\left(\frac{27}{2}\right)} \pm \sqrt{81} = \pm 9$$

2. Insert 2 numbers between -6 and $20\frac{1}{4}$, so that the sequence is a G.P.

   Solution

   Let the common ratio of the G.P. be r then consecutive terms of the G.P. are $-6, -6r, -6r^2, 20\frac{1}{4}$.

   $$\Rightarrow \frac{-6r}{-6} = \frac{20\frac{1}{4}}{-6r^2} \Rightarrow r^3 = -\frac{81}{24} = -\frac{27}{8} \Rightarrow r = -\frac{3}{2}$$

   Therefore, the 2 geometric means are $-6\left(-\frac{3}{2}\right)$ and $-6\left(-\frac{3}{2}\right)^2$ i.e. 9 and $-\frac{3}{2}$ in that order.

Module 17, Topic 16: Sequences

 Exercise 16:4

1. The first three terms of a geometric progression are $2x, x+6, 4x+1\frac{1}{2}$.
   Find (a) The two possible values of x.
   (b) The common ratio and the fourth term of that progression for which x is positive.

2. The second term of a G.P. is -6 and the fifth term is $20\frac{1}{4}$. Find the common ratio and the nth term.

3. The third term of a geometric sequence exceeds the second by $\frac{9}{14}$ and the second exceeds the first by $\frac{3}{7}$. Find the common ratio.

4. The first term of a geometric progression is $\frac{3}{5}$ and the fourth term is $9\frac{3}{8}$.
   Find the 12th term and the sum of the first 12 terms.

5. Find the geometric means of (a) 6 and 24 (b) 36 and $20\frac{1}{4}$
   (c) x^2 and x^4

6. Insert three numbers between the following so that the sequence is a G.P.
   (a) 2 and 32 (b) 16 and $\frac{1}{16}$

7. In Table 24:1, find the sum of the series in column A to the number of terms indicated in column B.

A	B
(a) $1+2+4+8+...$	10
(b) $4-8+16-32+...$	8
(c) $6+2+\frac{2}{3}+\frac{2}{9}+...$	7
(d) $1+3a+9a^2+27a^3+...$	10

8. In both an *A.P.* and a *G.P.* the first and fourth terms are 4 and $\frac{1}{2}$ respectively. Calculate
   (a) The common difference of the *A.P.*
   (b) The common ratio of the *G.P.*
   (c) The second and fifth terms of each progression.
   (d) The number of terms of the *A.P.*, which have a sum of $3\frac{1}{2}$.

Multiple Choice Exercise 16

1. The next 2 terms in the sequence 1, 2, 4, 7, 11, 16... are:
   [A] 17, 29   [B] 29, 24   [C] 22, 29   [D] 29, 40
2. The next term in the sequence 1, 4, 9, 16,...is:
   [A] 20   [B] 25   [C] 23   [D] 27
3. The next term in the sequence 2, 5, 11, 23, 47...is:
   [A] 95   [B] 93   [C] 71   [D] 27
4. In the sequence 1,3,7,15,31 the number that must be added to 31 to give the next term is:
   [A] 4   [B] 8   [C] 16   [D] 32
5. The number represented by * in the arithmetic progression 14, − 3, *, − 37 is:
   [A] 11   [B] −14   [C] 17   [D] − 20
6. The 4th term of an $A.P.$ whose first term is 2 and whose common difference is 0.5 is:
   [A] 3   [B] $\frac{7}{2}$ [C] $\frac{11}{2}$ [D] 5
7. The first term of an $A.P.$ is equal to twice the common difference d. In terms of d the 5th term of the $A.P.$ is:
   [A] $4d$ [B] $5d$ [C] $6d$ [D] $a + 5d$
8. The 9th term of the Arithmetic progression 18, 12, 6, 0,−6,... is:
   [A] −54   [B] −30   [C] 30   [D] 42
9. The eleventh term of an $A.P.$ is 25 and its first term is 3. Its common difference is:
   [A] $1\frac{9}{10}$ [B] $2\frac{1}{5}$ [C] $2\frac{4}{5}$ [D] $2\frac{1}{2}$
10. If the first term of an $A.P.$ is 4 and the 5th term is 12, then the mean of the first five terms is:
    [A] 4   [B] 6   [C] 8   [D] 10
11. The nth term U_n of the $A.P.$ 11, 4, −3... is:
    [A] $U_n = 19 + 7n$ [B] $U_n = 19 - 7n$
    [C] $U_n = 18 - 7n$ [D] $U_n = 18 + 7n$
12. The nth term Un of the sequence 4, 10, 16,... is:
    [A] $2(3n-1)$ [B] $2(2+3^{n-1})$ [C] $2^2 + 2$ [D] $2(3n+1)$
13. In an $A.P.$, the first term is 2, and the sum of the first and the 6th terms is $16\frac{1}{2}$. The 4th term is:

[A] $5\frac{1}{2}$ [B] $9\frac{1}{2}$ [C] 8 [D] 7

14. The sum of the 1st and 2nd terms of an A.P. is 4 and the 10th term is 19. The sum of the 5th and 6th terms is:
    [A] 11 [B] 22 [C] 21 [D] 20

15. It is observed that:
    If $1 + 3 + 5 + 7 + 9 + 11 + 13 + 15 = p^2$, then the value of p is:
    [A] 6 [B] 7 [C] 8 [D] 9
    $$1 + 3 = 2^2$$
    $$1 + 3 + 5 = 3^2$$
    $$1 + 3 + 5 + 7 = 4^2$$

16. The common ratio of a G.P. is 2. If the 5th term is greater than the first by 45, the 5th term will be:
    [A] 3 [B] 6 [C] 45 [D] 48

17. The fifth and seventh terms of the geometric progression
    $-2, -3, -\frac{9}{2}, -\frac{27}{4} \cdots$ are respectively:
    [A] $-\frac{81}{16}, -\frac{729}{32}$ [B] $-\frac{8}{81}, \frac{72}{18}$ [C] $-\frac{21}{8}, \frac{32}{618}$ [D] $-\frac{27}{16}, -\frac{79}{18}$

18. The 6th term of a geometric progression is $-\frac{2}{27}$ and the first term is 18. The common ratio is:
    [A] $-\frac{1}{2}$ [B] $-\frac{1}{3}$ [C] $\frac{1}{4}$ [D] 3

19. The common ratio of the G.P. log 3, log 9, log 81,... is:
    [A] 1 [B] 2 [C] 3 [D] 6

20. The 16th term of the G.P. 2, 6, 18,... is given by:
    [A] 2×3^{12} [B] 2×3^{13} [C] 2×3^{15} [D] 2×3^{16}

21. The sum of the first 5 terms of the G.P. 2, 6, 18,... is:
    [A] 121 [B] 243 [C] 242 [D] 130

22. If the second and fourth terms of a G.P. are 8 and 32 respectively, the sum of the first four terms will be:
    [A] 28 [B] 40 [C] 48 [D] 60

23. If the second and 5th terms of a G.P. are –6 and 48 respectively, the sum of the first four terms is:
    [A] –45 [B] –15 [C] 15 [D] 45

24. The nth term of a sequence is represented by $3 \times 2^{(2-n)}$. The first three terms of the sequence are respectively:

[A] $\frac{3}{2}, 3, 6$ [B] $6, 3, \frac{3}{2}$ [C] $\frac{3}{2}, 3, \frac{1}{3}$ [D] $\frac{2}{3}, 3, \frac{8}{3}$

25. The nth term of a sequence is $2^{(2n-1)}$. If $U_n = 2^9$, n is:
    [A] 3   [B] 4   [C] 5   [D] 6
26. The next two terms of the sequence 1,5,14,30,55,... are respectively:
    [A] 61,110   [B] 67,116   [C] 81,140   [D] 91,140
27. The nth term of a sequence is given by $(-1)^{(n-2)}$. The sum of the second and third terms is:
    [A] 0   [B] 1   [C] 2   [D] 6
28. The sum of the first n terms of a sequence is given by $S_n = 17n - 3n^2$. The fourth term of the sequence is:
    [A] 20   [B] -10 [C] -4 [D] 10
29. The nth term of a sequence is given as $\frac{n^2 + n}{2}$. The 7th term of the sequence is:
    [A] 36   [B] 28   [C] 21   [D] 14
30. The common difference in the sequence 10, 2, -6, -14, . . .is:
    [A] 10   [B] 8   [C] -16 [D] -8
31. The next three terms and the rule that describe the sequence 10, 20, 40, 80 . . . . are:
    [A] 82, 84, 86; start with 10 and add 2 repeatedly.
    [B] 90, 100, 110; start with 10 and add 10 repeatedly.
    [C] 320, 1280, 5120. Start with 10 and multiply by 4 repeatedly.
    [D] 160, 320, 640; start with 10 and multiply by 2 repeatedly.
32. The next three terms in the sequence 3,12,21,30,... are:
    [A] 40, 50, 60   [B] 38, 46, 54   [C] 39, 48, 57   [D] 36, 32, 39
33. The next four terms in the sequence $-4, -1, 2, 5$... are:
    [A] 5, 8, 11, 14   [B] 8, 11, 14, 17   [C] 3, 6, 9, 12   [D] 0, 8, 11, 14
34. If n points are marked on a circle, where n is a whole number greater than 1, we can draw $\frac{1}{2}n^2 - \frac{1}{2}n$ segments to connect these points. The number of segments that we can draw if 8 points are marked on the circle is:
    [A] 28 segments   [B] 32 segments
    [C] 60 segments   [D] 56 segments
35. If the pattern continues, the number of squares in the 8th diagram is:
    [A] 16 squares   [B] 18 squares   [C] 14 squares   [D] 15 squares

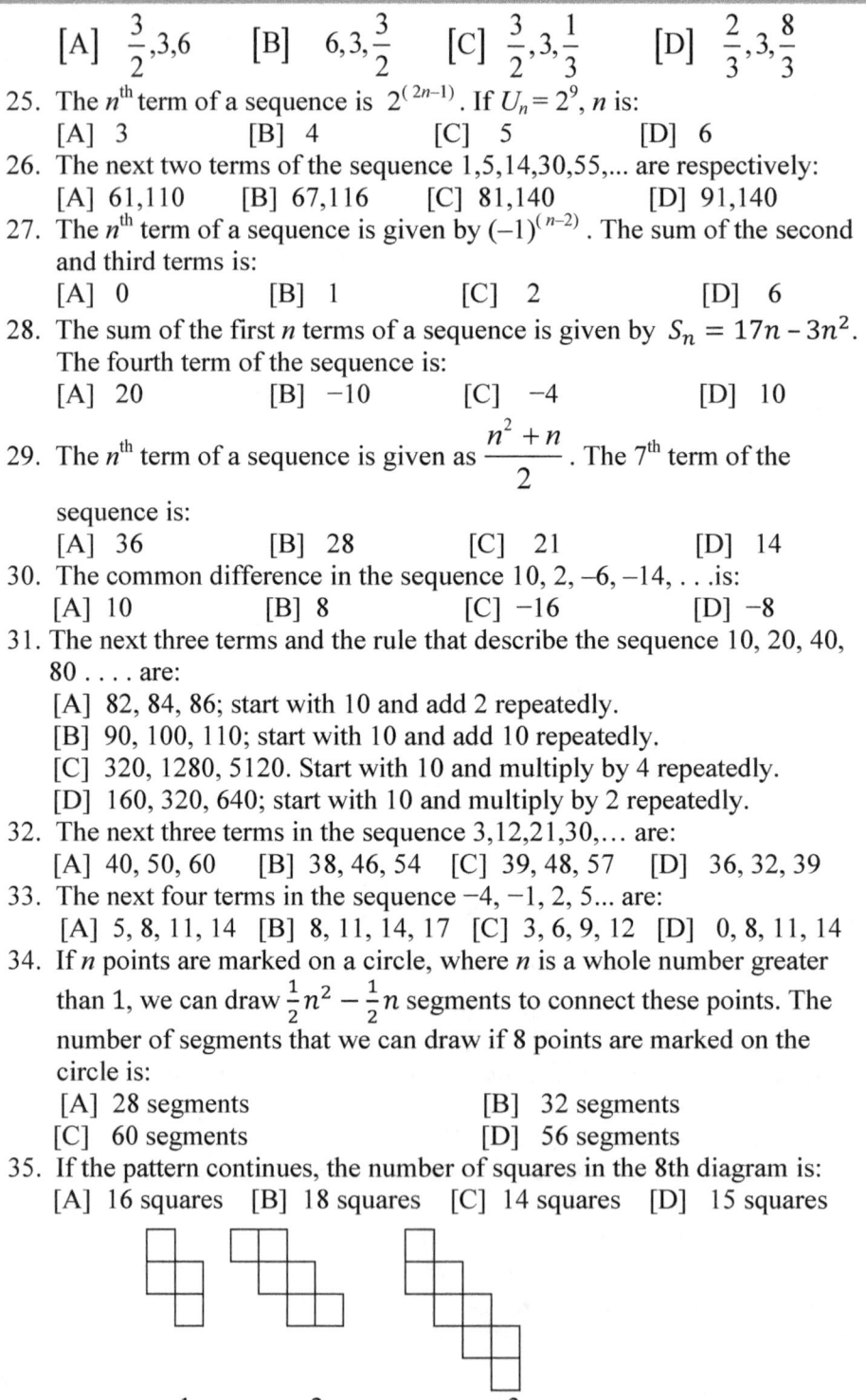

36. The number of dots in the ninth figure is:
    [A]   25 dots   [B]   27 dots   [C]   19 dots   [D]   26 dots

```
                  •              •
         •      • • •        • • • • •
         1        2              3
```

37. The diagram below represents a five by five square of black and white floor tiles. The number of black tiles, which we can add to the existing pattern to make it six by six, is:
    [A]   8   [B]   6   [C] 5   [D] 4

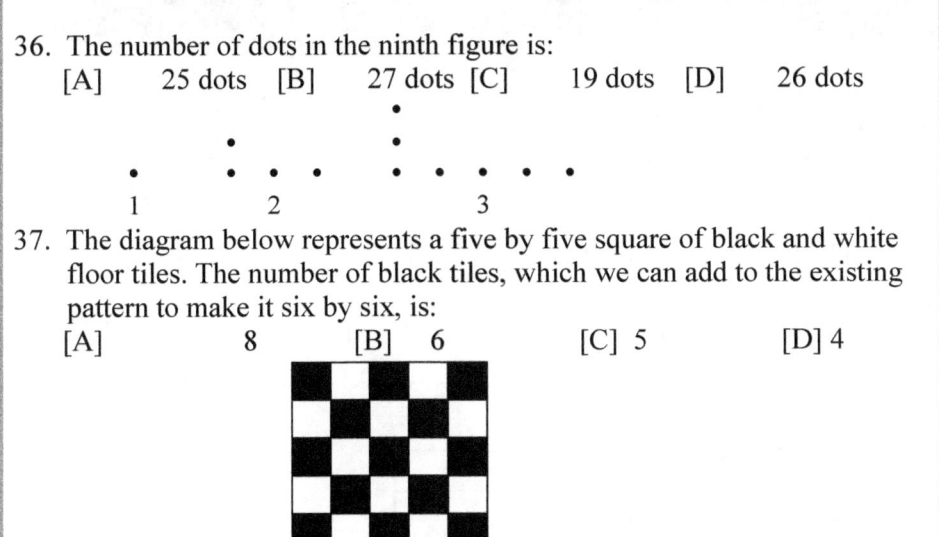

Answers to Structural Exercises

Exercise 1:1
1. (a) 1080 (b) 500 (c) 9500 (d) 5140
   (e) 22100 (f) 360 (g) 470 (h) 47400
2. (a) 3 (b) 12 (c) 4 (d) 6
3. (a) 1.1 (b) 14.6 (c) 0.6 (d) 3.9

Exercise 1:2
1. (a) 1,000 (b) 38,000 (c) 30 (d) 0.8 (e) 54 (f) 24
2. (a) 3 (b) 10 (c) 2.1 (d) 0.1 (e) 20 (f) 0.08

Exercise 1:3

1.
Number		Number of significant figures			
		1	2	3	4
a	0.0068398	0.007	0.0068	0.00684	0.006840
b	2.0068398	2	2.0	2.01	2.007
c	4.69768	5	4.7	4.70	4.698
d	1.006127	1	1.0	1.01	1.006

2.
Number		Number of Decimal places			
		1	2	3	4
a	0.0068398	0.0	0.01	0.007	0.0068
b	2.0068398	2.0	2.01	2.007	2.0068
c	4.69768	4.7	4.70	4.698	4.6977
d	1.006127	1.0	1.01	1.006	1.0061

3. (a) 0.004 (b) 0.457 (c) 0.505 4. (a) 14.90 (b) 23.11 (c) 6.04
5. (a) 0.0249 (b) 4.03 6. (a) 550 (b) 60 (c) 5400
7. (a) 0.009 (b) 5 (c) 0.2

Exercise 1:4
(a) 5×10^3 (b) 4.8×10^2 (c) 1.02×10^4 (d) 7×10^5 (e) 3.2×10^{-3}
(f) 7.3×10^{-5} (g) 9.25×10^{-1} (h) 1×10^{-3} (i) 5.6×10^{-1} (j) 3×10^{-5}
(k) 1.96×10^{-3} (l) 3.4×10^{-10}

Exercise 1:5
1. (a) 9.192×10^7 (b) 3.5×10^1 (c) 5.4×10^{-3} (d) 2×10^1 (e) 3.5×10^2
   (f) 6×10^{-2} (g) 4×10^4 (h) 9.687×10^2 (i) 3.66×10^{-2} (j) 1.33×10^0
   (k) 6×10^{-3} (l) 4.37×10^0

Answers to Structural Exercises

Exercise 1:6
1. 27499
2. (a) 2 mm (b) $\frac{1}{10}$ (c) 10 %
3. (a) 0.3 m (b) $\frac{1}{116}$ (c) 0.86 %
4. 10%

Exercise 2:1

1. (a) 5 (b) $\frac{13}{3}$ (c) $\frac{3\sqrt{2}}{2}$ (d) 6
2. (a) $2\sqrt{6}$ (b) $\frac{\sqrt{30}}{10}$ (c) $6\sqrt{2}$ (d) $1+\frac{3\sqrt{2}}{2}$
   (e) $\sqrt{3}$ (f) $\sqrt{c^3}$ (g) $\sqrt{7}$ (h) $2\sqrt{2}$
3. (a) $\frac{4}{7}$ (b) $\frac{41}{4}$ (c) $x=-5$ or $x=-6$ (d) $x=1$ or $x=-7$
4. (a) 4 (b) 12 (c) 1 (d) 5 (e) 444 (f) $\frac{3}{5}$
5. (a) $\frac{5}{11}$ (b) 6 (c) $\frac{\sqrt[3]{3}}{3}$ (d) 16
6. (a) $\frac{2\sqrt{3}+1}{11}$ (b) $\sqrt{2}+\sqrt{3}$ 7. (a) $\frac{\sqrt{14}}{2}$ or 1.87 (b) $\frac{2\sqrt{7}}{3}$ or 1.76

Exercise 2:2

1. $\frac{3}{2}$ 2. 2250 3. $\frac{8}{5}$ 4. $-\frac{31}{28}$ 5. $-\frac{17}{12}$ 6. $\frac{1}{12}$ 7. $\frac{1}{20}$
8. $x=-\frac{5}{4}$ 9. $x=-\frac{3}{5}$ 10. $x=-6$ 11. $x=\frac{5}{3}$
12. $x=5$ 13. $x=-\frac{4}{5}$ 14. $x=\frac{9}{5}$, $y=\frac{27}{5}$

Exercise 2:3

1. (a) −4 (b) $\frac{3}{2}$ (c) 2 (d) −1 (e) 2 (f) 6
2. (a) 1.8060 (b) 0.3010 (c) 0.1505 3. (i) 0.3891 (ii) − 0.1743 4. 2.130

Exercise 3:1

1. $\frac{10}{3}$ 2. ±2 3. $\begin{pmatrix} -2 & -2 \\ 2 & 2 \end{pmatrix}$ 4. $\begin{pmatrix} 11 & -7 \\ -16 & 10 \end{pmatrix}$, −2
5. −4 or 11 6. −1 or 3 7. 3 or 7

Exercise 3:2

(a) $\begin{pmatrix} 1 & 3 \\ 4 & 5 \end{pmatrix}$ (b) $\begin{pmatrix} 1 & -4 \\ -5 & 3 \end{pmatrix}$ (c) $\begin{pmatrix} 3 & -1 \\ 2 & 2 \end{pmatrix}$ (d) $\begin{pmatrix} 0 & -1 \\ -2 & 4 \end{pmatrix}$ (e) $\begin{pmatrix} 5 & -2 \\ -3 & 4 \end{pmatrix}$

(f) $\begin{pmatrix} 1 & 3 \\ 2 & -2 \end{pmatrix}$ (g) $\begin{pmatrix} 1 & 0 \\ 0 & 1 \end{pmatrix}$ (h) $\begin{pmatrix} 4 & -1 \\ -2 & \frac{1}{2} \end{pmatrix}$ (i) $\begin{pmatrix} 7 & -5 \\ -3 & 2 \end{pmatrix}$

Exercise 3:3

1. (a) $\begin{pmatrix} -\frac{1}{7} & -\frac{3}{7} \\ -\frac{4}{7} & \frac{5}{7} \end{pmatrix}$ (b) $\begin{pmatrix} -\frac{1}{17} & -\frac{3}{17} \\ -\frac{4}{17} & \frac{5}{17} \end{pmatrix}$ (c) $\begin{pmatrix} \frac{3}{8} & -\frac{1}{8} \\ \frac{1}{4} & \frac{1}{4} \end{pmatrix}$

(d) $\begin{pmatrix} 0 & \frac{1}{2} \\ 1 & -2 \end{pmatrix}$ (e) $\begin{pmatrix} \frac{5}{14} & -\frac{1}{7} \\ -\frac{3}{14} & \frac{2}{7} \end{pmatrix}$ (f) $\begin{pmatrix} -\frac{1}{8} & -\frac{3}{8} \\ -\frac{1}{4} & \frac{1}{4} \end{pmatrix}$

(g) $\begin{pmatrix} 1 & 0 \\ 0 & 1 \end{pmatrix}$ (h) No inverse (i) $\begin{pmatrix} -7 & 5 \\ 3 & -2 \end{pmatrix}$

2. $\begin{pmatrix} 1 & -1 \\ -2 & 3 \end{pmatrix}, \begin{pmatrix} 1 & 0 \\ 0 & 1 \end{pmatrix}$ 3. $\begin{pmatrix} \frac{5}{2} & -\frac{3}{2} \\ -\frac{1}{2} & \frac{1}{2} \end{pmatrix}, \begin{pmatrix} 1 & 0 \\ 0 & 1 \end{pmatrix}$

4. (i) $\begin{pmatrix} \frac{3}{14} & \frac{1}{7} \\ -\frac{1}{14} & \frac{2}{7} \end{pmatrix}$ (ii) $\begin{pmatrix} 1 & 0 \\ 0 & 1 \end{pmatrix}$ (iii) $\begin{pmatrix} \frac{1}{3} & 0 \\ 0 & \frac{1}{2} \end{pmatrix}$ (iv) $\begin{pmatrix} \frac{1}{8} & \frac{1}{4} \\ -\frac{3}{8} & \frac{1}{4} \end{pmatrix}$

Exercise 3:4

1. $p = 4, q = -2$ 2. $s = 7, t = 4$ 3. $x = 4, y = 3$
4. $m = \frac{1}{2}, y = \frac{1}{4}$ 5. $x = 3, y = 1$ 6. $u = 5, v = 2$

Answers to Structural Exercises

Exercise 3:5

1. (a) $\begin{array}{c} \\ A \\ B \\ C \\ D \end{array} \begin{array}{c} A\ B\ C\ D \\ \begin{pmatrix} 0 & 1 & 1 & 0 \\ 1 & 0 & 2 & 1 \\ 1 & 2 & 0 & 3 \\ 0 & 1 & 3 & 0 \end{pmatrix} \end{array}$ (b) $\begin{array}{c} \\ A \\ B \\ C \\ D \end{array} \begin{array}{c} s\ t\ u\ v\ w\ x\ y\ z \\ \begin{pmatrix} 1 & 0 & 1 & 0 & 0 & 0 & 0 & 0 \\ 1 & 1 & 0 & 0 & 0 & 1 & 1 & 0 \\ 0 & 1 & 1 & 1 & 1 & 0 & 1 & 1 \\ 0 & 0 & 0 & 1 & 1 & 1 & 0 & 1 \end{pmatrix} \end{array}$

2. $\begin{array}{c} \\ P \\ Q \\ R \end{array} \begin{array}{c} P\ Q\ R \\ \begin{pmatrix} 0 & 2 & 1 \\ 1 & 0 & 0 \\ 1 & 1 & 2 \end{pmatrix} \end{array}$

Exercise 4:1

1.

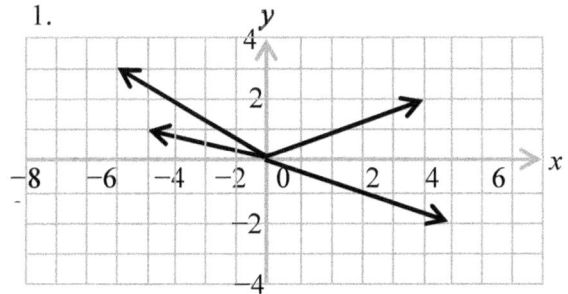

2. $\mathbf{AB} = \begin{pmatrix} 3 \\ -11 \end{pmatrix}$, $\mathbf{CD} = \begin{pmatrix} 7 \\ 6 \end{pmatrix}$,

$\mathbf{EF} = \begin{pmatrix} 0 \\ 8 \end{pmatrix}$, $\mathbf{GH} = \begin{pmatrix} 0 \\ 8 \end{pmatrix}$

3.

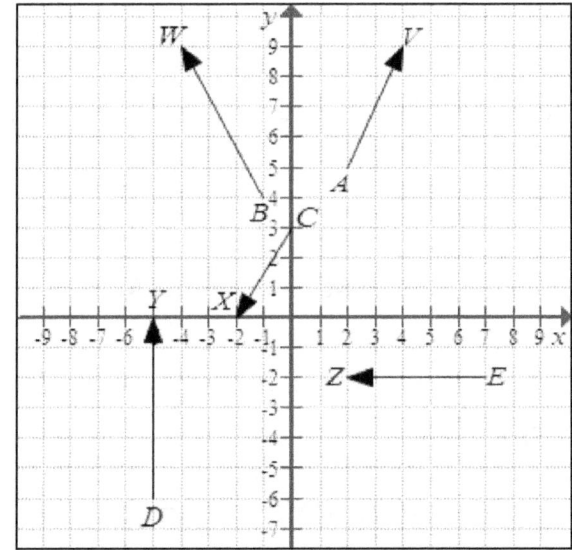

4.

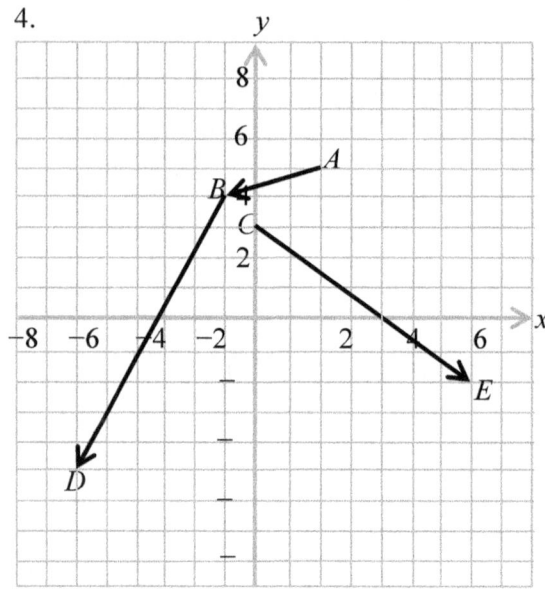

Exercise 4:2

1.

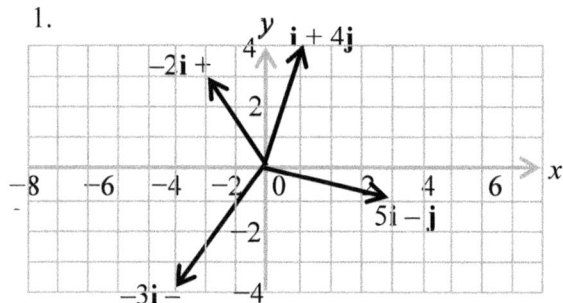

2. (i) $2\mathbf{i}+5\mathbf{j}$ (ii) $-5\mathbf{i}+3\mathbf{j}$ (iii) $-4\mathbf{i}-\mathbf{j}$ (iv) $6\mathbf{i}-2\mathbf{j}$

3. **OD** and **OE**

Exercise 4:3

1. $\mathbf{OD} = \frac{1}{2}(\mathbf{b}+\mathbf{c})$, $\mathbf{OG} = \frac{1}{3}(\mathbf{a}+2\mathbf{c})$ 2. $\mathbf{TM} = \frac{2}{3}\mathbf{r}+\frac{1}{2}\mathbf{p}, \mathbf{PM} = \mathbf{r}-\frac{1}{2}\mathbf{p}$

3. (a) $\mathbf{d}-\mathbf{a}$ (b) (i) $\mathbf{a}+\mathbf{b}$ (ii) $\mathbf{d}+\mathbf{c}, \mathbf{c} = \frac{1}{2}\mathbf{d}$ 4. (a) $\mathbf{a}+\mathbf{b}$ (b) (i) $\mathbf{a}-\mathbf{b}$

5. $\mathbf{PQ} = -3\mathbf{a}-\mathbf{b}, \mathbf{QR} = 3\mathbf{a}-2\mathbf{b}, \mathbf{RP} = 3\mathbf{b}$ 6. $\frac{1}{4}(\mathbf{a}+3\mathbf{b})$

7. (a)(i) $12\mathbf{i}-3\mathbf{j}$ (ii) $\mathbf{OL} = \frac{m}{m+n}(-3\mathbf{i}+6\mathbf{j}), \mathbf{OM} = \frac{m}{m+n}(9\mathbf{i}+3\mathbf{j})$

Answers to Structural Exercises

(iii) $\mathbf{LM} = \dfrac{m}{m+n}(12\mathbf{i} - 3\mathbf{j})$ (b) $\dfrac{m}{m+n}\mathbf{AB}$

(c) (i) 2:3   (ii) 17.5 square units   8. (a) $\mathbf{a} + \mathbf{b}$ (b) $\mathbf{a} - \mathbf{b}$

Exercise 4:4
1. (a) 5   (b) 90°   2. (a) $\sqrt{13}$ (b) 90°   3. $18\sqrt{2}$ 4. (a) 0   (b) 27
5. 120°   6. 0, perpendicular   7. 44°   8. 73.7°

Exercise 5:1
1. (2,–3)
2.

	(2,3)	(3,0)	(0,–5)	(3,3)	(–4,–6)
x-axis	(2,–3)	(3,0)	(0,5)	(3,–3)	(–4,6)
y-axis	(–2,3)	(–3,0)	(0,–5)	(–3,3)	(4,–6)
y = x	(3,2)	(0,3)	(–5,0)	(3,3)	(–6,–4)
y = –x	(–3,–2)	(0,–3)	(5,0)	(–3,–3)	(6,4)

3. (2, –4)

4.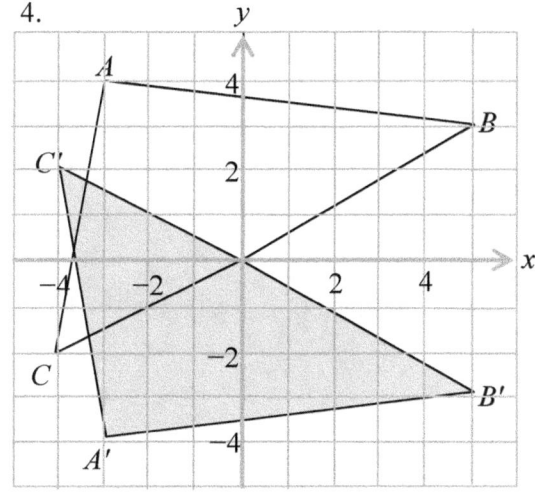

5.(i) $\left(\dfrac{3\sqrt{3}}{2} - 1, \sqrt{3} + \dfrac{3}{2}\right)$ (ii) $\left(\dfrac{3}{2}, \dfrac{3\sqrt{3}}{2}\right)$ (iii) $\left(-\dfrac{5\sqrt{3}}{2}, \dfrac{5}{2}\right)$ (iv) $\left(\dfrac{3\sqrt{3}}{2} - \dfrac{3}{2}, \dfrac{3\sqrt{3}}{2} + \dfrac{3}{2}\right)$

(iii) $\left(-\dfrac{5\sqrt{3}}{2}, \dfrac{5}{2}\right)$ (iv) $\left(\dfrac{3\sqrt{3}}{2} - \dfrac{3}{2}, \dfrac{3\sqrt{3}}{2} + \dfrac{3}{2}\right)$ (v) $\left(2 - 3\sqrt{3}, 3 - 2\sqrt{3}\right)$

6. (–7,7)
7. (a) (3,3)   (b) (2,4)   (c) (–1,2)   (d) (1, 5)   (e) (0, 1)   (f) (0,0)
   (g) (a+2,b+3)   (h) (3,4)   (i) (–9,18)   (j) (–2,6)

8.

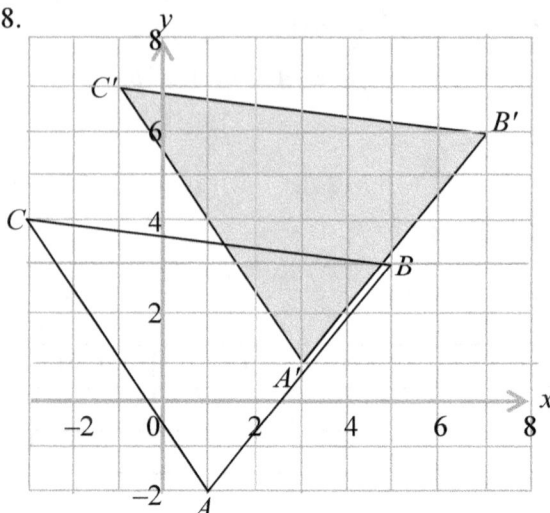

T_1 is the inverse of T. TT_1 is the identity transformation, because it leaves all points unchanged.

9. (a) (1,−7) (b) (2, −6) (c) (2, −7) (d) (0, −5) (e) (5, -9) (f) (−1, −10)

10.

The single transformation is a translation of 4.5 cm along AC.

11. $y = x^2 + 8x + 8$ 12. $(-1, 4)$
13. (a) (1,4)   (b) (−5,4)   (c) (−7,3)   (d) (−1,3)

Exercise 5:2

1. (a) $(-5, -3)$ (b) $(-10, -6)$ (c) $\left(-\dfrac{5}{2}, \dfrac{3}{2}\right)$ (d) $(-3, -5)$

2.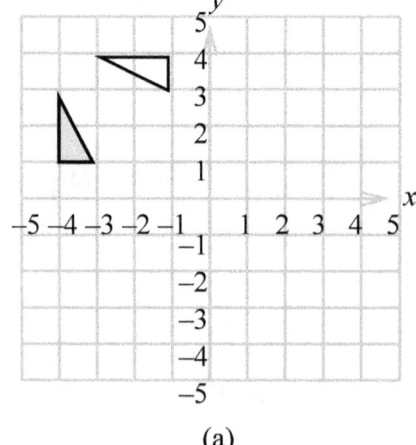

(a)   (b)

326

Answers to Structural Exercises

Exercise 5:3
1. (0,0), (2,2), (2,5), (0,3)   2. (0,0), (7,–5), (6,–4), (–1,1)

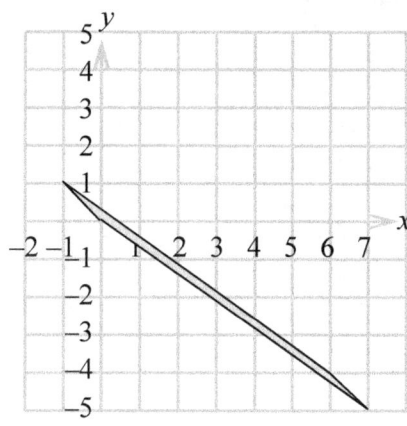

3. (a)

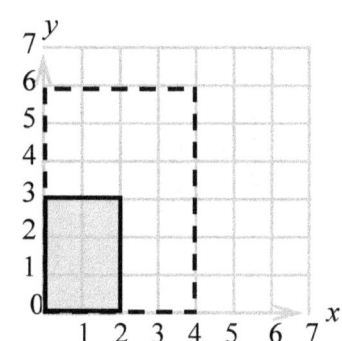

(b)

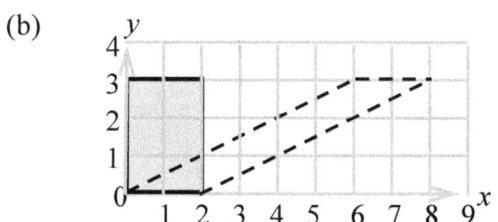

(c)

Exercise 5:4
1. **A** is an identity transformation; object and image are identical.
   B is a rotation about the origin through 90° in an anticlockwise sense.
   C is a rotation about the origin through 90° in an clockwise sense.
   D is a rotation about the origin through 180° in the clockwise or anticlockwise sense.
   E is a reflection in the line $y = x$.
   F is a reflection in the y-axis.
   G is a reflection in the line $y = -x$.
   H is a reflection in the x-axis.
   P is an enlargement scale factor 2, center (0,0).
   Q is an enlargement scale factor $\frac{1}{2}$, center (0,0).
   R is a rotation about the origin through 180° in the clockwise or anticlockwise sense followed by an enlargement scale factor $\frac{1}{2}$.
   S is a rotation about the origin through 180° in the clockwise or anticlockwise sense followed by an enlargement scale factor 2.
   T is a stretch, stretch factor 2 in the Oy direction
   U is a shear, shear factor 1 in the Ox direction with points on the x-axis invariant.

V is a shear, shear factor 2 in the Oy direction with points on the y-axis invariant.
W is a stretch, stretch factor 3 in the Ox direction.
X is a stretch, stretch factor 2 in the negative Ox direction.
Y is a stretch, stretch factor 3 in the Oy direction.

2.

A

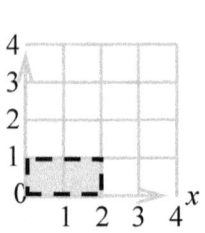

B

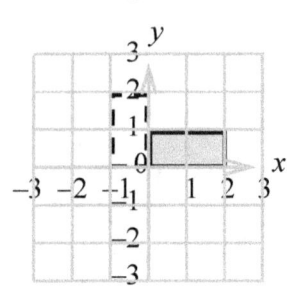

C

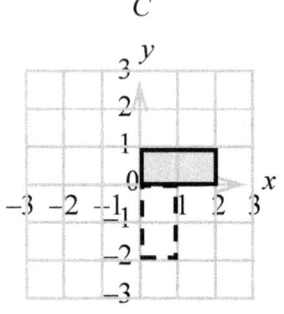

D

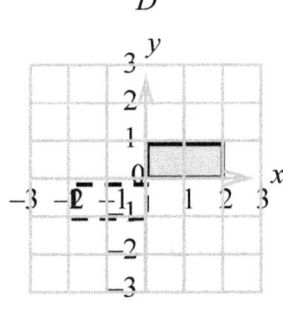

E

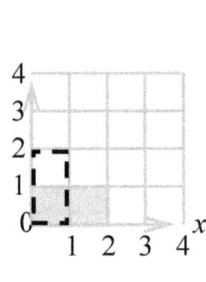

F

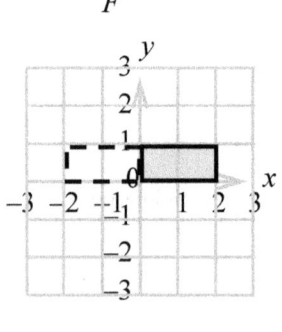

G

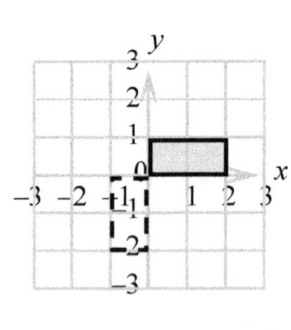

P

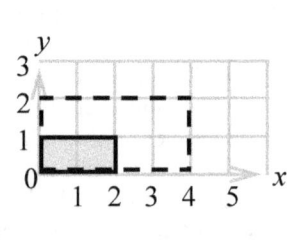

Q

R

S

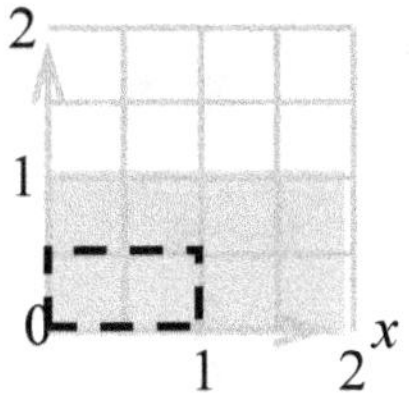

Answers to Structural Exercises

T

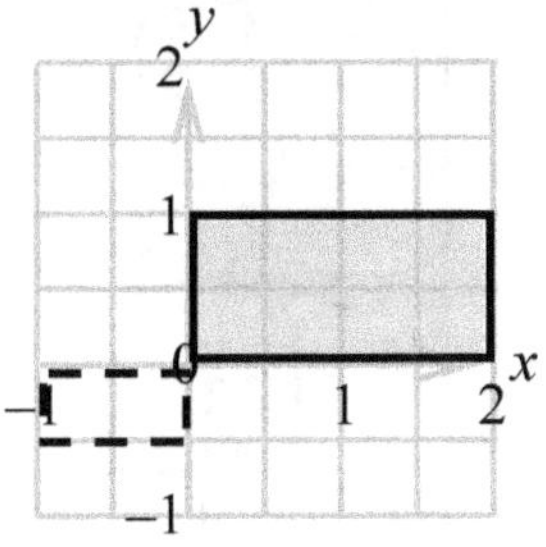

U

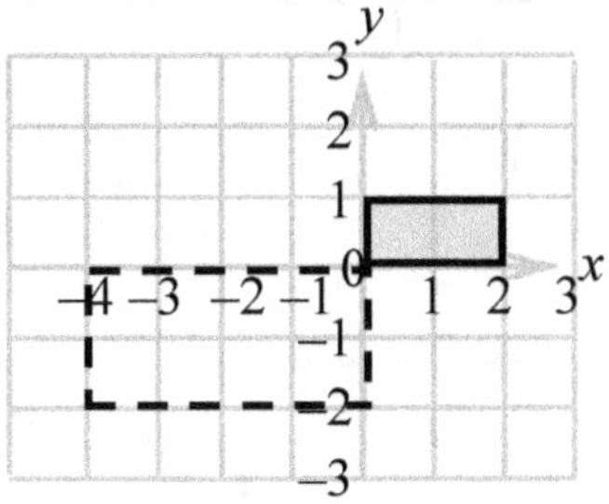

V

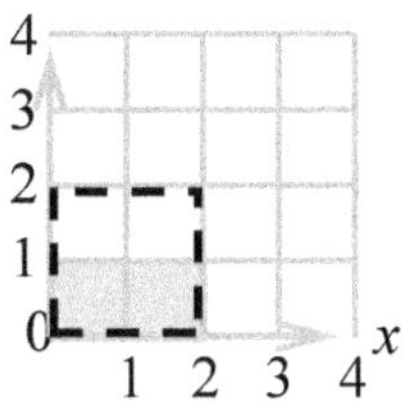

W

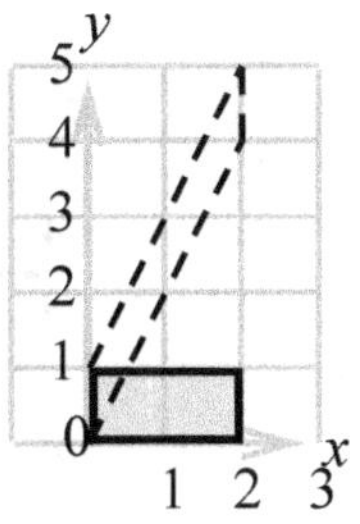

X

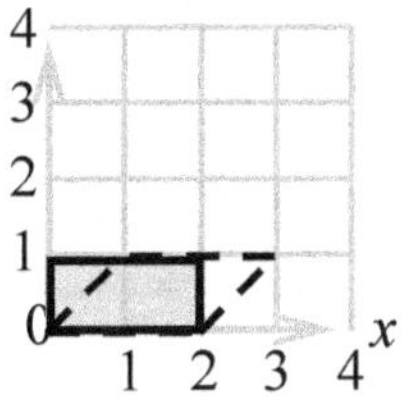

Y

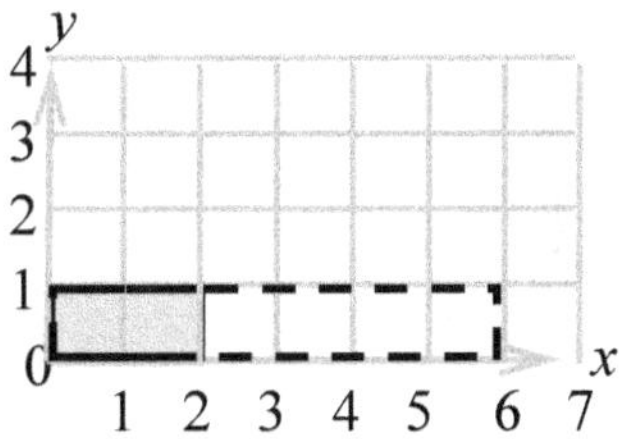

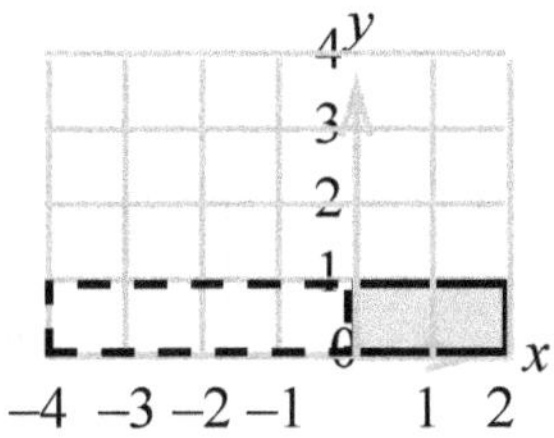

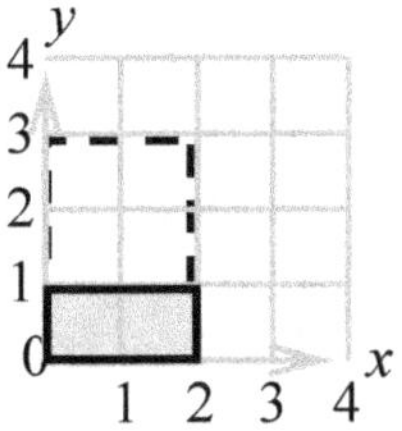

Exercise 5:5

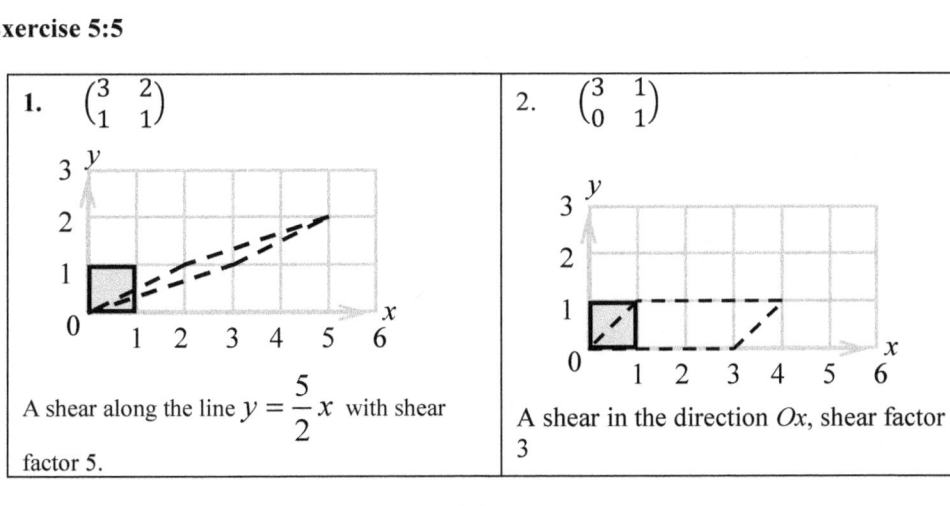

1. $\begin{pmatrix} 3 & 2 \\ 1 & 1 \end{pmatrix}$

   A shear along the line $y = \dfrac{5}{2}x$ with shear factor 5.

2. $\begin{pmatrix} 3 & 1 \\ 0 & 1 \end{pmatrix}$

   A shear in the direction Ox, shear factor 3

3. $\begin{pmatrix} 2 & 1 \\ 0 & 0 \end{pmatrix}$

4. $\begin{pmatrix} 2 & 0 \\ 0 & 2 \end{pmatrix}$

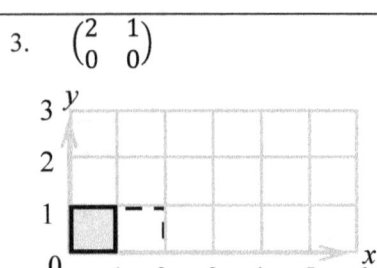

5. $\begin{pmatrix} 4 & 1 \\ 0 & 1 \end{pmatrix}$

6. $\begin{pmatrix} 1 & 1 \\ 0 & -1 \end{pmatrix}$

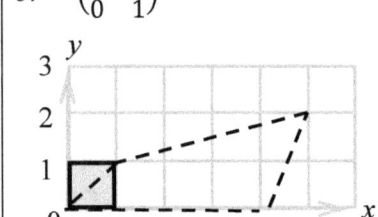

7. $\begin{pmatrix} 3 & 0 \\ 0 & -3 \end{pmatrix}$

8. $\begin{pmatrix} 1 & -1 \\ 0 & 1 \end{pmatrix}$

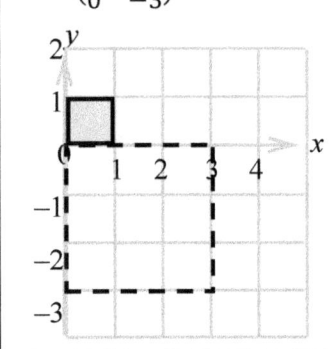
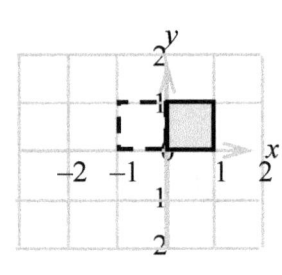

9. $\begin{pmatrix} 3 & 0 \\ 0 & 3 \end{pmatrix}$

10. $\begin{pmatrix} 1 & -2 \\ 0 & 1 \end{pmatrix}$

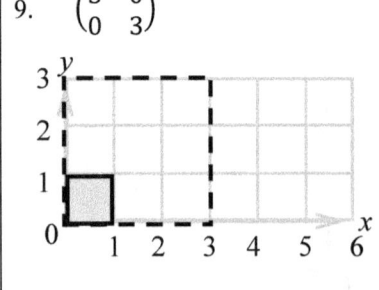
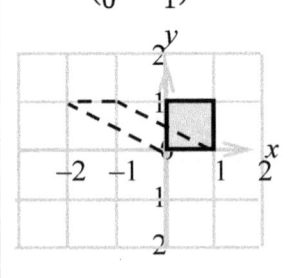

Answers to Structural Exercises

11. $\begin{pmatrix} 3 & 1 \\ 1 & 2 \end{pmatrix}$ 12. $\begin{pmatrix} 0 & 1 \\ \frac{1}{2} & 1 \end{pmatrix}$

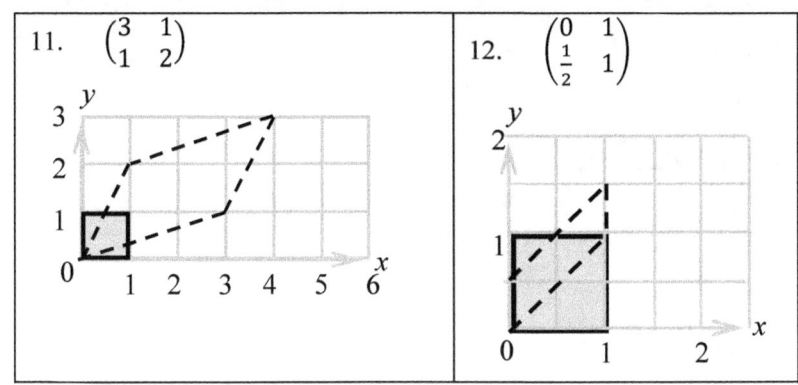

Exercise 5:6

1. (a) $\begin{pmatrix} 0 & -6 & -8 & -6 \\ 0 & -8 & 0 & -2 \end{pmatrix}$, 8:1   (b) $\begin{pmatrix} 8 & 4 & 0 & 2 \\ -2 & -8 & -2 & 0 \end{pmatrix}$, 8:1

2. (a) $\begin{pmatrix} 2 & 8 & 2 \\ 2 & 2 & 6 \end{pmatrix}$ (b) 8:1   3. 48 cm²   4. 0 un²

5.

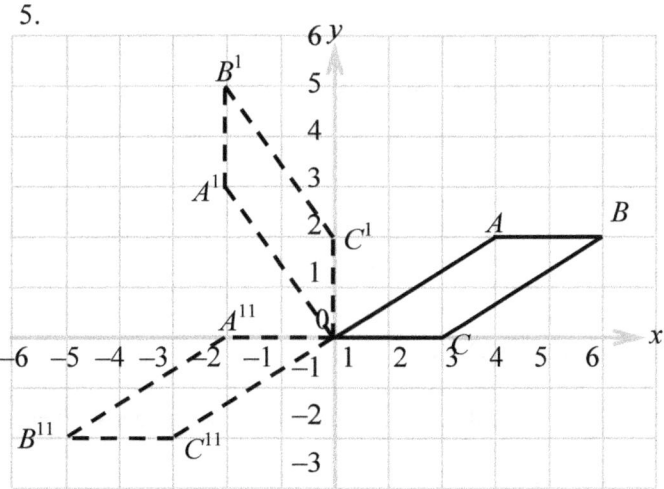

(d) T is a rotation anti-clockwise through 90° about (0,0).
T^2 is a rotation anti-clockwise through 180° about (0,0)
6. M represents a reflection in the x-axis.
$A^1(-1,-3)$, $B^1(-3,0)$, $C^1(-1,2)$ and $D^1(1,-1)$.
7. (a) $A^1(0,-1)$, $B^1(1,-2)$, $C^1(0,-4)$.

Competency Based Mathematics for Secondary Schools. Book 4

(b)
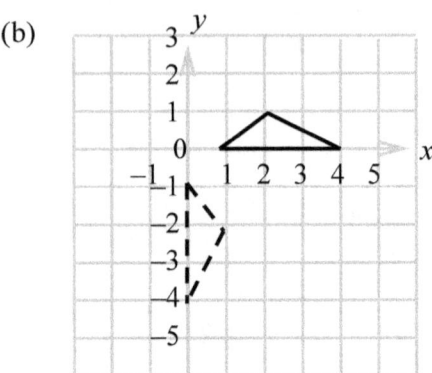

(c) T is a rotation clockwise through 90° about (0,0).

(d) $\mathbf{T}^{-1} = \begin{pmatrix} 3 & 1 \\ 1 & 2 \end{pmatrix}$. T^{-1} is a rotation anti-clockwise through 90° about (0,0).

8. (a) $A^1(-6,-4)$.
   (b) M is a rotation anti-clockwise through 90° about (0, 0).
9. $ABCD$ is a parallelogram.
   M represents a rotation anti-clockwise through 180° about (0,0), followed by an enlargement scale factor 2.
   $A''(0,8), B''(4,8), C''(12,4), D''(8,4)$
   Area of $ABCD : A''B''C''D'' = 16:1$

Exercise 6:1
1. (iii) 10 cm (v) 6.2 cm   2. (e) 5.4 cm (f) 90°___3. (c) 10 cm
4. (f) 14.1 cm (g) 23°   5. (c) 10 cm

Exercise 6:2
1. mediator or perpendicular bisector.
3. (a) $YX = 9$ cm
   (b) The locus of P is a straight line perpendicular to YZ and distant 9 cm from Y.

Exercise 7:1
1. (i) (a) $\dfrac{5}{12}$ (b) $\dfrac{12}{13}$ (c) $\dfrac{5}{13}$ (d) $\dfrac{12}{5}$ (e) $\dfrac{5}{13}$ (f) $\dfrac{12}{13}$

(ii) (a) $\dfrac{9}{40}$ (b) $\dfrac{40}{41}$ (c) $\dfrac{9}{41}$ (d) $\dfrac{40}{9}$ (e) $\dfrac{9}{41}$ (f) $\dfrac{40}{41}$

2. (a) $\dfrac{4}{3}$ (b) $\dfrac{4}{5}$ (c) $\dfrac{3}{5}$ (d) $\dfrac{8}{15}$ (e) $\dfrac{8}{17}$ (f) $\dfrac{15}{17}$

(ii) (a) $\dfrac{3}{4}$ (b) $\dfrac{4}{5}$ (c) $\dfrac{3}{5}$ (d) $\dfrac{15}{8}$ (e) $\dfrac{8}{17}$ (f) $\dfrac{15}{17}$

3. (a) $\dfrac{7}{25}$ (b) $\dfrac{24}{25}$ 4. (i) $\dfrac{\sqrt{3}}{2}$ (ii) $\dfrac{1}{2}$ (iii) $\dfrac{\sqrt{3}}{3}$ 5. (i) $\dfrac{\sqrt{2}}{2}$ (ii) $\dfrac{\sqrt{2}}{2}$ (iii) 1

Answers to Structural Exercises

6. $\sec x = \dfrac{\sqrt{n^2 - m^2}}{n}$, $\cot x = \dfrac{\sqrt{n^2 - m^2}}{m}$ 7. (a) $\dfrac{3}{5}$ (b) $\dfrac{4}{3}$

8. (a) $\dfrac{4}{5}$ (b) $\dfrac{3}{5}$ (c) $\dfrac{4}{3}$ (d) $\dfrac{3}{5}$ (e) $\dfrac{4}{5}$ (f) $\dfrac{3}{4}$
   (g) $\dfrac{5}{13}$ (h) $\dfrac{12}{13}$ (i) $\dfrac{5}{12}$ (j) $\dfrac{12}{13}$ (k) $\dfrac{5}{13}$ (l) $\dfrac{12}{5}$

Exercise 7:2

1. (a) 1.7883   (b) 1.0497   (c) 1.0125   (d) 1.6243   (e) 1.7263
   (f) 2.2583   (g) 0.9004   (h) 1.9542   (i) 0.1358
2. (a) 38°   (b) 16°12'   (c) 67°46'   (d) 9°40'   (e) 81°44'
   (f) 22°23'   (g) 42°42'   (h) 13°50'
3. (a) 34°   (b) 19°12'   (c) 76°46'   (d) 4°40'   (e) 61°44'
   (f) 27°23'   (g) 31°42'   (h) 84°50'
4. (a) 37°   (b) 22°12'   (c) 62°46'   (d) 8°40'   (e) 63°44'
   (f) 24°23'   (g) 33°42'   (h) 82°50'

Exercise 7:3

1. (a) $\dfrac{1}{2}$ (b) $\dfrac{\sqrt{3}}{3}$ (c) $\sqrt{3}$ (d) -1 (e) $\dfrac{\sqrt{2}}{2}$ (f) $\sqrt{3}$
2. (a) -0.1736 (b) 1.4281 (c) -1.4281 (d) -0.9848 (e) 0.342 (f) -0.9848
3. (a) $-\dfrac{2\sqrt{3}}{3}$ (b) $\sqrt{3}$ (c) $-\dfrac{\sqrt{3}}{3}$ (d) ∞ (e) $-\sqrt{2}$ (f) 2
4. (a) 1.0154 (b) 0.7002 (c) -0.7002 (d) -5.7588 (e) -1.0642 (f) 5.7588

Exercise 7:4

1. (a) $\dfrac{1}{2}$ (b) $-\dfrac{\sqrt{3}}{3}$ (c) $\sqrt{3}$ (d) 1   (e) $-\dfrac{\sqrt{2}}{2}$ (f) $-\dfrac{\sqrt{3}}{2}$
2. (a) -0.1736 (b) 1 (c) 1.4281 (d) 0.9848 (e) -0.3420 (f) -0.9848
3. (a) -0.1736 (b) 1 (c) 1.4281 (d) 0.9848 (e) -0.3420 (f) -0.9848
4. (a) $\dfrac{2\sqrt{3}}{3}$ (b) $-\sqrt{3}$ (c) $\dfrac{\sqrt{3}}{3}$ (d) ∞ (e) $-\sqrt{2}$ (f) -2
5. (a) -1.0154 (b) -0.7002 (c) 0.7002
   (d) -5.7588 (e) -1.0642 (f) -5.7588
6. (a) -1.0154 (b) -0.7002 (c) 0.7002
   (d) -5.7588 (e) -1.0642 (f) -5.7588

Exercise 7:5

1. (a) $\dfrac{\pi}{6}$ (b) $\dfrac{2\pi}{3}$ (c) $\dfrac{4\pi}{3}$ (d) $\dfrac{5\pi}{3}$ (e) $\dfrac{\pi}{8}$
2. (a) 270°   (b) 225°   (c) 210°   (d) 15°   (e) 300°

Exercise 7:7
1. $x = 0$ 2. $x = 0, x = 60$ 3. $45°$ 4. 90 5. 90

Exercise 8:1
1. (a) $AB = 15.59$ cm, $BC = 9$ cm   (b) $AB = 15.05$ cm, $BC = 17.04$ cm
2. (a) $90°, 22.06°, 67.4°$ (b) $90°, 28.1°, 61.9°$
   (c) $90°, 16.3°, 73.7°$ (d) $90°, 79.6°, 10.4°$ 3. 8 m   4. 3 m
5. 11 m   6. (a) 6 m   (b) $120°$ (c)   7. (a) 1124 m   (b) 2782 m

Exercise 8:2
1. (a) $\sqrt{7}$, $\angle X = 41.4°, \angle Y = 48.6°$ (b) $\sqrt{1649}$, $\angle X = 52°, \angle Y = 38°$
   (c) 9, $\angle X = 53.1°, \angle Y = 36.9°$ (d) $\sqrt{111}$, $\angle X = 31.8°, \angle Y = 58.2°$
   (e) 3, $\angle X = 36.9°, \angle Y = 53.1°$ (f) $3\sqrt{231}$, $\angle X = 62.9°, \angle Y = 27.1°$
2. $40°$ 3. $3\sqrt{2}$ cm, $\angle X = \angle Z = 45°$

Exercise 8:3
1. 6.57 cm   2. $4\sqrt{2}$ cm, $\angle ABC = 38°, \angle BAC = BCA = 70.5°$
3. (a) $\sqrt{91}$ cm, $\angle A = \angle C = 72.5°, \angle B = 34.9°$
   (b) $\sqrt{7}$ cm, $\angle A = \angle C = 41.4°, \angle B = 97.2°$
   (c) $\dfrac{\sqrt{171}}{2}$ cm, $\angle A = \angle C = 69.1°, \angle B = 41.8°$
4. (a) 21 cm   (b) 79 cm   5. $\angle BAC = 36.4°, \angle ABC = ACB = 71.8°$

Exercise 8:4
1. (a) 18 m   (b) 47 m   2. (a) $13\sqrt{3}$ m   (b) 26 m   3. 8 m
4. (a) 69 m   (b) 57 m   (c) 224 m
5. (a) 23 m   (b) 39 m   (c) 4 m   6. 82 m
7. (a) 32 m   (b) 35 m   8. 61.4 m

Exercise 8:5
1. (a) 5 km   (b) $76°$ 2. (a) $\sqrt{13}$ km   (b) $73.3°$ 3. (a) 71 km   (b) 92 km
4. $BC = 7.8$ cm, $AC = 4.8$ cm   2. 6.7 km, $061°$
5. $\sqrt{19}$ km, $36.6°$ 6. (a) 185.2 km   (b) $071°$

Exercise 8:6
1. $AC = 4.8$ km, $BC = 7.8$ km   2. Distance $= 6.7$ km, bearing $= 61.1°$
3. Distance $= 4.4$ km, bearing $= 36.7°$ 4. Distance $= 185.2$ km, bearing $= 71.1°$
5. (a) $120°$ (b) (i) 15.6 km   (ii) $76.3°$

Answers to Structural Exercises

Exercise 9:1
1. (i) (a) chord (b) radius (c) diameter (d) minor arc (e) secant (f) tangent (g) major arc
   (ii) (a) semi-circle (b) minor sector (c) minor segment
2. (a) Minor arc (b) radius (c) isosceles (d) chord (e) Minor sector
3. (a) Minor segment (b) Minor arc (c) radius (d) diameter (e) Chord (f) Minor sector

Exercise 9:2
1. 38.5 cm^2 2. 154 cm^2 3. 3.5 cm 4. 28 cm 5. 22 cm 6. 88 cm 7. 7 cm 8. 1886.5 cm^2 9. 132 m 10. 7 cm 11. 110 m^2 12. 115.5 cm^2

Exercise 9:3
1. 3.5 cm 2. 85.9 cm 3. 27.2 4. 62.2 cm 5. 24 cm 6. 30° 7. 16 cm 8. 10.4 cm^2 9. 89.8 cm^2 10. 20 cm, 102 cm 11. 111 cm^2 12. 22200 m^2
13. (a) $P = \pi x + 2x + 2y$ (b) $x = \dfrac{P - 2y}{\pi + 2}$ 14. (a) 7 cm (b) 10.5 cm^2
15. 68.2 cm^2 16. 70.3 cm^2 17. 195.3 cm^2 18. 115.5 cm^2
19. (a) 12.57 cm, (b) 0.86 cm^2

Exercise 9:4
1. 7 cm, 1 cm 2. 60 cm^2, 7.06 cm 3. 8.3 cm 4. 48° 5. 37°
6. (a) 27° (b) 126° 7. $x = 73°, y = 92°$ 8. $x = 36°, y = 72°$
9. 36° 10. (a) 90° (b) 75° (c) 150° (d) 60°

Exercise 9:5
1. $\angle BTA = 50°$, $\angle CBT = 90°$, $\angle DTM = 50°$
   (a) $\angle ABT = \angle BTD = 90°$ (alt. $\angle$s btw ▯ lines)
   (b) $\angle CBT = 90°$
2. (i) 25° (ii) 65° (iii) 65° (iv) 12.7 cm (v) 16.6 cm (vi) 105.1 cm
3. $x = 130°$, $y = 30°$, $z = 50°$ 4.(a) 60° (b) 130° (d) $DE = 5.8$ cm, $DB = 11.5$ cm
5. 57° 6. $a = 75°, b = 15°, c = 75°$ 7. $x = 122°, y = 26°$
8. $x = 78°, y = 53°, z = 49°$

Exercise 9:6
1. 9 cm 2.(a) 4.2 cm (b) 24° 3. 3 cm 4. 8 cm 5. $AB = 14$ cm, $EF = 16.5$ cm
6. $r = 25$ cm, length of tangent $= 32$ cm 7. 3 cm
8. Radius $= 3.5$ cm, Area $= 38.5$ cm^2

Exercise 10:1
1. (a) 5544 cm^2, 38808 cm^3 (b) 1386 cm^2, 4850 cm^3 2. 1400 cm^2 3. 1:343 4. 1767 cm^3 5. 10,000 6. (a) 56.6 cm^2 (b) 56.5 cm^2 (c) 84.9 cm^2

Exercise 10:2
1. (a) 12962.5 km (b) 6146 km (c) 5810.8 km
2. (a) 18438.1 km (b) 4246.3 km (c) 10057.1 km 3. (a) 25.2° (b) 90°

Exercise 10:3
1. (a) 5:25 p.m. (b) 11:36 p.m. (c) 2:20 p.m. (d) 11:10 a.m.
2. (a) 20:34 (b) 5:56 (c) 12:45 (d) 00:45
3. (a) 9 hours 30 minutes (b) 14 hours 30 minutes (c) 19 hour 2 minutes

(d) 10 hours 42 minutes
4.(a) ante meridian (between midnight and midday)
  (b) post meridian (between midday and midnight)
5. 3 a.m.   6. 20 hours 30 minutes   7. (a) 8:40 a.m.   (b) Wednesday

Exercise 11:1
(a) $E = \{1, 2, 3\}$, $O = \emptyset$ or $\{\ \}$
(b) $E = \{1, 3\}$, $O = \{2, 4\}$
(c) $E = O = \emptyset$ or $\{\ \}$, $O = \{1, 2, 3, 4\}$
(d) $E = \{1, 2\}$, $O = \{3, 4\}$
(e) $E = \{3, 5\}$, $O = \{1, 2, 4, 6\}$
(f) $E = \{3, 5\}$, $O = \{1, 2, 4, 6\}$
(g) $E = \{3, 6\}$, $O = \{1, 2, 4, 5, 7, 8\}$
(h) $E = \{1, 3, 5, 6\}$, $O = \{2, 4\}$
(i) $E = \{1, 3, 4, 6\}$, $O = \{2, 5, 8\}$

Exercise 11:2
(a) traversable because there are exactly 2 odd vertices.
(b) traversable because all vertices are even.
(c) not traversable because there are 4 odd vertices and 4 is greater than 2.

Exercise 11:3
1. (a)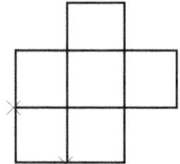

   The beginning point can be either of the two odd vertices; the endpoint will be the other odd vertex.
   (b)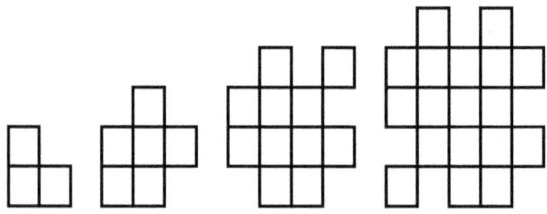

   Remove 1 square from the 2 × 2 grid; remove 3 squares from the 3 × 3 grid; remove 5 squares from the 4 × 4 grid; remove 7 squares from the 5 × 5 grid.
   (c) The number of squares to remove from a 12 × 12 grid is the 11th odd number, 21.
   (d) The number of squares to remove from an $n \times n$ grid is the $(n-1)^{th}$ odd number, $2n - 3$.

2.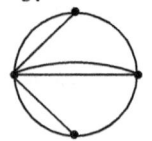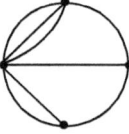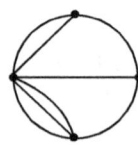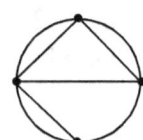

Answers to Structural Exercises

3. (a) Yes

(b)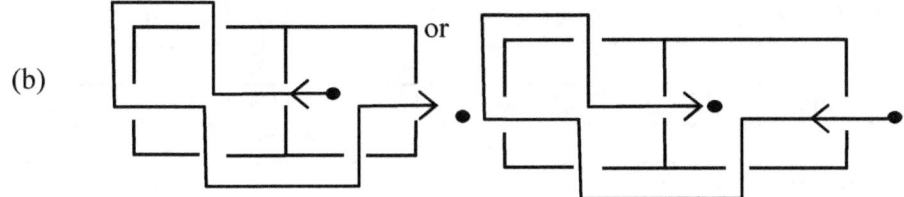

Exercise 11:4
1. (a) {1,2} and {2,6}, {2,3} and {3,4}, {4,6} and {4,5}, {4,5} and {5,6}, {4,6} and {5,6}, {4,6} and {2,6}, {1,2} and {1,3}, {2,3} and {2,6}.
   (b) (1,2) and (2,3), (6,2) and (2,3), (6,4) and (4,3).
   (c) (1,2), (2,3), (4,3), (6,4), (6,2).
   (d) 1 and 2, 2 and 3, 3 and 4, 4 and 5, 4 and 6, 5 and 6, 2 and 6.
   (e) 1 and {1,2}, 6 and {2,6}, 2 and {2,3}, 3 and {3,4}, 4 and {4,6}, 4 and {4,5}, 5 and {5,6}, 2 and {1,2} etc.
2. $G = (V, E)$ where $V = \{1, 2, 3, 4, 5, 6\}$ and
   $E = \{(1,2), (2,3),(4,3),(6,4), (6,2), \{4,5\},\{5,6\}\}$.
3.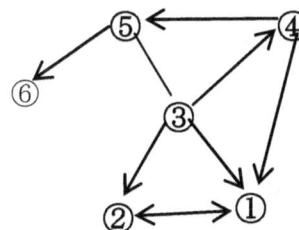

Exercise 11:5
1.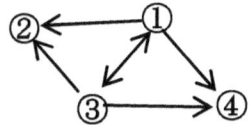

2. $D = (V, A)$, where $V = \{1, 2, 3,4,5\}$ and
   $A = \{(1, 2), (2, 1), (2, 4), (3, 2),(3,4), (4,1), (4,5),(5,3), (5,4)\}$.

3. 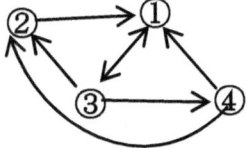   4.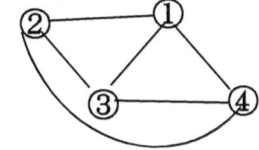

5. $G = (V, E)$ where $V = \{1, 2, 3, 4,5,6\}$ and
   $E = \{\{1,2\},\{1,3\},\{1,5\},\{2,3\},\{2,4\},\{3,4\}, \{4,5\}, \{5,6\}\}$.
6. (a) 6 hours   (b) 1650 FCFA   7. Answers vary.
8. Trees :   (b), (c), (f). $n(E) = n(V) - 1$
   Forest :   (a), (d). Combination of trees.
   Neither : (e) $n(E) \neq n(V) - 1$

Exercise 12:1

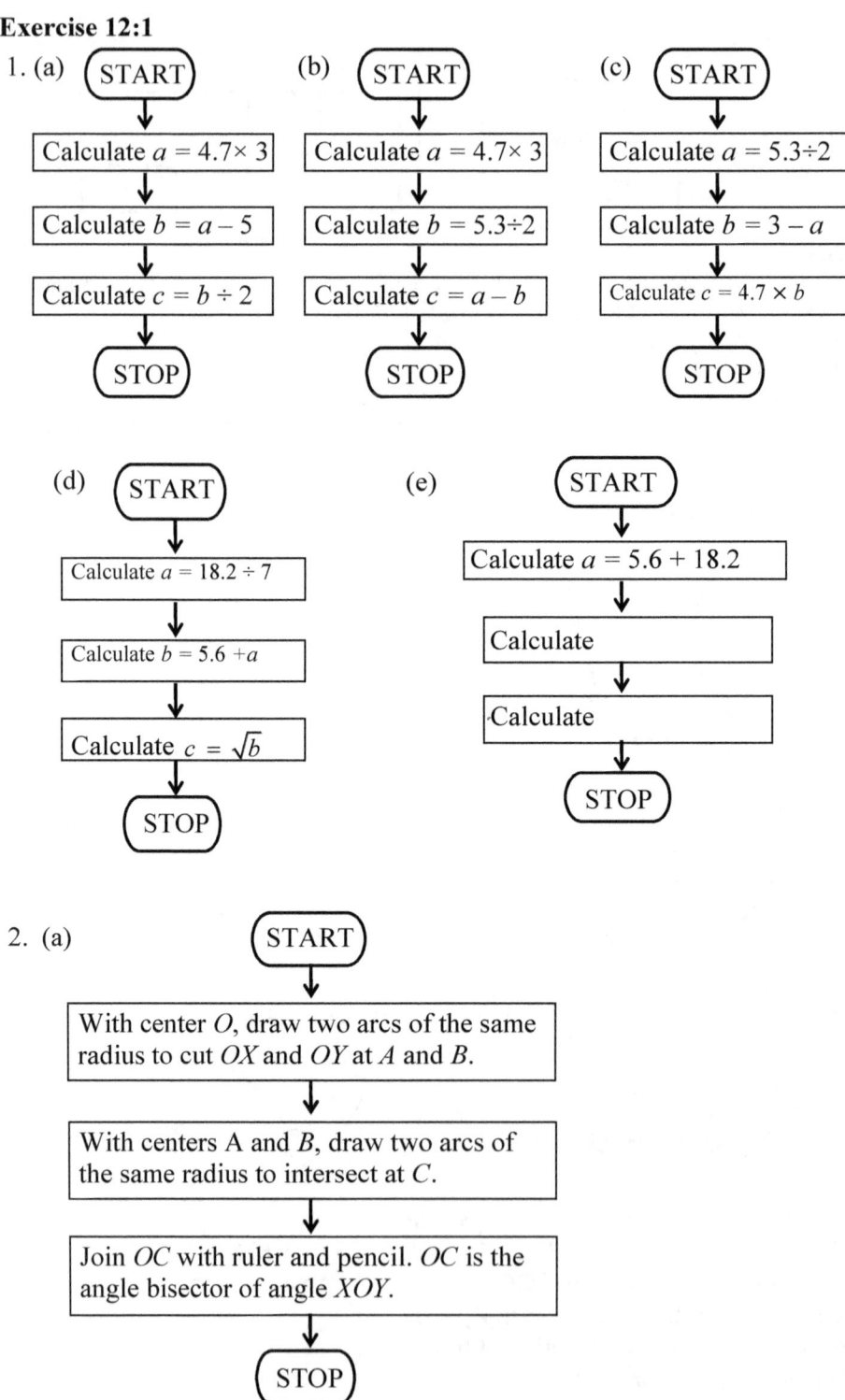

Answers to Structural Exercises

(b)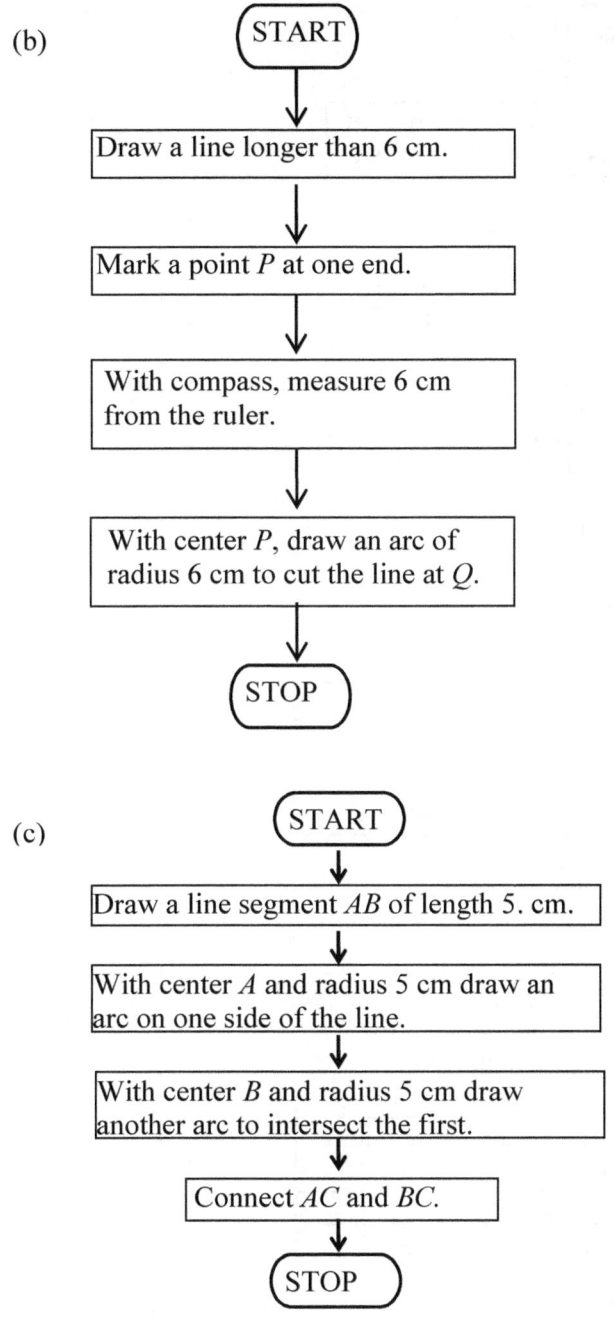

(c)

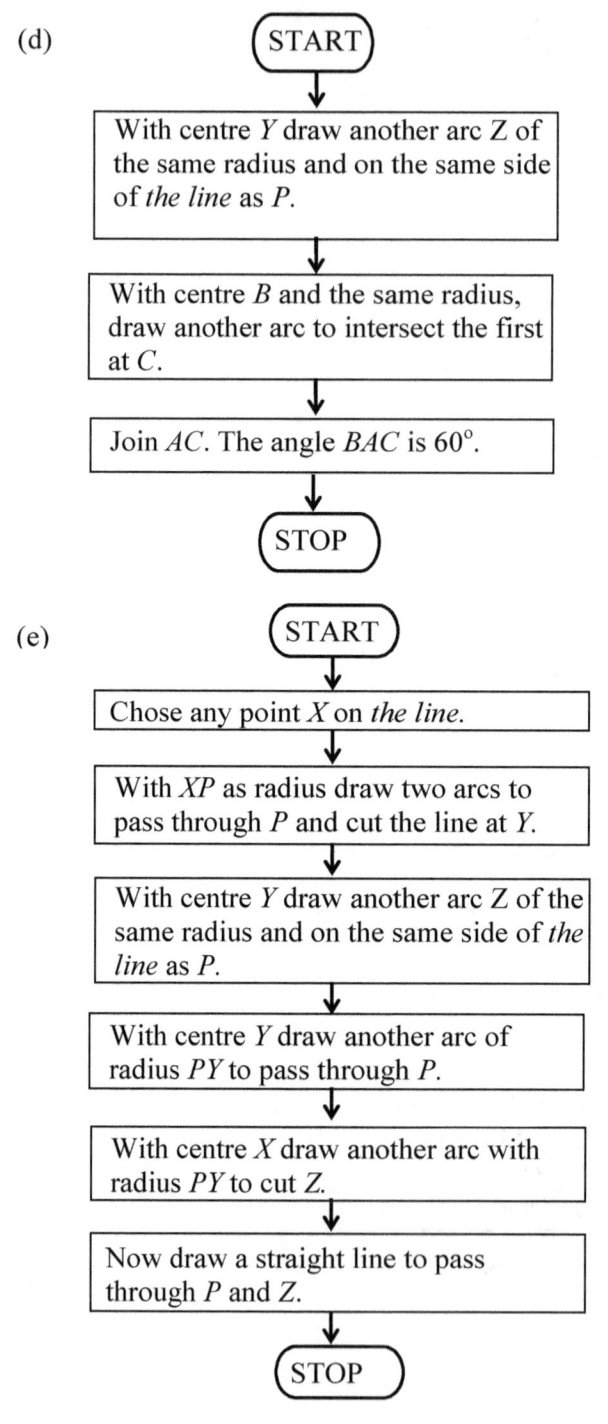

3. (a) $7 \times 14 - 12 \div 4 + 5$ (b) $13 - 12 \div 3 + 2 \times 8 - 5$ (c) $2x^2 + 3x - 5$

Answers to Structural Exercises

4. (a)

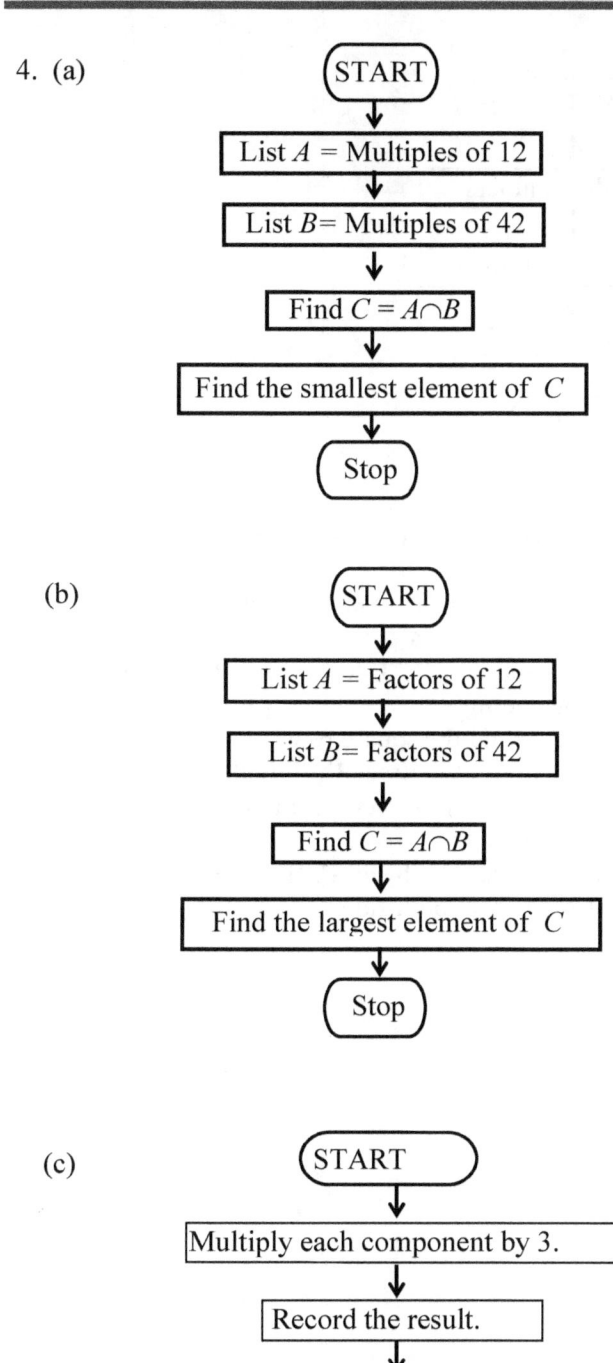

(b)

(c)

Exercise 12:2

1.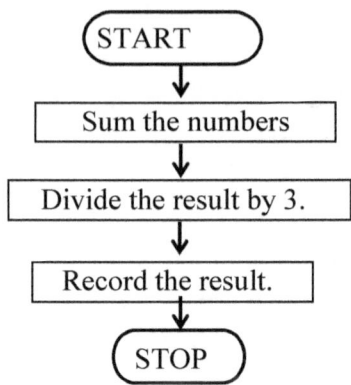

The instructions in the boxes between start and stop for question 2 to 11 are as stated.

2. 'List all the elements in both sets A and B', 'Separate the elements by commas', 'Enclose the elements in braces'.
3. 'Find two factors of –35 whose sum is –2', 'Decompose –2x in terms of these factors', 'Factorise by grouping'.
4. 'Write $A = \pi r^2$', 'Divide by π', 'Find the square root'.
5. 'Add like components'.   6. 'Subtract 3', 'Divide by 2'.
7. 'Add 9', 'Divide by 7'.   8. 'Subtract 5', 'Subtract 2'.
9. 'Multiply by 2', 'Subtract 5', 'Divide by 5'.
10. 'Add 5 to both sides', 'Subtract x from both sides', 'Divide both sides by 2'.
11. 'Add 6 to both sides', 'Subtract x from both sides',

Answers to Structural Exercises

12.

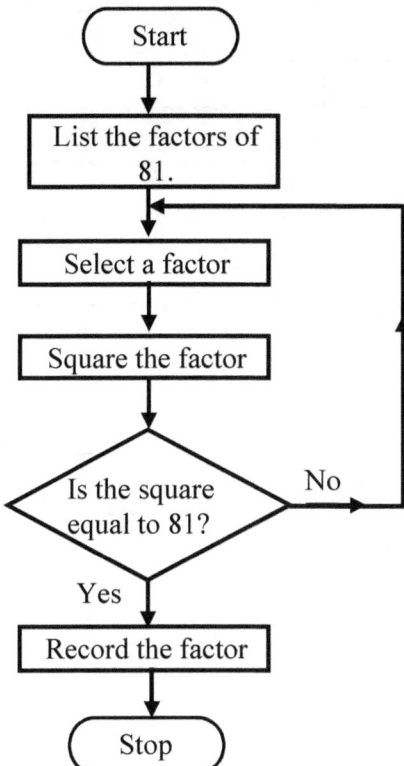

Exercise 12:3

1.

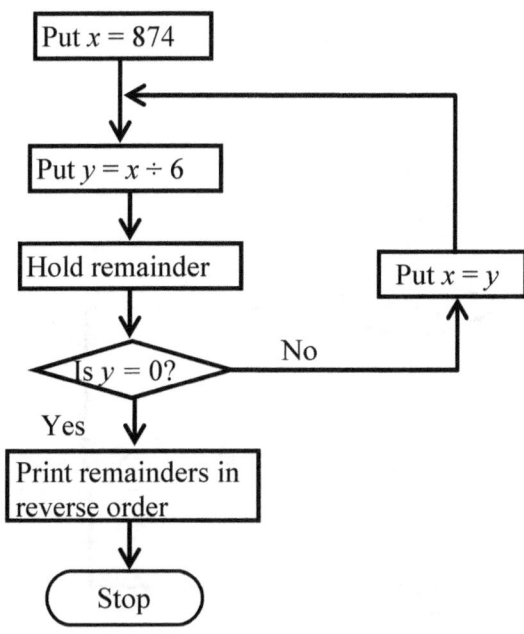

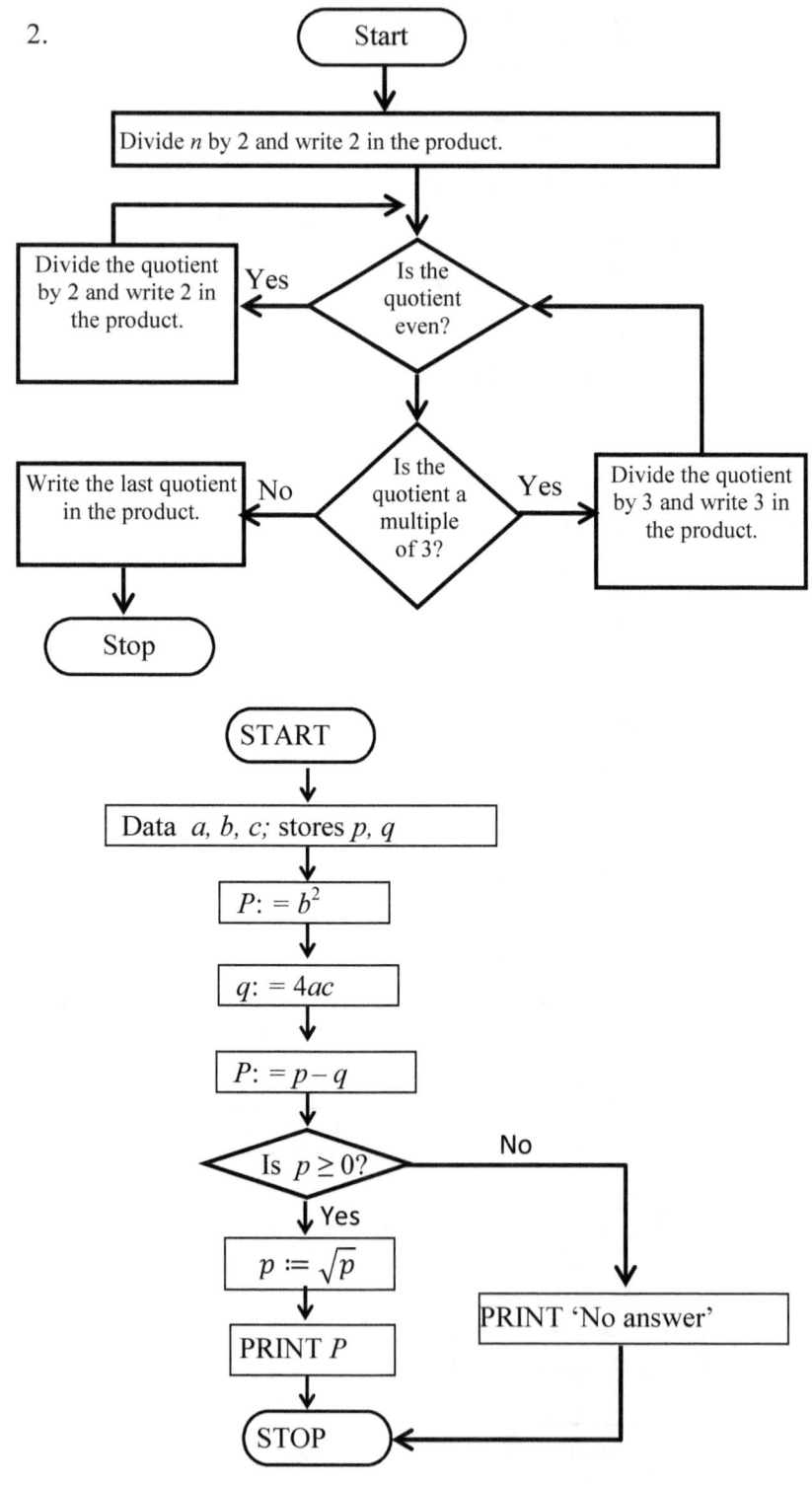

Answers to Structural Exercises

4.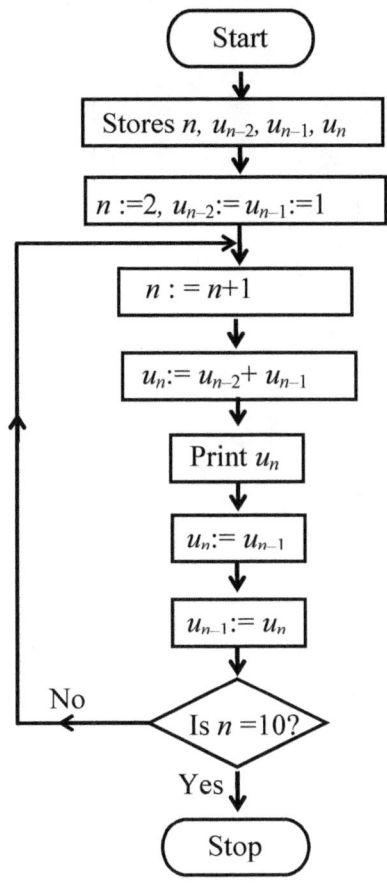

5. 1, 3, 4, 7, 11, 18, 29, 47,...
6. (a) (i) 1, 3, 6, 10, 15, 21, 28, ...   (ii) 1, 9, 36, 100,...
   (iii) The cubes 1, 8, 27, 64, 125,...
   (b) The sequences in (a) (i) and (a) (ii) are the triangular numbers and the squares of the triangular numbers respectively.
7. (i) 2, 2, 3   (ii) 3, 5, 5   (iii) 7, 7   The prime factors are printed.
8. (i) 3   (ii) $AX^2 + BX + C$

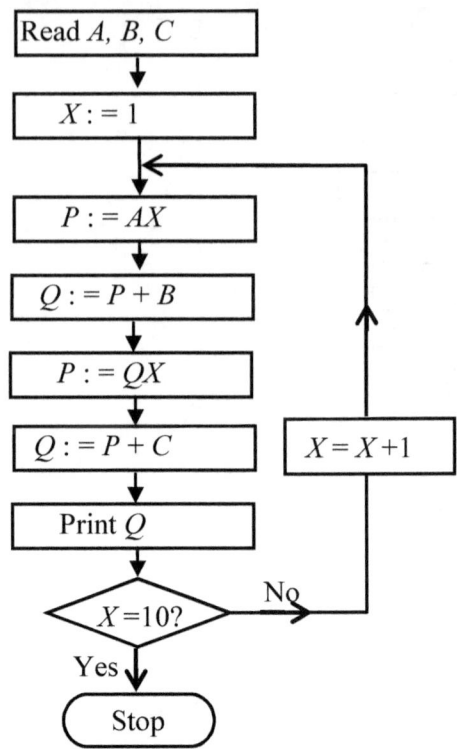

9.

A	B
12	30
12	18
12	6
6	6
0	6

The program fails because it did not take care of the case when $B = 0$. We can amend the program as in the figure below.

Answers to Structural Exercises

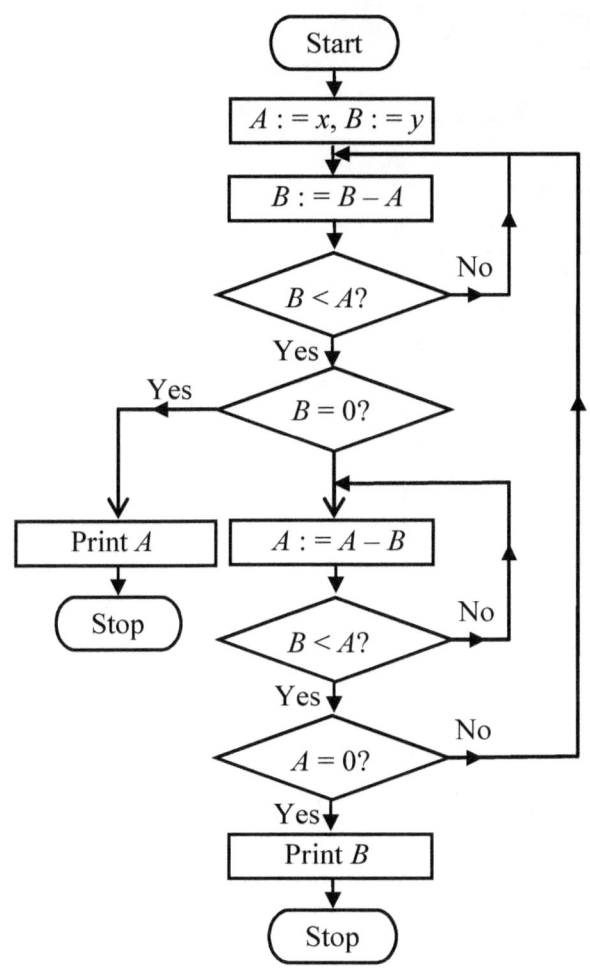

10.

A	B	C
5	−4	2
3.4	1.6	10
3.02	0.38	6.8

$$\frac{x^2+9}{2x}$$

11. (a) $1+2+3+4+\ldots+20$ (b) $A := A+2$ (for $A+1$)

12.

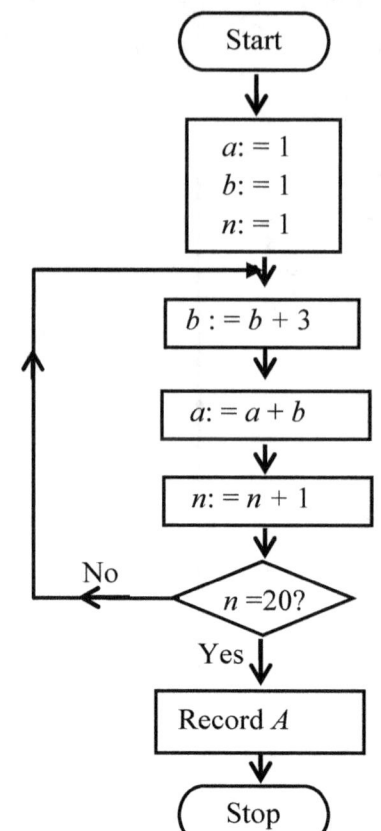

13.

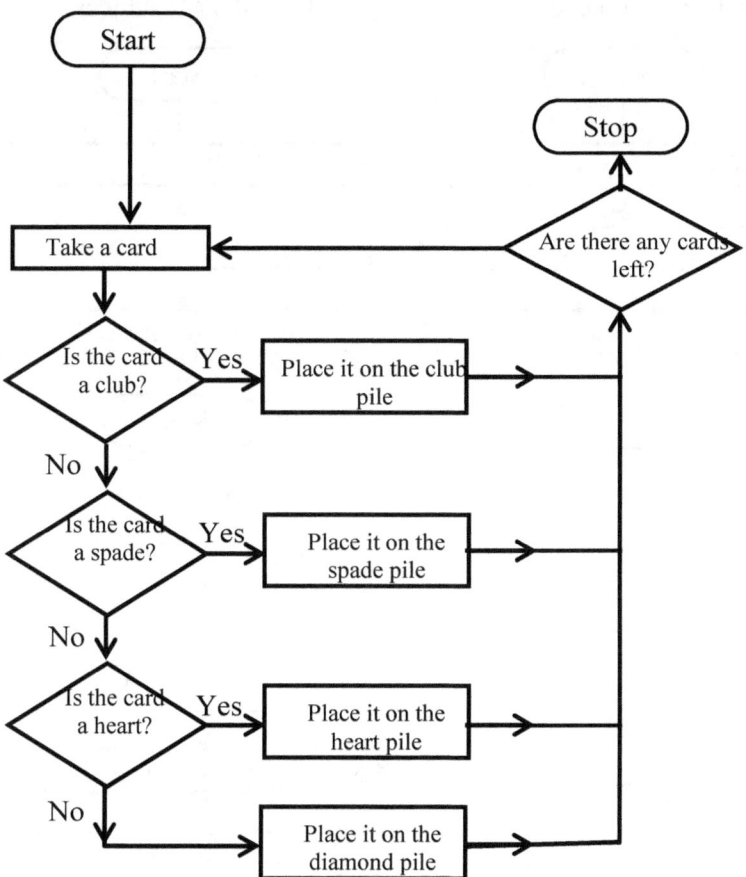

14. To make *C* a subject   To make *F* a subject

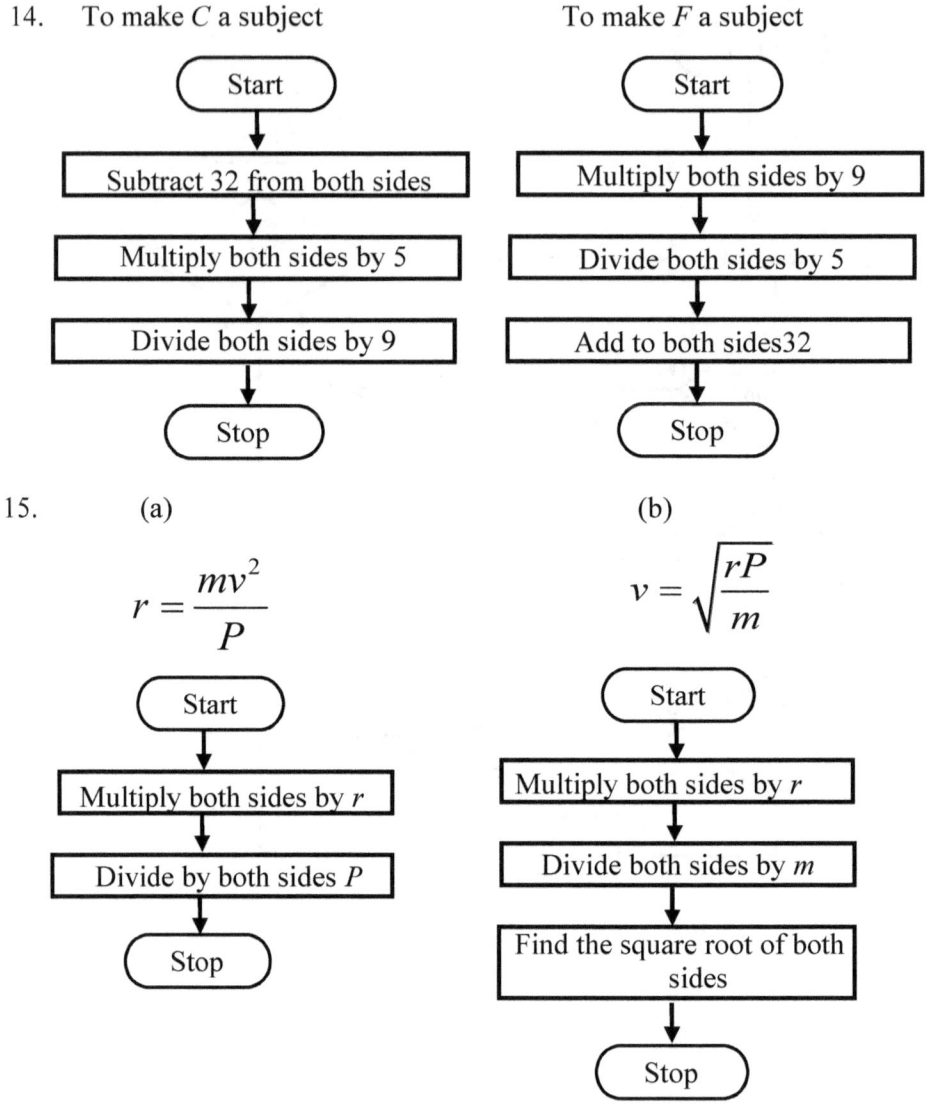

15. (a) $r = \dfrac{mv^2}{P}$ (b) $v = \sqrt{\dfrac{rP}{m}}$

16.

$b = \sqrt{r^2 - a^2}$

$r = \sqrt{a^2 + b^2}$

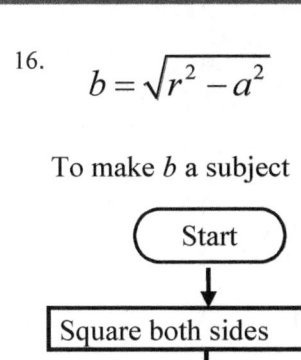

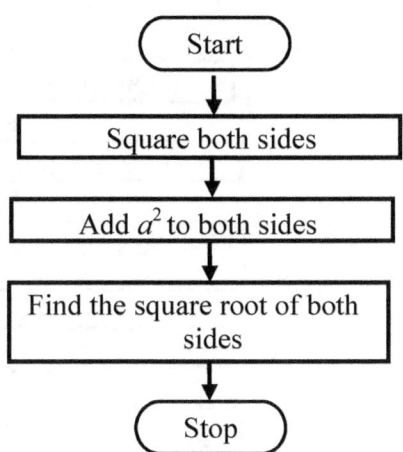

17. To prove that $ax^2 + bx + c = 0$

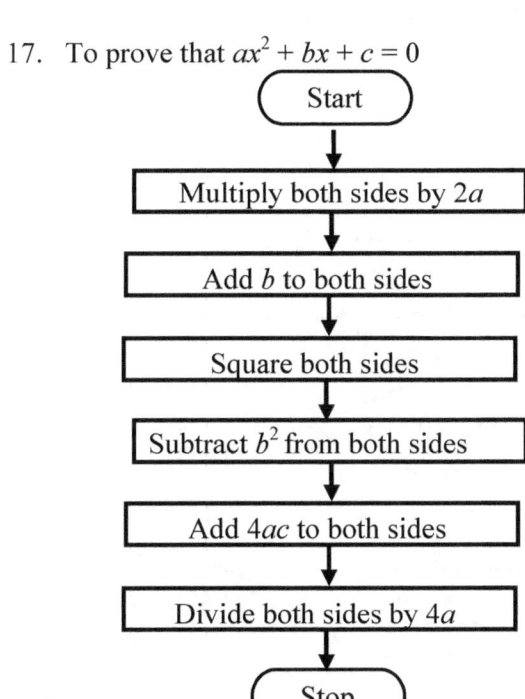

To prove that $x = \dfrac{\sqrt{b^2 - 4ac} - b}{2a}$

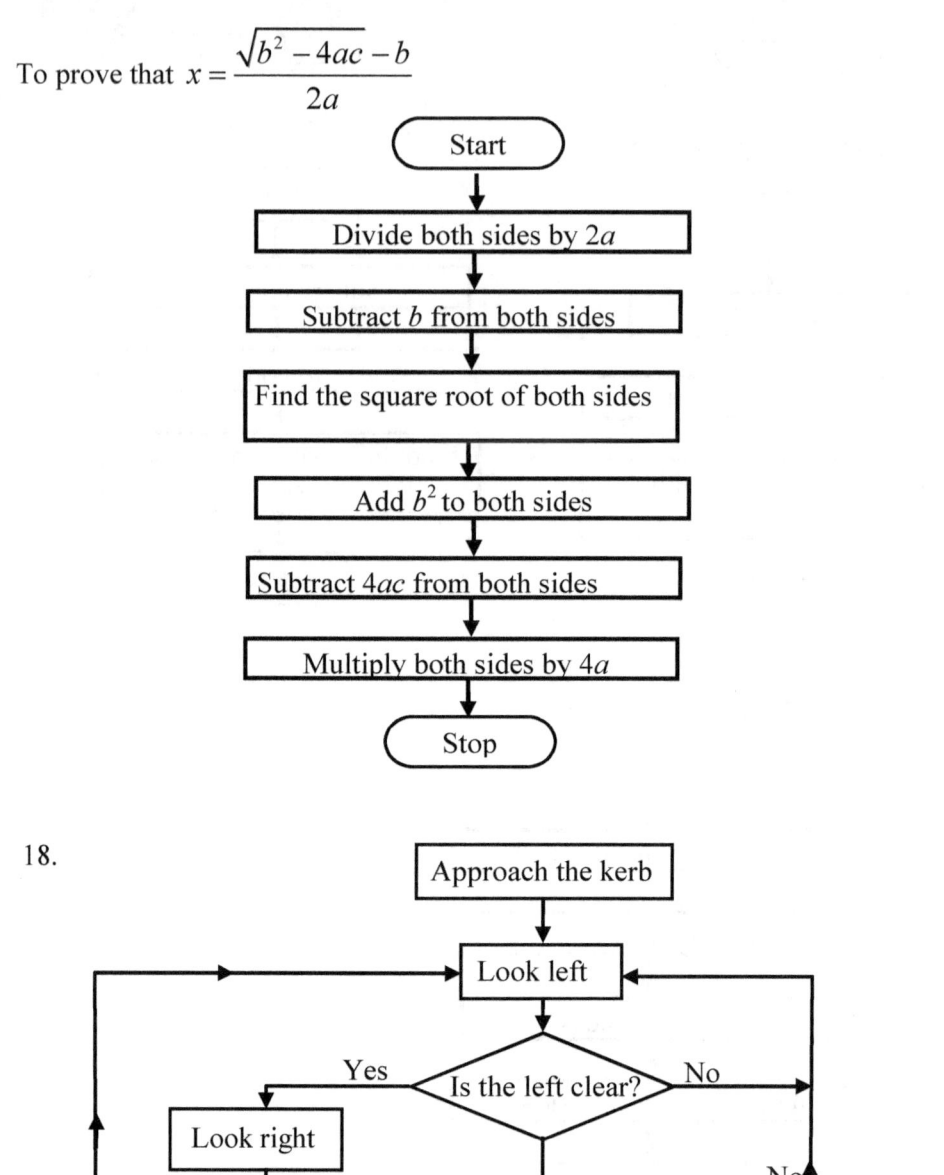

18.

$P(C) = \left(\dfrac{1}{2}\right)\left(\dfrac{1}{3}\right)$

Answers to Structural Exercises

Exercise 13:1

1. (a) $p^2 - 20p + 100$ (b) $y^2 - 14y + 49$ (c) $x^2 + 5x + \dfrac{25}{4}$
   (d) $u^2 + \dfrac{4}{3}u + \dfrac{4}{9}$ (e) $x^2 + 18x + 81$ (f) $x^2 - 4x + 4$

2. (a) $(t+6)^2$ (b) $(x-9)^2$ (c) $\left(y+\dfrac{7}{2}\right)^2$ (d) $\left(u-\dfrac{11}{2}\right)^2$
   (e) $(x+10)^2$ (f) $(m-15)^2$

Exercise 13:2

1. $(x+3)(x-2)$ 2. $(p+1)(p+11)$ 3. $(y-4)(y-3)$
4. $(x+8)(x-2)$ 5. $(x-5)(x+3)$ 6. $(5x-3)(2x+1)$
7. $(4a+5)(a-2)$ 8. $(5-6y)(1-2y)$ 9. $(3+2x)(1-x)$
10. $(5-3x)(x+5y)$ 11. $(x+4y)(x-3y)$ 12. $(2x-3y)(x+5y)$
13. $(6k-x)(k+3x)$ 14. $(2p-q)(6p-5q)$

Exercise 13:3

1. $x=3$ or $x=-15$ 2. $u=-19$ or $u=-1$ 3. $p=-8$ or $p=1$
4. $y=13$ or $y=5$ 5. $a=4$ or $a=7$ 6. $x=33$ or $x=-3$
7. $x=-16$ or $x=2$ 8. $x=-19$ or $x=1$

Exercise 13:4

1. $\dfrac{5\pm\sqrt{57}}{4}$ 2. $\dfrac{-7\pm\sqrt{29}}{2}$ 3. $\dfrac{-5\pm\sqrt{17}}{4}$ 4. $-3\pm\sqrt{19}$ 5. $\dfrac{2\pm\sqrt{10}}{3}$
6. $\dfrac{5\pm\sqrt{7}}{5}$ 7. $\dfrac{5\pm\sqrt{5}}{5}$ 8. $\dfrac{-1\pm\sqrt{73}}{18}$

Exercise 13:5

1. $u=-2$ or $p=-4$ 2. $a=8$ or $a=-3$ 3. $x=\dfrac{1}{3}$ or $x=-7$
4. $m=-\dfrac{5}{2}$ or $m=-1$ 5. $p=2$ or $p=-10$ 6. $x=\dfrac{7}{2}$ or $x=1$
7. $x=\dfrac{3}{2}$ or $x=-4$ 8. $a=-\dfrac{1}{2}$ or $a=10$

353

Competency Based Mathematics for Secondary Schools. Book 4

Exercise 13:6
1. $x = \pm 8$
2. $y = 0$ or $y = 7$
3. $a = \pm \dfrac{1}{2}$
4. $p = 0$ or $p = \dfrac{1}{9}$
5. $x = 0$ or $x = 4$
6. $y = \pm 3$
7. $b = 0$ or $b = 6$
8. $p = \pm 5\sqrt{2}$

Exercise 13:7
1. -3 and -2 or 3 and 2
2. -3 and $-3^2 = 9$ or 4 and $4^2 = 16$
3. 15, 135 FCFA
4. 15 m by 16.5 m
5. $n = -13$ or $n = 12$
6. 3 and 9
7. 16
8. 12 m by 2 m
9. $h = 8$ m, $b = 12$ m
10. 8
11. 11 and 3
12. 12 rows

Exercise 13:8
1. $x = 4, y = -\dfrac{3}{2}$ or $x = 3, y = -2$
2. $x = \dfrac{4}{3}, y = -\dfrac{4}{3}$ or $x = 4, y = 28$
3. $x = 3, y = 4$ or $x = 4, y = 3$
4. $x = \dfrac{1}{2}, y = \pm 1$
5. $x = 1, y = 3$ or $x = -\dfrac{95}{41}, y = \dfrac{21}{41}$
6. $x = \dfrac{27}{4}, y = \dfrac{49}{6}$ or $x = -4, y = 1$
6. $x = \dfrac{27}{4}, y = \dfrac{49}{6}$ or $x = -4, y = 1$
7. $x = -2, y = 1$ or $x = -\dfrac{3}{11}, y = -\dfrac{62}{33}$
8. $x = 1, y = 2$ or $x = -\dfrac{3}{2}, y = -\dfrac{1}{2}$
9. $x = 2, y = \pm 3$
10. $x = 1, y = 2$ or $x = \dfrac{1}{9}, y = \dfrac{22}{9}$

Exercise 14:1
1. (a) $(x-2), (x-1)(x+1)$ (b) $(2x+1), (x-1)(x+1)$
2. (a) $x = \dfrac{1}{2}$ (b) $x = 1, -1$ or $\dfrac{1}{6}$
3. -8
4. $k = -7, f(x) = (x+3)(x-2)(x-1)$
5. $a = 3, b = -5, c = 8$
6. $x^2 - 7x - 6$
7. $a = 3, b = 12$
8. $k = 2$, roots are $x = 2$ and $x = 3$
9. $k = -7$, $(x+3)(x-1)(x-2)$
10. $k = 2, f(x) = (x+1)(x+2)(x-1)$
11. Remainder $= 5$, required number $= -5$

Exercise 14:2
1. $(x+4)(x^2-4x+16)$
2. $(x-3)(x^2+3x+9)$
3. $(x-2y)(x^2+2xy+4y^2)$
4. $(2x+5y)(4x^2 - 10xy + 25y^2)$
5. $(4x-3y)(16x^2 - 12xy + 9y^2)$

Exercise 15:1
1. (a)   (b)   (c)   (d)   (e)

Answers to Structural Exercises

2. P is closed. Both 17 and 19 are included.
   Q is closed. Both 1 and 3 are included
   R is half opened, half closed, −1 is excluded but 2 is included.
   S is half opened, half closed, −2 is included but 2 is excluded.
   T is closed. Both 2 and 5 are included.
   U is opened. Both −2 and 0 are excluded.
   V is opened. Both 5 and 8 are excluded.
   W is opened. Both −7 and −2 are excluded.

3.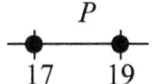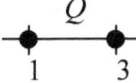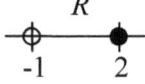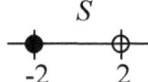

   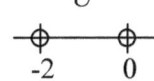 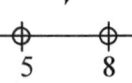 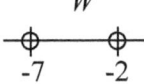

4. (a) (−2,2],]−2,2]   (b) [98,101], 98 ≤ x ≤ 101
   (c) (2,5), 2 < x < 5,]2,5[(d) [1,3), [1,3[
   (e) (−6, −1), −6 < x < −1   (f)]14, 19], 14 < x ≤ 19
   (g) [0, 7[, 0 ≤ x < 7   (h) [−3, 1), −3 ≤ x < 1
   (i) (11, 13] , 11 < x ≤ 13

5. (a) closed   (b) Half open, half closed   (c) Half open, half closed
   (d) Opened   (e) Closed   (f) Open   (g) Closed

Exercise 15:2

1. $x+8 \leq 14$ 2. $\dfrac{40}{n} > 4$ 3. $3(x+7) \leq 27$ 4. $4m \geq 20$ 5. $\dfrac{3}{5}y - 10 < 12$

6. $3x - \dfrac{1}{2}x < x+6$ 7. $2(m+7) \leq m+12$ 8. $3(p+5) < 4p+2$

9. $30 - 5n \geq 4$ 10. $4x - 60,000 \leq 600,000$

Exercise 15:3

1. $x < 2$ 2. $x > 9$ 3. $x \geq 4$ 4. $y \geq 4$ 5. $x \leq \dfrac{16}{5}$

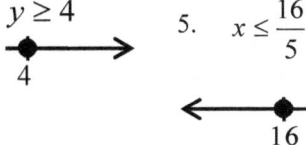

6. $n > 4$ 7. $x < 40$ 8. $x > 5$ 9. $y \geq 35$ 10. $t < 10$

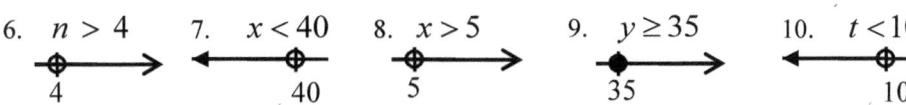

355

11. $u < 9$

12. $x \leq 10$

13. $x > \dfrac{11}{2}$

14. $x \geq 3$

15. $y \leq \dfrac{3}{2}$

16. $x < 5$

17. $x \leq 15$

18. $p < \dfrac{5}{2}$

19. $m < 2$

20. $x \leq 7$

21. $x \geq 5$

22. $x \leq 1$

23. $-\dfrac{13}{4} \leq x \leq 0$

24. $-1 \leq w \leq 8$

25. $-1 < x < 2$

26. $-1 < x < 2$

27. $-1 \leq x \leq 2$

Exercise 15:4

1. $-4 < x < -2$

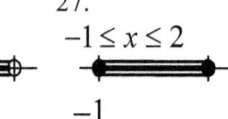

2. $a < -3$ or $a > 9$

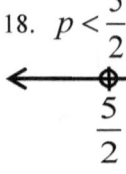

3. $x \leq -7$ or $x > \dfrac{1}{3}$

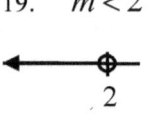

4. $m \leq -\dfrac{5}{2}$ or $m \leq 1$

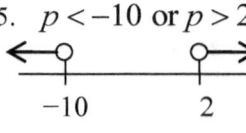

5. $p < -10$ or $p > 2$

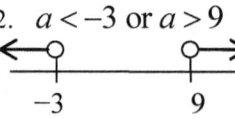

6. $x < 1$ or $x > \dfrac{7}{2}$

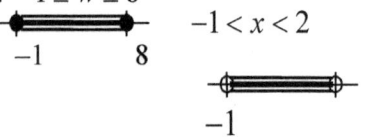

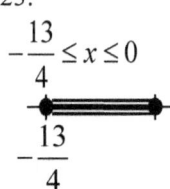

7. $-4 < x < \dfrac{3}{2}$ 8. $-\dfrac{1}{3} \le a \le 10$

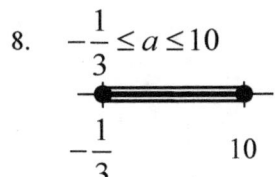

Exercise 15:5
1. Imaginary   2. Real and distinct   3. Imaginary   4. Imaginary
5. Real and distinct   6. Real and equal   7. Imaginary   8. Real and distinct
9. Real and equal   10. Imaginary   11. Real and distinct   12. Real and equal

Exercise 15:6
1. 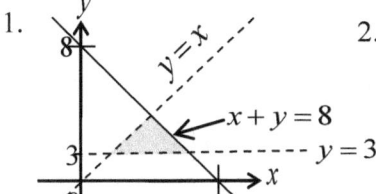   2. (a) 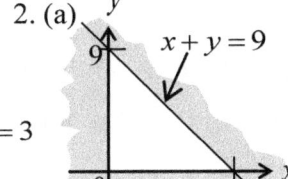   (b)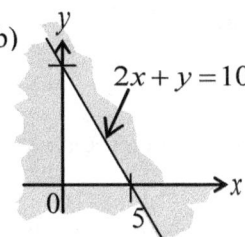

(c) 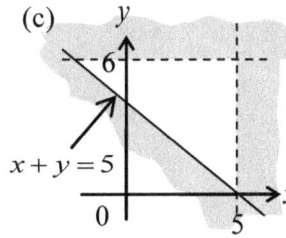   (d) 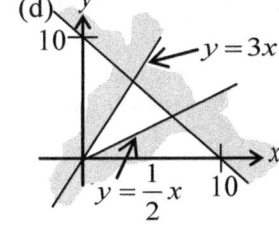   (e)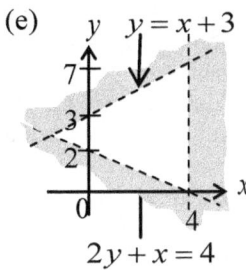

(f) 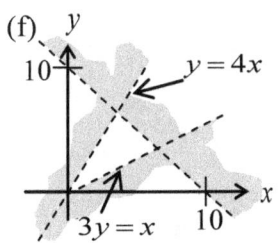   (g) 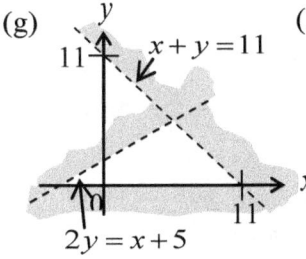   (h)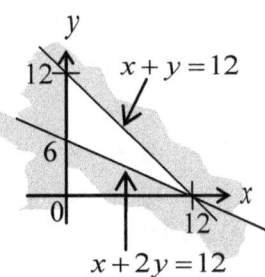

(i)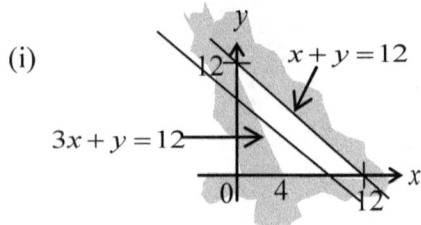

3. (a) $2 \leq x \leq 7$ (b) $-2 < x < 5$ (c) $-2 \leq y < 4$
   (d) $x \leq 0, x+y < 10, y > x$ (e) $y \leq -2, x < 8, y < x$
   (f) $y \geq 0, 3y < 2x+12, y \geq -2x+4$

Exercise 16:1
(1) Even numbers; 10,12,14 (2) Odd multiples of 3; 27,33,39
(3) Consecutive square numbers; 25,36,49 (4) Adding 3 to preceding term; 20,23 26
(5) Dividing preceding term by 2; 10,5, $\dfrac{5}{2}$ (6) Adding 1,2,3,4...; 11,16,22
(7) Dividing preceding term by 3; 1, $\dfrac{1}{3}, \dfrac{1}{9}$ (8) Adding 3 to preceding term; 1, $\dfrac{1}{3}, \dfrac{1}{9}$
(9) Adding 10 to preceding term; 42, 52, 62
(10) Subtracting 4 from preceding term; 25, 21, 17
(11) Adding 10 and subtracting 3 alternately; 17, 27, 24
(12) Subtracting 4 from preceding term; $-16, -20, -24$
(13) Consecutive even numbers to the preceding term; 33, 45, 59
(14) Adding 4 to preceding term; 19, 23, 27

Exercise 16:2
1. (a) $3(2^{n-1})$ (b) $3n+2$ (c) n^3 (d) $10(4^{n-1})$ (e) 2^{1-n} (f) $\dfrac{n}{n+1}$
2. (a) 5,12,21 (b) 2,6,12 (c) $\dfrac{11}{10}, \dfrac{7}{10}, \dfrac{19}{30}$ (d) $-4, -6, -6$ (e) 6,12,24
3. 2,6,12,20; $U_n = n(n+1)$ 4. 4,6,8; $U_n = 2(n+1)$
5. $S_1 = 1, S_2 = 5, S_3 = 14, S_4 = 30, S_5 = 55$
   $U_1 = 1, U_2 = 4, U_3 = 9, U_4 = 16, U_5 = 25$ 6. 1,3,6; $U_n = n$
7. (a) 31, 41; adding consecutive even numbers
(b) $\dfrac{1}{720}, \dfrac{1}{4840}$ Multiply the preceding term by $\dfrac{1}{2}, \dfrac{1}{3}, \dfrac{1}{4}, \dfrac{1}{5}, \dfrac{1}{6}$, respectively.
8. 8,3,0,-1,0 9. 16 10. (a) 5,11,17 (b) 456
11. (a) $-10, -6, -2, 2, 6$ (b) 110

Answers to Structural Exercises

Exercise 16:3

1. 145 2. $a = -27, U_{28} = 27$ 3. 24 4. $5, \dfrac{7}{3}, -\dfrac{1}{3}, -3 \cdots$

5. (a) 167 (b) $\dfrac{14}{83}$ (c) 11

6. 13, 17, 21, 25 7. 210 8. 10

9. 16, 13, 10, 7, 4 or 4, 7, 10, 13, 16 10. 12 11. $\dfrac{3a+b}{4}, \dfrac{a+b}{2}, \dfrac{a+3b}{4}$

Exercise 16:4

1. (a) $-\dfrac{12}{7}, 3$ (b) $r = \dfrac{3}{2}, U_4 = \dfrac{81}{4}$ 2. $r = -\dfrac{3}{2}, U_n = 4\left(-\dfrac{3}{2}\right)^{n-1}$ 3. $\dfrac{3}{2}$

4. $U_{12} = \dfrac{29296875}{2048}, S_{12} = \dfrac{244136529}{102040}$ 5. (a) ± 12 (b) ± 27 (c) $\pm x^3$

6. (a) 4, 8, 16 (b) $4, 1, \dfrac{1}{4}$ 7. (a) 1023 (b) -340 (c) $\dfrac{2186}{243}$ (d) $\dfrac{(3a)^{10}-1}{3a-1}$

8. (a) $-\dfrac{7}{6}$ (b) $\dfrac{1}{2}$ (c) A.P.: $U_2 = \dfrac{17}{6}, U_5 = -\dfrac{2}{3}$,

G.P.: $U_2 = 2, U_5 = \dfrac{1}{4}$ (d) 7

www.ingramcontent.com/pod-product-compliance
Lightning Source LLC
Chambersburg PA
CBHW071410180526
45170CB00001B/43